BASIC OPTICAL ENGINEERING

FOR ENGINEERS AND SCIENTISTS

BASIC OPTICAL ENGINEERING

FOR ENGINEERS AND SCIENTISTS

HAIYIN SUN

SPIE PRESS
Bellingham, Washington USA

Library of Congress Cataloging-in-Publication Data

Names: Sun, Haiyin, author
Title: Basic optical engineering for engineers and scientists / Haiyin Sun.
Description: Bellingham, Washington : SPIE Press, [2019] | Includes bibliographical references and index.
Identifiers: LCCN 2018028913 (print) | LCCN 2018051262 (ebook) | ISBN 9781510622067 (pdf) | ISBN 9781510622074 (epub) | ISBN 9781510622081 (mobi) | ISBN 9781510622050 (softcover)
Subjects: LCSH: Optical engineering.
Classification: LCC TA1520 (ebook) | LCC TA1520 .H345 2019 (print) | DDC 621.36–dc23
LC record available at https://lccn.loc.gov/2018028913

Published by

SPIE
P.O. Box 10
Bellingham, Washington 98227-0010 USA
Phone: +1 360.676.3290
Fax: +1 360.647.1445
Email: books@spie.org
Web: http://spie.org

Printed in the United States of America.
First Printing.
For updates to this book, visit http://spie.org and type "PM294" in the search field.

SPIE.

Contents

Preface

Optical engineering is increasingly used in industrial, scientific, medical, and military applications, among others. Engineers and scientists need to have certain knowledge in order to effectively apply optics to their projects; however, most of them did not major in optical engineering. Their busy schedules do not allow them to spend a lot time to study the details of optical engineering. Their interests are often limited to quickly gaining the most basic concepts and identifying the right optical components, devices, instruments, and approaches for their applications. What they need is an information source that can promptly provide them the very practical knowledge about a specific part of optical engineering.

In the past several decades, about ten notable books on optical engineering have been published. Several recently published books are available, e.g., Refs. 1–5. The first three approach the topic in a gradual way with detailed explanations and are excellent textbooks for students majoring in optical engineering, as well as for engineers and scientists who have enough time and interest to dig deep into the subject. Fischer[4] covers optical system design with practical design examples and also addresses basic optics and optical components. Smith[5] is a comprehensive book that covers many technical details about classical optical engineering. Both Fischer and Smith are great references for optical engineers and scientists. However, none of these books discuss lasers, laser optics and devices, and fiber optics, even though lasers and optical fibers are now conventional optical components. Smith devotes a section to optical fibers but treats optical fiber as a bulk glass waveguide; most of the important features of optical fibers are not covered.

Basic Optical Engineering for Engineers and Scientists introduces the very practical parts of optical engineering to address the needs of those engineers and scientists who are not specialized in the subject but need to quickly learn something about it for their projects. The text briefly introduces the most basic optics, but most of it describes various optical components, optical devices and systems, lasers, laser optics and devices, optical fibers, opto-electrical devices, optical designs, and optical assemblies. The performance specifications of some optical components that represent current technologies are also provided.

This book tries to avoid detailed manual numerical calculations and raytracings because these techniques are complex, time consuming, and often inaccurate. Instead, computers and optical software are used to perform these tasks because computers and optical software are widely available and the results provided by programs are much more accurate than the results obtained manually. There are no proofs or problem-solving exercises. These features are intended to let readers find the content relevant to their interests and get the results they need. This book tries to cover most areas of modern optical engineering but not in depth.

Haiyin Sun
October 2018

References

1. K. Kasunic, *Optical Systems Engineering*, McGraw-Hill Professional, New York (2011).
2. C. A. DiMarzio, *Optics for Engineers*, CRC Press, Boca Raton, FL (2011).
3. B. Walker, *Optical Engineering Fundamentals*, 2nd Ed., SPIE Press, Bellingham, WA (2009).
4. R. Fischer, *Optical System Design*, 2nd Ed., McGraw-Hill Education, New York (2008).
5. W. Smith, *Optical Engineering*, 4th Ed., McGraw-Hill Education, New York (2007).

Chapter 1
Geometrical Optics

1.1 General Comments

A light wave is an electromagnetic wave, and the wavelength that optics studies ranges from the ultraviolet (~0.2 μm) to the middle infrared (~10 μm). The spatial scales involved in most optical applications are much larger than the light wavelength. In these cases, a light wave can be approximately described by a bundle of straight optical rays. The science of studying optical rays that travel through optical media is called "geometrical optics" and is the most widely used field of optics. If the spatial scales involved are not much larger than the light wavelength, the wave nature of light must be considered; the science of studying the wave nature of light is called "wave optics."

The technique to trace an optical ray through various optical media is called "sequential raytracing" and is the main way to study geometrical optics. This chapter briefly introduces basic geometrical optics using a sequential raytracing technique. (Smith[1] is a widely cited optical-engineering reference book, and Hecht[2] is a popular optics textbook. Both are recommended here as additional information sources.)

1.2 Snell's Law

When an optical ray travels from one optical medium to another that has a different refractive index, the ray is split in two by the interface of the two media. One ray is reflected back to the first medium. Another ray is refracted and enters the second medium, as shown in Fig. 1.1. The equation describing such a refraction is called Snell's law:[3]

$$n_1 \sin(\theta_1) = n_2 \sin(\theta_2), \tag{1.1}$$

where θ_1 is the angle between the incident ray and the normal of the interface of the media at the point where the ray hits, θ_2 is the angle between the refracted ray and the same normal of the interface, and n_1 and n_2 are the refractive indices of the two media, respectively.

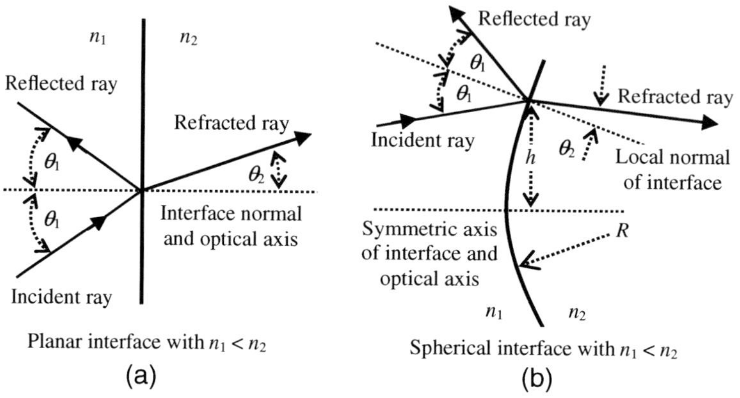

Figure 1.1 (a) An optical ray is incident on a planar interface of two optical media with refractive index n_1 and n_2, respectively, and $n_1 < n_2$. (b) An optical ray is incident on a convex interface of two media with refractive index n_1 and n_2, respectively, and $n_1 < n_2$. The symmetric axis of the interface is the optical axis. The incident angle varies as the ray height h varies. For $h \ll R$, where R is the interface radius of curvature, the interface can be considered planar.

Figure 1.1(a) illustrates a planar interface. The normal of the interface is the normal of the point at which the ray is reflected and refracted, and it is the optical axis. The reflection angle always equals the incident angle, which is the "reflection law" and will be discussed in Section 1.10.1. For $n_1 < n_2$, $\theta_1 > \theta_2$, and vice versa, according to Snell's law.

Figure 1.1(b) illustrates a curved interface. The symmetric axis of the interface is the optical axis. The incident ray height h is the vertical distance between the point on the interface where the ray hits and the optical axis. The figure shows that the incident angle θ_1 is a function of h and the interface radius of curvature R.

Snell's law is the foundation of geometrical optics, which is why every optics book mentions it.

1.3 Total Internal Reflection

Equation (1.1) shows that if $n_1\sin(\theta_1)/n_2 > 1$, there is no solution for θ_2 because the maximum possible value of $\sin(\theta_2)$ is 1. Thus, no ray will be refracted and enter the second medium. Instead, the incident ray will be totally reflected at the interface. This phenomenon is called total internal reflection (TIR).[4]

For example, if medium 1 is N-BK7 glass with an index $n_1 \approx 1.52$ and medium 2 is air with an index $n_2 \approx 1.0$, then condition $n_1\sin(\theta_1)/n_2 \geq 1$ leads to $\theta_1 \geq 41.1°$. Any incident ray with incident angle >41.1° will be totally reflected (41.1° is the critical angle in this case). The condition for TIR to happen is $n_1 > n_2$. A different ratio of n_1/n_2 would have a different critical angle.

Note that when a ray is incident on an interface of two media with an angle smaller than the TIR angle, the refracted ray will carry most of the energy of the incident ray, and the reflected ray will carry a relatively much smaller portion of the energy. The reflected energy can be calculated using wave optics theory, as will be discussed in Section 2.3. In many applications, the reflected energy is unwanted and is reduced by applying an antireflection (AR) coating on the interface.

1.4 Paraxial Approximation

Before the invention of computers, tracing a ray using Snell's law [Eq. (1.1)] through several curved optical interfaces required a lot of calculation and drawing. Efforts were taken to simplify the situation whenever possible. When $\theta_1 < 1$, the calculation can be simplified by expanding the sine function in Snell's law into a series:

$$\begin{aligned}
\theta_2 &= \sin^{-1}\left[\frac{n_1}{n_2}\sin(\theta_1)\right] \\
&\approx \frac{n_1}{n_2}\sin(\theta_1) + \frac{1}{6}\frac{n_1^{\,3}}{n_2^{\,3}}\sin(\theta_1)^3 + \frac{3}{40}\frac{n_1^{\,5}}{n_2^{\,5}}\sin(\theta_1)^5 + \ldots \\
&\approx \frac{n_1}{n_2}\left(\theta_1 - \frac{1}{6}\theta_1^{\,3} + \frac{1}{120}\theta_1^{\,5}\ldots\right) + \frac{1}{6}\frac{n_1^{\,3}}{n_2^{\,3}}\left(\theta_1 - \frac{1}{6}\theta_1^{\,3} + \frac{1}{120}\theta_1^{\,5}\ldots\right)^3 \\
&\quad + \frac{3}{40}\frac{n_1^{\,5}}{n_2^{\,5}}\left(\theta_1 - \frac{1}{6}\theta_1^{\,3} + \frac{1}{120}\theta_1^{\,5}\ldots\right)^5 + \ldots \\
&\approx \frac{n_1}{n_2}\theta_1 + \frac{1}{6}\left(\frac{n_1^{\,3}}{n_2^{\,3}} - \frac{n_1}{n_2}\right)\theta^3{}_1 + \frac{1}{4}\left(\frac{1}{30}\frac{n_1}{n_2} + \frac{1}{3}\frac{n_1^{\,3}}{n_2^{\,3}} + \frac{3}{10}\frac{n_1^{\,5}}{n_2^{\,5}}\right)\theta_1^{\,5}\ldots.
\end{aligned} \tag{1.2}$$

The calculation accuracy requirement determines how many terms to keep in Eq. (1.2).

"Paraxial approximation"[5] means $\theta_1 \ll 1$ so that only the first-order terms of θ_1, $\theta_2 = (n_1/n_2)\theta_1$ must be kept for a fairly accurate calculation. Paraxial approximation is an ambiguous concept; there is no simple line to determine whether a θ_1 value is small enough to qualify for paraxial approximation. Rather, it depends on how accurate the calculation result must be. Figure 1.1(b) shows that $h \ll R$ leads to $\theta_1 \ll 1$. If the terms of the third order or higher of θ_1 are used, the calculation will be quite complex.

With the calculation power of modern computers, paraxial approximation is not as important as before. However, paraxial approximation is still useful for qualitative and manual analysis, and it is still used in much of the literature, partially due to tradition.

1.5 Lenses

1.5.1 Lens types

Lenses are the most commonly used optical component and can be separated into two categories: positive and negative.

Any lenses with a central thickness larger than their edge thickness are positive lenses. A positive lens can convergently refract rays that pass through the lens, i.e., focus the rays passing through the lens.

Any lenses with a central thickness smaller than their edge thickness are negative lenses. A negative lens can divergently refract the rays passing through the lens.

Figure 1.2 shows several shapes of positive lenses and negative lenses. The symmetric axis of a lens is its optical axis. The functionality of a lens can be determined by tracing rays through the lens using Snell's law.

1.5.2 Positive lenses

Figure 1.3 shows how raytracing is performed through an equi-convex positive lens to analyze the lens function. The first ray usually traced is parallel to the optical axis of the lens, traced left to right through the lens, as indicated by

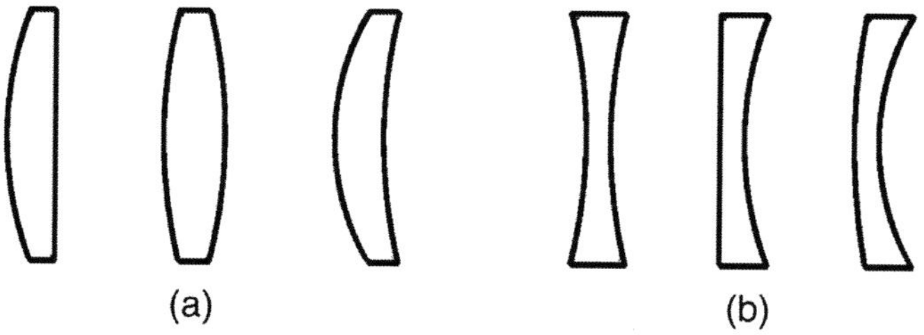

Figure 1.2 Three shapes of a (a) positive lens and (b) negative lens.

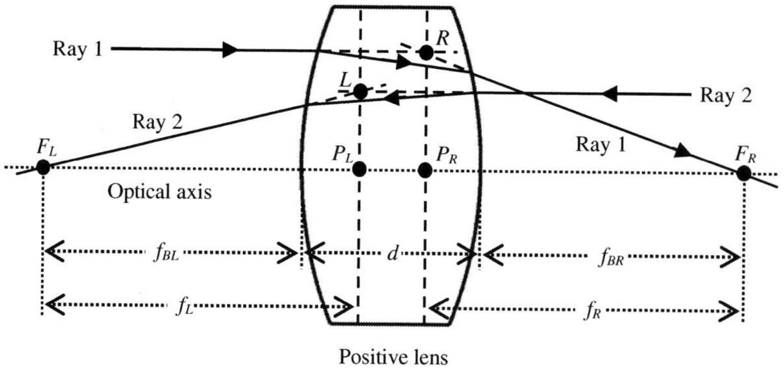

Figure 1.3 Two rays traced through an equi-convex positive lens. The rays are convergently refracted. The two cross-points of the rays and the optical axis are the two focal points of the lens, marked F_L and F_R, respectively.

Ray 1 in Fig. 1.3. The left surface of the lens convergently refracts the ray, and the right surface of the lens further convergently refracts the ray. As the ray travels forward, it eventually crosses the optical axis at point F_R, which is the right focal point of the lens. The ray is said to be focused.

The second ray traced is parallel to the optical axis of the lens and is traced from right to left through the lens, as marked by Ray 2 in Fig. 1.3. The right surface of the lens convergently refracts the ray, and the left surface of the lens further convergently refracts the ray. The ray eventually crosses the optical axis at point F_L, which is the left focal point of the lens. The ray is also focused.

For Ray 1, the forward extension of the incident ray and the backward extension of the exit ray meet at point R. For Ray 2, the forward extension of the incident ray and the backward extension of the exit ray meet at point L. Points L and R determine the axial locations of the two principal planes, which are shown by the two vertical dashed lines. The cross-points of the two principal planes and the optical axis are the two principal points, marked P_L and P_R, respectively, in Fig. 1.3.

The axial distance between the left (right) principal plane and the left (right) focal points is the focal length denoted by f_L (f_R), as marked in Fig. 1.3. f_L always equals f_R. Lens thickness d is defined as the axial distance between the two vertices of the lens surfaces.

The "back focal length" is the axial distance between the left (right) focal point and the vertex of the lens left (right) surface, as marked by f_{BL} and f_{BR}, respectively, in Fig. 1.3. f_{BL} equals f_{BR} only when the two surfaces of the lens have the same shape but opposite orientations. The equi-convex lens shown here is such a lens.

Positive lenses of any shape can focus rays, and their focal lengths are, by definition, positive.

1.5.3 Negative lenses

Figure 1.4 shows how rays are traced through an equal-concave negative lens. Again, the first ray traced is parallel to the optical axis of the lens and is traced from left to right through the lens, as marked by Ray 1 in Fig. 1.4. The left surface of the lens divergently refracts the ray, and the right surface of the lens further divergently refracts the ray. The ray never crosses the optical axis as it travels forward. However, a virtual ray that is the leftward extension of the exit Ray 1 can be conceived, as shown by the dashed line in Fig. 1.4. This virtual ray will cross the optical axis at point F_L, which is the left focal point of the lens. Such a focal point is a virtual focal point; the ray is never actually focused.

Similarly, the second ray traced is parallel to the optical axis of the lens and is traced from right to left through the lens, as marked by Ray 2 in Fig. 1.4. The right-side surface of the lens divergently refracts the ray, and the left-side surface of the lens further divergently refracts the ray. The ray

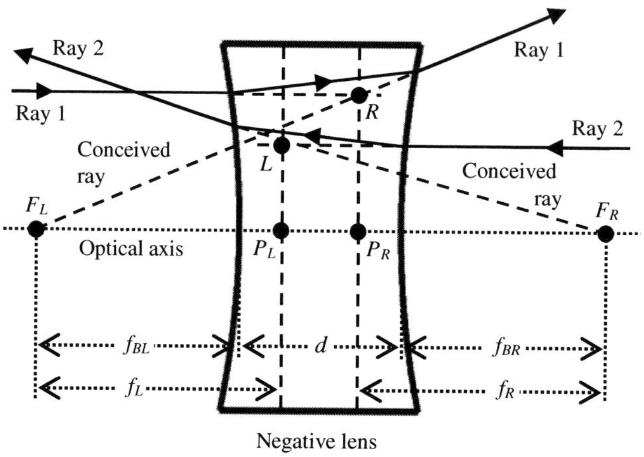

Figure 1.4 Two rays traced through an equal-concave negative lens. The rays are divergently refracted. The two cross-points L and R of the backward extension of the exit rays and the optical axis are the two focal points of the lens, marked by F_L and F_R, respectively. Points L and R determine the axial positions of the two principal points P_L and P_R.

never crosses the optical axis. The conceived virtual ray that is extended rightward from the exit Ray 2 will cross the optical axis at point F_R, which is the right-side focal point of the lens. Such a focal point is also a virtual focal point since the ray is never focused.

The two principal points and planes of this negative lens can be found by using the same raytracing technique as described earlier for the equi-convex lens. The focal length and back focal length of a negative lens are measured the same way as for the equi-convex positive lens. However, the focal length of any negative lens is defined as negative. No negative lens will ever focus rays.

1.5.4 Cardinal points

Any lens has three pairs of cardinal points:[6] a pair of principal points, a pair of focal points, and a pair of nodal points. The principal point pair P_L and P_R and the focal point pair f_L and f_R were discussed earlier and marked in Figs. 1.3 and 1.4, respectively. This subsection further discusses the optical meaning of the principal points and planes.

In Figs. 1.3 and 1.4, if the lens is viewed from the right side, the refracted Ray 1 appears to be emitted from point R, and P_R is the optical position of the lens when viewed from the right side. If the lens is viewed from the left side, the refracted Ray 2 appears to be emitted from point L, and P_L is the optical position of the lens when viewed from that side.

Figure 1.5 shows the positions of the two principal planes for various shapes of lenses. The principal plane positions of some lenses can be outside

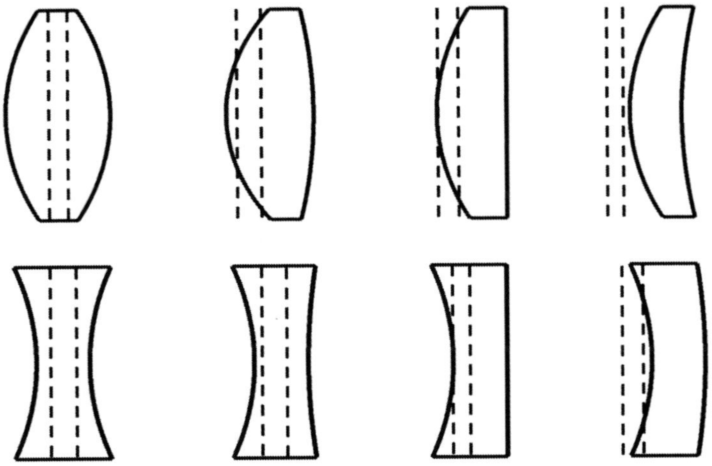

Figure 1.5 Positions of the two principal planes for various shapes of lenses.

the lenses. If the distance between the two principal points is much smaller than the focal length, the two principal points can be considered to coincide, and the lens is called a "thin lens."

The raytracings shown in Figs. 1.3 and 1.4 did not consider the effects of spherical aberrations and are therefore only approximations. The actual positions of the two principal points and the focal points, and thus the focal length, vary as the incident ray height changes.

Figure 1.6 shows accurate raytracing diagrams obtained using the optical design software Zemax. The diagrams illustrate how the position of the principal point changes as the incident ray height changes. The positions of the principal point in the paraxial approximation (i.e., the incident ray height approaches zero), marked by the larger dots in Fig. 1.6, are usually used. The positions of the focal points and the value of focal length usually used are also paraxial approximation values.

1.5.5 Nodal points

Every lens has a pair of nodal points. In most cases, including the case shown in Fig. 1.7, the medium at the left and right side of the lens are the same, such as air. Then the two nodal points coincide with the two principal points. Nodal points have two unique properties:

1. A ray aimed at one of the nodal points will be refracted by the lens such that it appears to come from the other nodal point and have the same angle with respect to the optical axis.
2. The right (left) focal point position of a lens is not shifted when the lens is rotated about its right (left) nodal point.

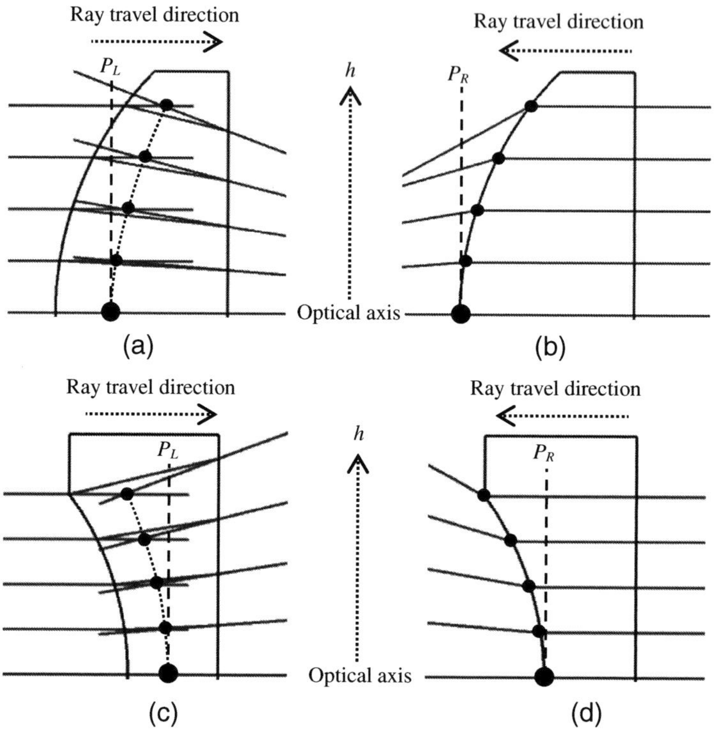

Figure 1.6 Zemax-generated accurate raytracing diagram of the principal point positions as a function of incident ray height h. The principal points are marked by dots. (a) and (b) Rays are traced right and left, respectively, to determine the positions of the two principal points for a positive lens with convex-planar surfaces. (c) and (d) Rays are traced right and left, respectively, to determine the positions of the two principal points for a negative lens with concave-planar surfaces.

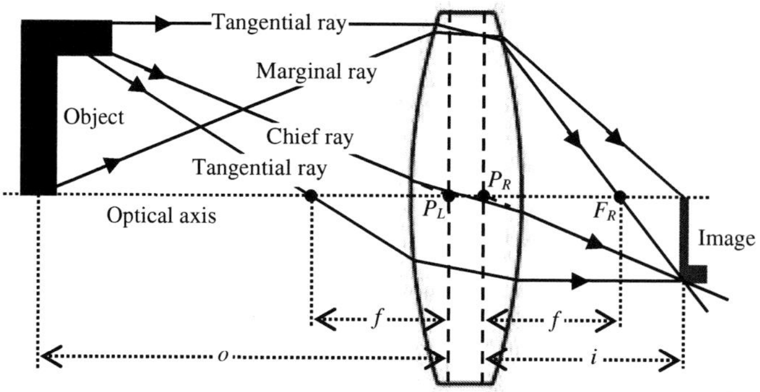

Figure 1.7 An equi-convex positive lens illustrates the nodal points, tangential plane and rays, and the formation of an image. Since the object is placed outside the focal point of the lens, the image formed is a smaller, inverse, and real image. The image is thinner than the object because the axial magnification (see Section 1.9) is less than 1.

Figure 1.7 illustrates the first property of the nodal points by tracing the chief ray through the lens. (Figures 14.11 and 14.12 in Section 14.6 demonstrate the second property of nodal points with two Zemax-generated, accurate raytracing diagrams.)

1.6 Rays and Planes

Raytracing is a geometrical optics technique widely used in the design and analysis of optical imaging systems. For a given object and lens, the approximate location, size, and orientation of the image formed by the lens can be found by tracing a few rays from the object through the lens. Although raytracing is now mainly performed by computers and optical software, familiarity with raytracing is still helpful.

To effectively trace rays, some special rays and planes must first be defined. Each of them has a special name and meaning. Figures 1.7 and 1.8 use the simplest one-lens example to illustrate these rays and planes.

1.6.1 Tangential planes and tangential rays

For a given lens and object, two orthogonal planes and several types of rays are defined. The tangential plane is defined by the optical axis and the object point from which the ray originated. In Fig. 1.7, the tangential plane is the plane of the page. A tangential plane is also called a meridional plane.

A ray that travels in the tangential plane is a tangential ray or meridional ray. All four rays shown in Fig. 1.7 are tangential rays or meridional rays. There are different types of tangential rays, such as chief and marginal rays.

The chief ray travels from the top of the object through the center of the aperture stop (see Section 1.7). In Fig. 1.7, the lens aperture is the aperture stop. A chief ray is also called a principal ray and is frequently used in raytracing.

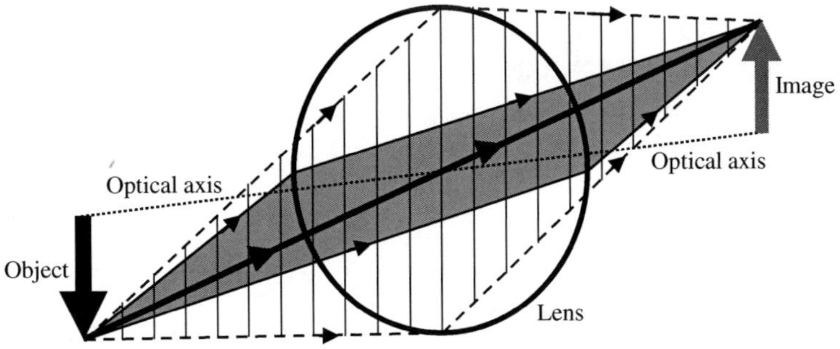

Figure 1.8 Sagittal plane and sagittal rays. The plane marked by the thin vertical lines is the tangential plane. Dashed lines indicate tangential rays. The shaded plane is the sagittal plane. All of the rays in the sagittal plane are sagittal rays, drawn with solid lines. The intersection line of the two planes is the chief ray, denoted by the thick solid line.

A marginal ray travels from the point where the object and the optical axis cross the edge of the aperture stop. In Fig. 1.7, the lens edge is the aperture stop edge. Tracing at least two tangential rays from an object point through a lens can determine the location of the image point, as shown in the figure.

1.6.2 Sagittal planes, sagittal rays, and skew rays

There are an infinite number of planes that are perpendicular to the tangential plane. Among these planes, only one plane contains the chief ray; such a plane is called the sagittal plane, as shown in Fig. 1.8.

Rays traveling in the sagittal plane are sagittal rays. The chief ray is the intersecting line of the tangential and sagittal planes. Technically, it is also a sagittal ray, but it is usually considered as a tangential ray. Sagittal rays are also called transverse rays. Sagittal rays are rarely used in manual raytracing because it is difficult to conceive and draw rays perpendicular to the plane of the page. However, modern computers and optical software can easily trace sagittal rays.

Skew rays are those rays that neither travel in the tangential plane nor cross the optical axis anywhere, and they are not parallel to the optical axis. Sagittal rays, except the chief ray, are special skew rays.

1.7 Stops and Pupils

1.7.1 Definitions

Stops are important optical components in an optical system. Pupils are the images of stops. Most image lenses have an aperture stop and a field stop. The aperture stop sets the largest cone angle the lens imposes on the object. The field stop sets the largest field angle the lens can see. These stops can significantly affect the lens characteristics and should be understood. The stops can be either lens aperture edges or some other structures.

Most image lenses also have an entrance pupil and an exit pupil. These two pupils are the images of the aperture stop formed by the lens at the object side and image side, respectively. Manually locating the stops and pupils and determining their sizes can be complex, whereas computer and optical design software can easily perform these tasks. Three examples are included here to explain the process.

1.7.2 Example 1: a schematic microscope

Figure 1.9 includes a microscope that consists of an objective and an eyepiece to illustrate the concepts of an aperture stop, field stop, entrance pupil, and exit pupil. In Fig. 1.9(a), the aperture of the objective restrains the cone angle of the microscope imposing on the object and is the aperture stop. The aperture of the eyepiece restrains the maximum allowed field angle. For example, the rays from the solid square on the object plane are completely blocked by the edge of the

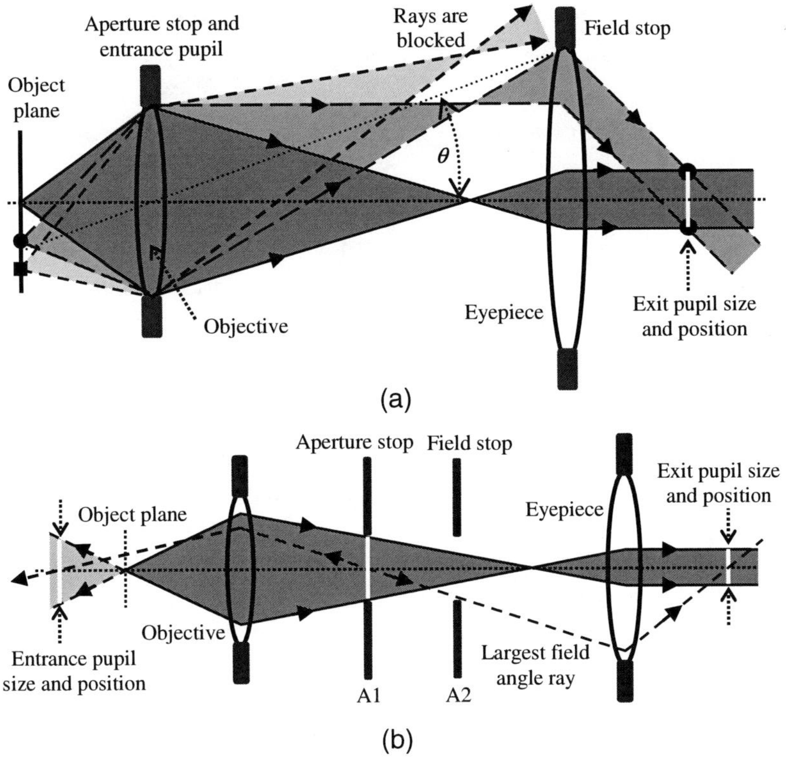

Figure 1.9 Schematics of two microscopes: (a) The microscope consists of an objective and an eyepiece. The aperture of the objective is the aperture stop and the entrance pupil. The aperture of the eyepiece is the field stop. The image of the objective aperture at the image space (right side of the microscope) is the exit pupil. (b) A similar microscope with two additional apertures A1 and A2 in the optical path. The image of A1 at the image space is the exit pupil.

eyepiece, and all of the rays from the solid dot on the object plane can travel through the eyepiece. The aperture of the eyepiece is the field stop. The largest field angle allowed by this field aperture is θ, marked in Fig. 1.9(a).

The entrance pupil of a lens is defined as the image of the aperture stop at the object space. In Fig. 1.9(a), there is no lens at the left side of the aperture stop. The aperture stop itself is the entrance pupil.

The exit pupil of a lens is defined as the image of the aperture stop at the image space. In Fig. 1.9(a), the exit pupil position and size can be found by tracing four rays rightward. The two rays from the top of the aperture stop (objective) meet at the point marked by a dot. The two rays from the bottom of the aperture stop meet at the point marked by another dot. These two dots determine the exit pupil location and size, as shown by the white vertical line linking the two dots.

Most lenses that generate an image for direct viewing have their exit pupil outside of the lenses. If a viewer places an eyeball at the exit pupil, they only need to rotate their eyeball to view the entire scene; otherwise, they must move their head to view the entire scene.

1.7.3 Example 2: a schematic microscope with two additional apertures

For other lenses with different structures, the aperture stop and field stop locations and sizes can be very different from those shown in Fig. 1.9(a). Figure 1.9(b) shows a microscope similar to the one in Fig. 1.9(a) but with two additional apertures A1 and A2 in the optical path. Aperture A1, rather than the aperture of the objective, restrains the cone angle of rays from the object. A1 is the aperture stop.

A more standard approach to determine the field stop and the pupils traces rays from the center of the aperture stop both right and left, as shown by the dashed line in Fig. 1.9(b). Aperture A2, rather than the aperture of the eyepiece, restrains the largest field angle. A2 is the field stop.

By tracing the dashed-line ray left, the ray will cross the optical axis. The cross-point is the location of the entrance pupil. The size of the ray bundle at the entrance pupil location is the entrance pupil size, as marked by the white line.

By tracing this dashed-line ray right, it will cross the optical axis again. The cross-point is the location of the exit pupil. The size of the ray bundle at the exit pupil location is the exit pupil size, as marked by the white line.

In real lenses, the entrance and exit pupils may have severe aberrations caused by lens aberrations. The pupils can be curved. The locations of the pupils can be far away or at infinity. The pupil sizes can be large or even infinite. Minimizing pupil aberration is one of the optical design goals.

1.7.4 Example 3: a real double Gauss lens

Figure 1.10 presents a Zemax-generated raytracing diagram of a real double Gauss lens to further illustrate stops, pupils, and some other lens parameters. Two notes to make here:

1. Both the aperture stop and field stop are real physical components. While both the entrance and exit pupils are images, either real or virtual, they can appear anywhere or disappear (no image can be formed).
2. The positions and sizes of stops and pupils obtained here by using manual raytracings are not accurate but are sufficient for illustration. Any optical design software can provide much more accurate results.

Most real-image lenses use a size-adjustable iris to limit the amount of light that passes through the lens or adjust the *F*-number of the lens. This iris is placed at the aperture stop and is marked in Figs. 1.10(a–c).

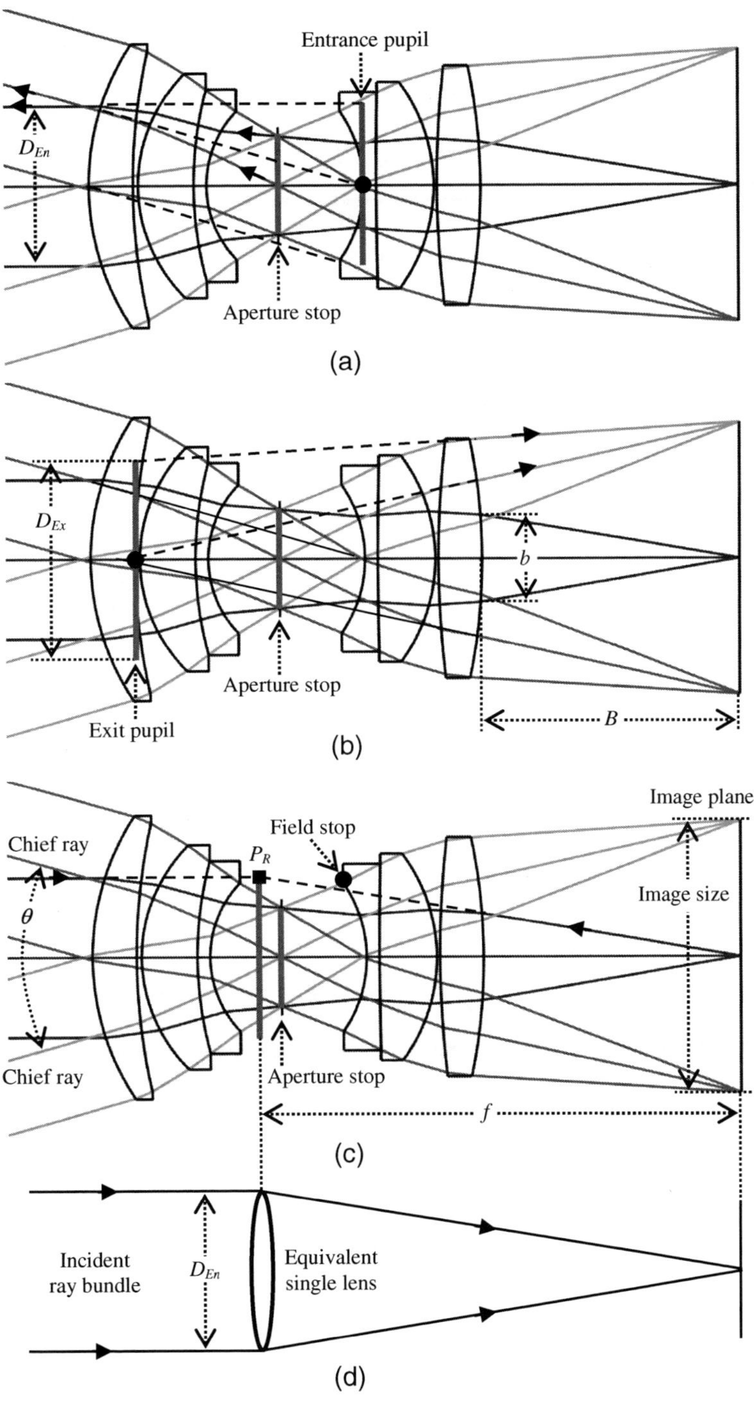

Figure 1.10 (a)–(c) Zemax-generated raytracing diagram of a double Gauss lens. (d) A single-lens model for the double Gauss lens.

In Fig. 1.10(a), where a ray is traced from the center of the aperture stop leftward, as indicated by the arrow, the ray never crosses the optical axis. However, the backward extension of the ray marked by the dashed line crosses the optical axis at the point marked by the black dot. This point is the location of the entrance pupil. If a ray were traced from the top of the aperture stop leftward, as indicated by another arrow, the ray again never crosses the optical axis. However, the backward extension of the ray marked by the dashed line passes through the plane where the entrance pupil is located and determines the entrance pupil size D_{En}.

In Fig. 1.10(b), where a ray is traced from the center of the aperture stop rightward, as indicated by the arrow, the ray never crosses the optical axis. However, the backward extension of the ray marked by the dashed line crosses the optical axis at the point marked by the black dot. This point is the location of the exit pupil. If a ray were traced from the top of the aperture stop rightward, as indicated by the arrow, the ray again never crosses the optical axis, but the backward extension of the ray marked by the dashed line passes through the plane where the exit pupil is located and determines the exit pupil size D_{Ex}. It is not rare that the exit pupil is located at the left side of the entrance pupil.

The field stop is relatively easy to find, as marked by the black dot in Fig. 1.10(c), because this point limits the largest field angle.

The right principal plane of this double Gauss lens can be found by extending forward the incident ray and extending backward the focused ray, as indicated and marked by the arrows and dashed lines, respectively, in Fig. 1.10(c). The cross-point of these two extended rays, marked by the solid black square, determines the principal-plane location and size. The distance between the principal plane and the image plane is the focal length f. (The determination of the left principal plane is omitted.)

Figures 1.10 also illustrate a few other lens parameters. The back working distance B and image space F-number B/b are marked in Fig. 1.10(b). The angle between the two chief rays that have the largest incident angles is the maximum field angle θ of this lens, as marked in Fig. 10(c); the image size and plane are also marked.

1.7.5 Single-lens model of a complex lens

This double Gauss lens and any other complex lenses can be approximately represented by a single lens model, as shown in Fig. 1.10(d). This single lens is located at the right principal plane location of the double Gauss lens, and its size is the same as the entrance pupil size of the Gauss lens. The focal length, the image plane location, and the size of this single lens are the same as those of the double Gauss lens.

The single-lens model can provide insight into complex lenses and significantly simplify the qualitative analysis.

1.8 Analytical Modeling of a Lens and Rays

1.8.1 Some comments

Manually tracing rays through a lens takes a lot of effort and is rarely performed. People either use computers and optical design software to trace rays for accurate results or solve analytical equations for approximate results. Analytical modeling is a simple and approximate tool that has a unique advantage and is still widely used today for fast estimations.

A few equations about lens focal lengths and how a lens manipulates rays have been derived for analytical modeling. This section discusses the derivation and applications of these equations. Note that it is more important to be able to skillfully use these equations than to derive them. Readers who do not have enough time and/or interest in the mathematical details can skip the derivation process without affecting their ability to work through the book.

1.8.2 Rules of sign for lens focal length and surface radius of curvature

Before proceeding, the rules of sign for the lens focal length and surface radius of curvature must be clearly stated, using the lens shown in Fig. 1.11 as an example:

1. Any positive lenses that focus rays have a positive focal length.
2. Any negative lenses that divert rays have a negative focal length.
3. When the vertex of a lens surface is to the left of the rest of the surface, the surface radius of curvature is positive. According to this rule, R_1 in Fig. 1.11 is positive.
4. When the vertex of a lens surface is to the right of the rest of the surface, the surface radius of curvature is negative. According to this rule, R_2 in Fig. 1.11 is negative.

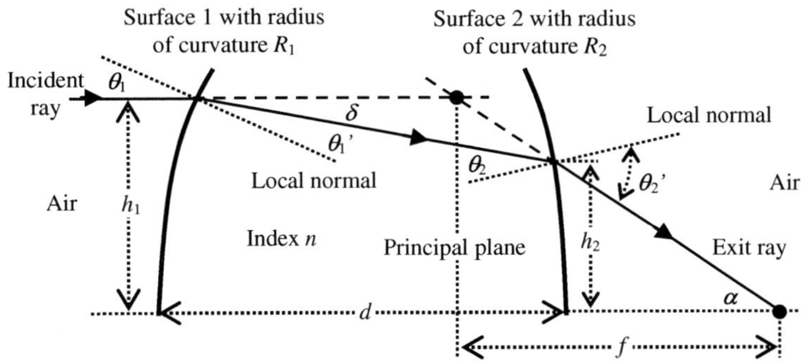

Figure 1.11 Derivation of the paraxial focal length for a thick lens.

1.8.3 Derivation of the paraxial focal length for a lens

The lens in Fig. 1.11 has a central thickness d, two surface radii of curvature R_1 and R_2, respectively, and a lens material index of n. The following process derives the paraxial focal length of this lens. Consider a ray parallel to the optical axis of the lens incident on surface 1 of the lens with ray height h_1 and incident angle θ_1, as shown in Fig. 1.11; the paraxial condition leads to

$$\theta_1 = \frac{h_1}{R_1}. \tag{1.3}$$

According to Snell's law, with paraxial approximation and Eq. (1.3), the ray refracted by surface 1 has an angle θ_1', given by

$$\begin{aligned}\theta_1' &= \frac{\theta_1}{n}\\ &= \frac{h_1}{R_1 n}.\end{aligned} \tag{1.4}$$

The difference between θ_1 and θ_1' is given by Eqs. (1.3) and (1.4):

$$\begin{aligned}\delta &= \theta_1 - \theta_1'\\ &= \frac{h_1}{R_1}\left(1 - \frac{1}{n}\right).\end{aligned} \tag{1.5}$$

The height of the refracted ray hitting surface 2 is found by using Eq. (1.5) and the paraxial condition

$$\begin{aligned}h_2 &= h_1 - d\delta\\ &= h_1\left[1 - \frac{d}{R_1}\left(1 - \frac{1}{n}\right)\right].\end{aligned} \tag{1.6}$$

The incident angle of this ray on surface 2 is

$$\begin{aligned}\theta_2 &= \frac{h_2}{-R_2} + \delta\\ &= h_1\left[\frac{1}{-R_2} - \frac{d}{R_1(-R_2)}\left(1 - \frac{1}{n}\right) + \frac{1}{R_1}\left(1 - \frac{1}{n}\right)\right],\end{aligned} \tag{1.7}$$

where the negative sign of R_2 is from the rule of sign stated earlier. According to Snell's law, the ray refracted by surface 2 has an angle θ_2', given by

$$\theta_2' = n\theta_2. \tag{1.8}$$

This ray crosses the optical axis with angle α, which is given with paraxial approximation by

$$\begin{aligned}\alpha &= \theta_2' - \frac{h_2}{-R_2}\\ &= n\theta_2 + \frac{h_2}{R_2}\\ &= nh_1\left[\frac{1}{R_2} + \frac{d}{R_1R_2}\left(1-\frac{1}{n}\right) + \frac{1}{R_1}\left(1-\frac{1}{n}\right)\right] + \frac{h_1}{R_2}\left[1-\frac{d}{R_1}\left(1-\frac{1}{n}\right)\right]\\ &= h_1(n-1)\left[\frac{1}{R_1} - \frac{1}{R_2} + \frac{(n-1)d}{nR_1R_2}\right].\end{aligned} \tag{1.9}$$

The last step to derive Eq. (1.9) requires some rearranging and cleaning up of the equation. The paraxial focal length f is related to h_1 and α by the following equation, as shown in Fig. 1.11:

$$\begin{aligned}\frac{1}{f} &= \frac{\alpha}{h_1}\\ &= (n-1)\left[\frac{1}{R_1} - \frac{1}{R_2} + \frac{(n-1)d}{nR_1R_2}\right].\end{aligned} \tag{1.10}$$

Equation (1.10) is the final result: the widely used paraxial focal length of a thick lens.

For any lens, once the two surface radii of curvature R_1 and R_2, the central thickness d, and the material index n are known, the paraxial focal length f of the lens can be easily calculated using Eq. (1.10).

When $d \ll (R_1R_2)^{0.5}$, Eq. (1.10) reduces to the widely used thin lens form

$$\frac{1}{f} \approx (n-1)\left(\frac{1}{R_1} - \frac{1}{R_2}\right). \tag{1.11}$$

1.8.4 Derivation of the thin lens equation for a thin lens

A thin lens is a simplified concept. A thin lens has properties such that its central thickness is much less than its focal length and the two principal planes of the lens can be considered to coincide. The thin lens equation is an analytical tool to analyze the effects of a lens on rays that travel through the lens. The derivation process of the thin lens equation for one lens is explained in this subsection.

Consider a positive lens with focal length f that focuses rays from an object with height h and a distance o from the lens, as shown in Fig. 1.12.

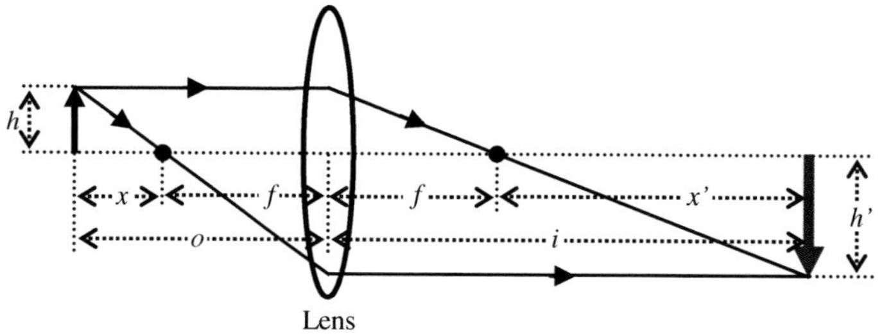

Figure 1.12 Derivation of the thin lens equation. *f* is the focal length of the lens, and *o* and *i* are the object and image distances, respectively. The lens position is its center position.

An image with height h' is formed by the lens at a distance i from the lens. Equations (1.12) and (1.13) can be written using some triangular geometrics:

$$\frac{h}{x} = \frac{h'}{f}, \tag{1.12}$$

$$\frac{h}{f} = \frac{h'}{x'}. \tag{1.13}$$

Equation (1.14) can be found by combining Eqs. (1.12) and (1.13) to eliminate h and h':

$$xx' = f^2. \tag{1.14}$$

Equation (1.14) is one form of the thin lens equation, but a more popular form can be found by inserting $x = o - f$ and $x' = i - f$ into Eq. (1.14); the result after some rearrangements is

$$\frac{1}{o} + \frac{1}{i} = \frac{1}{f}, \tag{1.15}$$

which is the widely used thin lens equation.

The object is conventionally placed to the left of the lens, and the rays travel rightward from the object through the lens to form an image at the right of the lens a distance i away from the lens. In such a case, $o > 0$ by definition. By definition, $i > 0$ means the image is on the right of the lens, and $i < 0$ means a virtual image is formed to the left of the lens. f is the focal length of the lens; $f > 0$ and $f < 0$ mean a positive lens and a negative lens, respectively. The thin lens equation can be used to conveniently analyze the location and size of an image formed by a lens. The results are accurate enough for many applications.

1.8.5 Derivation of the focal length of two thin lenses

When there are two thin lenses with focal length f_1 and f_2, respectively, and a small distance $d \ll f_1$ and $d \ll f_2$ between them, the focal length of the two lenses combined can be derived by the following process.

Consider a ray emitted by an on-axis object point located a distance o_1 from Lens 1, as shown in Fig. 1.13. If there is no Lens 2, this ray will be focused by Lens 1 at a point marked by the grey dot with a distance of i_1 from Lens 1. For Lens 1, the thin lens equation [Eq. (1.15)] takes the form

$$\frac{1}{o_1} + \frac{1}{i_1} = \frac{1}{f_1}$$

or

$$i_1 = \frac{f_1 o_1}{o_1 - f_1}. \tag{1.16}$$

For Lens 2, the grey point is a virtual object point with a distance

$$o_2 = i_1 - d \tag{1.17}$$

from Lens 2. The thin lens equation for Lens 2 takes the form

$$\frac{1}{-o_2} + \frac{1}{i_2} = \frac{1}{f_2}, \tag{1.18}$$

where the negative sign of o_2 comes from the fact that the virtual object point is located to the right of Lens 2. Equations (1.17) and (1.18) are combined to eliminate o_2 and solve for i_1:

$$i_1 = \frac{i_2 f_2}{f_2 - i_2} + d. \tag{1.19}$$

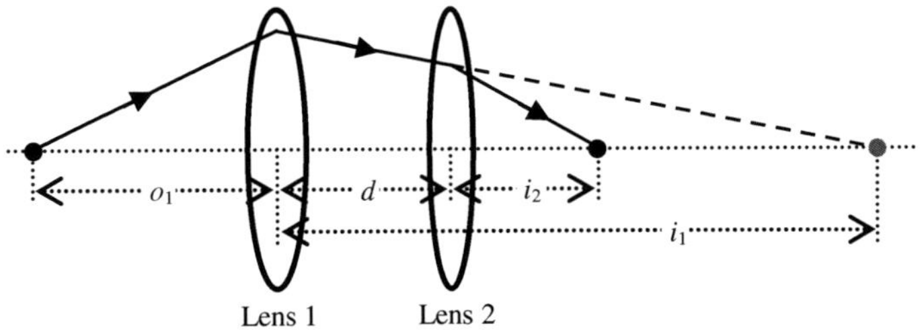

Figure 1.13 Derivation of the focal length of two thin lenses with distance d between them.

Equations (1.16) and (1.19) are combined to illuminate i_1:

$$\frac{i_2 f_2}{f_2 - i_2} + d = \frac{o_1 f_1}{o_1 - f_1}. \tag{1.20}$$

In the case of $o_1 \to \infty$, i_2 becomes the right focal length f_r of the two lenses. With some rearrangements, Eq. (1.20) reduces to

$$\frac{1}{f_r} = \frac{f_1 + f_2 - d}{f_2(f_1 - d)}. \tag{1.21}$$

For $d \to 0$, Eq. (1.21) reduces to the well-known form

$$\frac{1}{f_r} = \frac{1}{f_1} + \frac{1}{f_2}. \tag{1.22}$$

In the case of $i_2 \to \infty$, o_1 becomes the left focal length f_l of the two lenses. With some rearrangements, Eq. (1.20) reduces to

$$\frac{1}{f_l} = \frac{f_1 + f_2 - d}{f_1(f_2 - d)}. \tag{1.23}$$

For $d \to 0$, Eq. (1.23) also reduces to the well-known Eq. (1.22).

1.8.6 Application examples of the thin lens equation for a positive lens

Five cases about a positive lens are discussed in this section with the help of Fig. 1.14, where the focal point is marked by F, and the focal length f is not marked. In all of these cases, the object is on the lens optical axis, the rays are symmetric about the optical axis, and the image point is on the optical axis. These case studies are particularly useful for analyzing a laser beam propagating through a lens.

- Case 1: A convergent ray bundle is incident on the lens, as shown in Fig. 1.14(a). No point light source can emit such a ray bundle, and therefore there is no object point. However, the forward extensions of the incident rays cross the optical axis of the lens at a point marked by the open square. This point is a virtual object point. In such a case, $o < 0$. The focused spot of the rays is the image point. From Eq. (1.15) it can be found that $i < f$, and the rays will be focused at a point inside the focal length, as marked by the solid square in Fig. 1.14(a).
- Case 2: $o \to \infty$, the rays are from infinity and parallel to the lens optical axis when they reach the lens, as shown in Fig. 1.14(b). Equation (1.15) shows that $i = f$. The rays are focused at the right focal point of the lens.

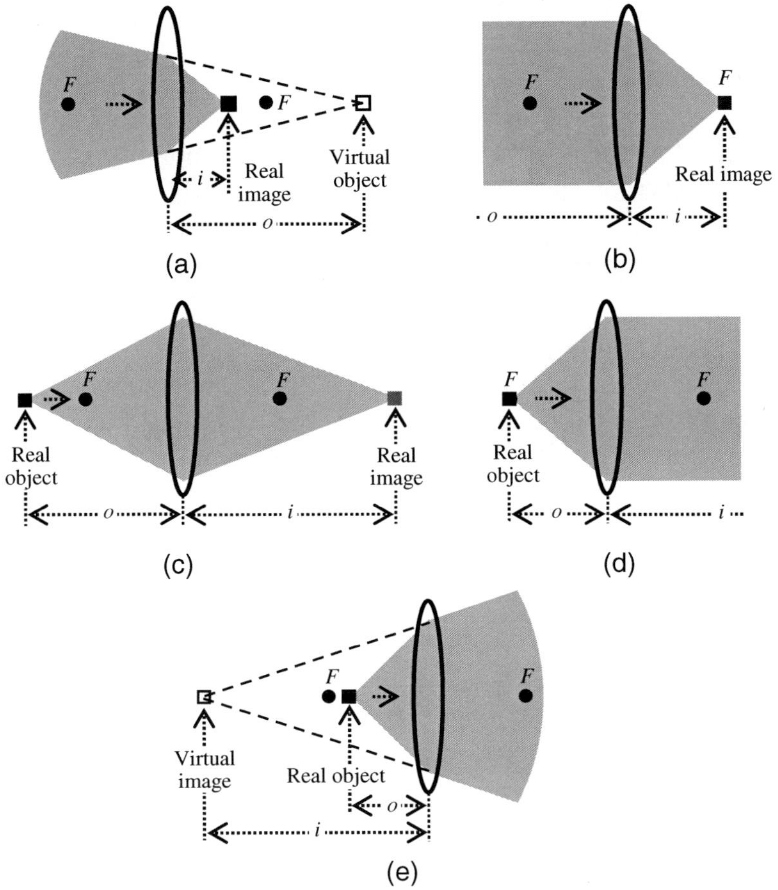

Figure 1.14 Illustration of the thin lens equation [Eq. (1.15)] for a positive lens with focal points marked by *F* and focal length $f > 0$. (a) When the ray bundle is convergently incident on the lens, the forward extension of the incident rays crosses the optical axis at a point marked by the open square. (b) When the object point is infinitely away, the rays are parallel to the optical axis as they reach the lens and will be focused by the lens on its focal point. (c) When the object point is outside the focal length, the rays will be focused by the lens at a point beyond the focal point. (d) When the object point is at a focal point, the rays will be collimated by the lens, which means the image is at infinity. (e) When the object point is inside the focal length, the rays passing through the lens will still be divergent. A virtual image marked by the grey open square can be found by extending the rays backward.

- Case 3: A real object is marked by the solid square shown in Fig. 1.14(c). The object distance is $o > f$. According to Eq. (1.15), $i > f$. The rays are focused at a point beyond the right focal point, as marked by the grey solid square.
- Case 4: A real object marked by the solid square is at the focal point, that is $o = f$, and $i \rightarrow \infty$, according to Eq. (1.15). The situation is shown in Fig. 1.14(d). The image appears at infinity, which means rays are collimated.

- Case 5: The real object marked by the solid square is inside the focal length, as shown in Fig. 1.14(e), i.e., $o < f$ and $i < 0$, according to Eq. (1.15). The rays passing through the lens are divergent. No real image is formed. But the backward extensions of the rays cross the optical axis at a point marked by the grey open square; this point is the virtual image of the object. Since the virtual image is to the left of the lens, $i < 0$.

1.8.7 Application examples of the thin lens equation for a negative lens

Four cases about a negative lens are discussed in this subsection with the help of Fig. 1.15, where the focal point is marked by F, and the focal length f is not marked. In all these cases, the object is on the lens optical axis, the rays are symmetric about the optical axis, and the image point is on the optical axis. Note that for a negative lens, the focal length f in Eq. (1.15) is negative.

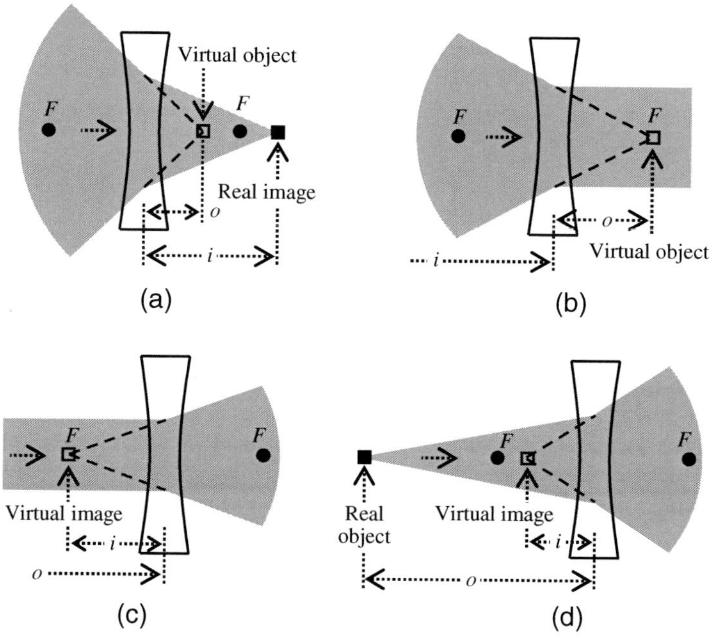

Figure 1.15 Illustration of thin lens equation [Eq. (1.15)] for a negative lens with focal points marked by F and focal length $f < 0$. (a) When a convergent ray bundle is incident on the lens, the forward extensions of the incident rays cross the optical axis at a point marked by the open square, which is a virtual object point. This virtual object point is inside the focal length, rays are focused. (b) Similar to the situation shown in (a), but the virtual object is at the focal point, and the rays are collimated. (c) When the object point is infinitely away, the rays passing through the lens are divergent. The backward extensions of the rays will cross the optical axis on the focal point and form the virtual image. (d) When the object point is not infinitely away, the extension of the exit rays will cross the optical axis at a point inside the focal length. This point is the virtual image.

- Case 1: A convergent ray bundle is incident on the lens, as shown in Fig. 1.15(a). No point light source can emit such a ray bundle, and therefore there is no object point. However, the forward extensions of the incident rays cross the optical axis of the lens at a point marked by the open square. This point is a virtual object point. In such a case, $o < 0$. For $o < 0$ and $f < 0$, i can be either positive or negative, based on Eq. (1.15). In Fig. 1.15(a), the virtual object point is inside the focal length; it can be found from Eq. (1.15) that $i > 0$. The rays will be focused at a point to the right of the lens marked by a solid square.
- Case 2: Similar to Case 1, but the virtual object point is at the focal point, as shown in Fig. 1.15(b). Equation (1.15) leads to $o = f$ and $i \to \infty$. The rays are collimated by the lens.
- Case 3: The object is at infinity, i.e., $o \to \infty$. Then, $i = f < 0$ according to Eq. (1.15). $i < 0$ means that there is a virtual image at the left focal point of the lens, as shown by the backward-extended rays and the grey open square in Fig. 1.15(c).
- Case 4: The object is at finite $o > 0$. Then, $0 > i > f$, according to Eq. (1.15). $i < 0$ means that there is a virtual image to the left of the lens, and $i > f$ means that $|i| < |f|$ since f is negative; the virtual-image point position is inside the focal plane, as shown by the backward-extended rays and the grey open square in Fig. 1.15(d).

1.9 Lateral and Axial Magnifications of Lenses

So far, the object being considered is a point. For a real object of a certain size, the image formed by a lens often has a size and orientation different than those of the object because the lens has a lateral magnification M_L and axial magnification M_A. Note that all of the lens imaging drawings in this section neglect spherical aberration to avoid unnecessary complexity.

1.9.1 Definition of lateral magnifications and axial magnifications

In Fig. 1.12, the lateral magnification M_L is defined as the ratio of the image height and object height. Based on the left side of Fig. 1.12 and some triangular geometry, M_L can be found to be

$$\begin{aligned} M_L &= -\frac{h'}{h} \\ &= \frac{f}{f - o}. \end{aligned} \tag{1.24}$$

Equations (1.15) and (1.24) are combined to eliminate f, and M_L is found to be

$$M_L = -\frac{i}{o}. \tag{1.25}$$

There are two notes to make about M_L:

1. A negative M_L means that the image has an orientation opposite to the object, which is the case shown in Fig. 1.12. Since h and h' in Fig. 1.2 are defined as positive values, a negative sign must be placed in front of M_L in Eq. (1.24).
2. Equation (1.24) states that for a given lens, f is fixed, and changing o can change the value of M_L.

Lateral magnification is also called transverse magnification. Axial magnification is defined by $M_A = \Delta i/\Delta o$, which can be found by differentiating Eq. (1.15). The result is

$$\begin{aligned} M_A &= \frac{\Delta i}{\Delta o} \\ &= -\frac{i^2}{o^2} \\ &= -M_L^2. \end{aligned} \tag{1.26}$$

Since the value of M_L^2 is always positive, M_A is always negative, which means that when an object is axially extended towards the lens, the corresponding image is extended away from the lens, as shown in Fig. 1.16. Axial magnification is also called longitudinal magnification.

1.9.2 Schematic examples of an image formed by a positive lens

Figure 1.16 is basically a re-plot of Fig. 1.7 but with the intention to illustrate the lateral and axial magnifications. The object is placed outside the focal length of this positive lens. The image found by some raytracings has an orientation opposite that of the object and has a height approximately half the object height. Thus, $M_L = -h'/h \approx -0.5$, and $M_A = -M_L^2 \approx -0.25$. Figure 1.16

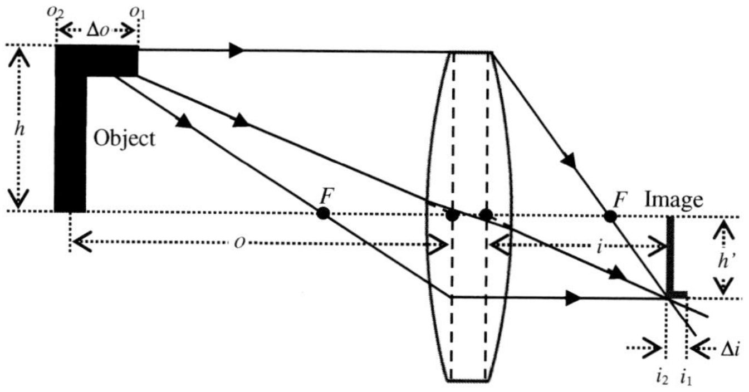

Figure 1.16 A positive lens forms a real and negative image of an object placed outside the focal length. Here, $M_L = h'/h \approx -0.5$ and $M_A = -M_L^2 \approx -0.25$. The image is thinner than the object.

shows that $o_2 > o_1$, while $i_2 < i_1$, which explains the negative value of M_A. The image is drawn approximately to the right proportion and is thinner than the object because M_A and M_L have different values.

Figure 1.17 shows another example of a positive lens forming an image. The object is placed inside the focal length of the lens. The image formed has the same orientation as the object and has a height about three times larger than the object height. This h' in Fig. 1.17 is defined as negative (opposite to h' in Fig. 1.16), so $M_L = -h'/h \approx 3$, and $M_A = -M_L^2 \approx -9$. Again, the image is much thicker than the object because M_A and M_L have different values.

1.9.3 Schematic examples of an image formed by a negative lens

Figure 1.18 shows an example of a negative lens forming an image. The object is placed outside the focal length of the lens. The image formed has the same orientation as the object and has a height approximately one-third of the object height: $M_L = -h'/h \approx 1/3$, and $M_A = -M_L^2 \approx -1/9$. The image is much thinner than the object because M_A and M_L have different values.

Figure 1.19 shows another example of a negative lens forming an image. The object is placed right at the focal point of the lens. The image formed has the same orientation as the object and has a height approximately half the object height. Thus, $M_L = -h'/h \approx 1/2$, and $M_A = -M_L^2 \approx -1/4$. The image is thinner than the object because M_A and M_L have different values.

There are two unique phenomena about negative lenses that have a negative focal length:

1. $f < 0$ and $f - o < 0$ (since the object is at the left of the lens $o > 0$) produce an M_L that is always positive, according to Eq. (1.24) and shown in Figs. (1.18) and (1.19).

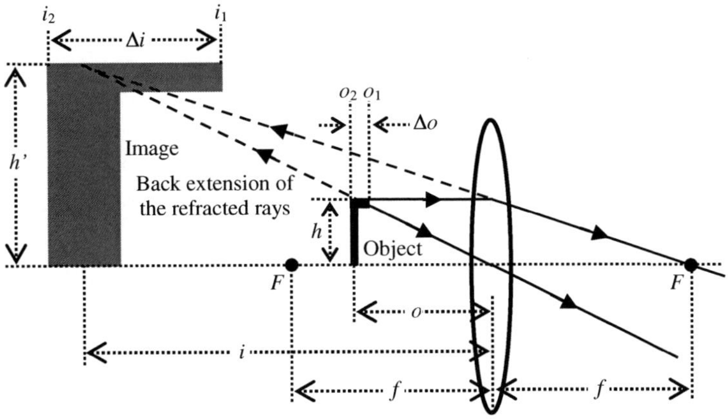

Figure 1.17 A positive lens forms a virtual and positive image of an object placed inside the focal length. Here, $M_L = -h'/h \approx 3$ and $M_A = -M_L^2 \approx -9$. The image is much thicker than the object.

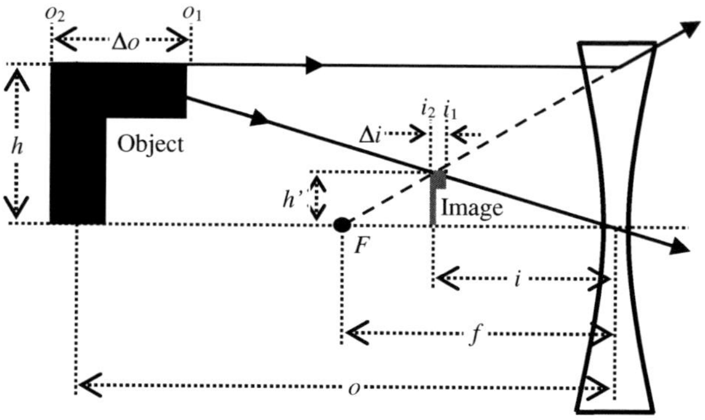

Figure 1.18 A negative lens forms a virtual and positive image of an object placed outside the focal length. In this setup, $M_L = -h'/h \approx 1/3$, and $M_A = -M_L^2 \approx -1/9$. The image is much thinner than the object.

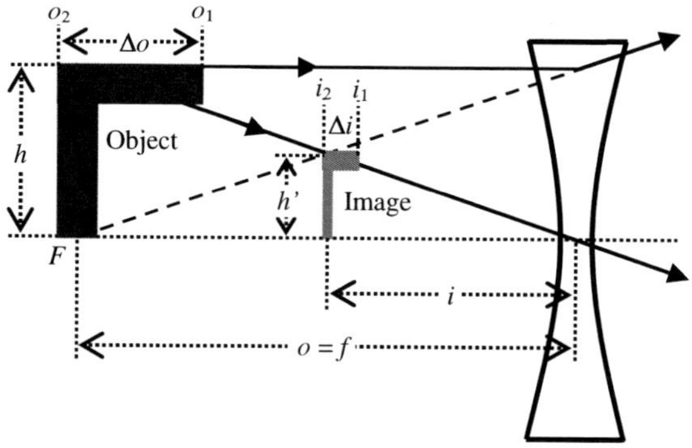

Figure 1.19 A negative lens forms a virtual and positive image of an object placed at the focal point. Here, $M_L = -h'/h \approx 1/2$, and $M_A = -M_L^2 \approx -1/4$. The image is thinner than the object.

2. $i = fo/(o-f)$ is always negative, and the images formed by a negative lens are always virtual images at the left of the lens, as shown in Figs. (1.18) and (1.19).

1.9.4 Distortion of a 3D image formed by a perfect lens

Figures (1.16)–(1.19) are drawn with the assumption that the lateral and axial magnifications are constant for a given lens and object. This is true only when the axial size of the object $\Delta o = o_2 - o_1$ meets the condition $\Delta o \ll f$, so that o and i can be treated as constant numbers, and the resulting lateral and axial magnifications are constants too. However, in some cases, $\Delta o \sim f$, and the different axial portion of the object has different o values and different magnifications, both lateral and axial.

Figure 1.20 illustrates such a case. The front position of the object is at $o_2 = 2f$. According to Eqs. (1.15), (1.25), and (1.26), $i_2 = 2f$, $M_L = -1$, and $M_A = -1$, respectively. The image of the front part of the object is just an inverse of the object, nothing else. The back position of the object is at

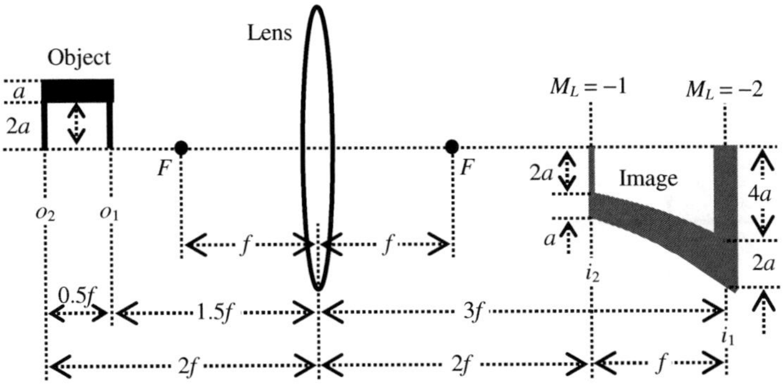

Figure 1.20 A detailed analysis and illustration of image distortion caused by lateral and axial magnifications varying as a function of the object position, even though the lens is perfect.

$o_1 = 1.5f$, which leads to $i_1 = 3f$, $M_L = -2$, and $M_L = -4$. The back part of the object has much larger magnification than the front part. The image formed in such a case will have an approximate size and shape; as shown in Fig. 1.20, the distortion is severe.

The images shown in Figs. (1.16)–(1.19) should contain severe distortions too, but these distortions are neglected for the purpose of simplification. The distortion discussed here is not caused by defects or aberrations of the lens. Even a perfect lens has such a distortion.

1.10 Mirrors

Mirrors are probably the second-most-used optical components next to lenses. Mirrors are similar and different to lenses in several aspects.

1.10.1 Reflection law

When a ray is incident on a mirror, most of the energy of the ray is reflected. The reflection law states that

1. The incident ray, the reflected ray, and the normal of the reflection surface at the point of incidence lie in the same plane.
2. The angle that the incident ray makes with the normal equals the angle that the reflected ray makes to the same normal.
3. The reflected ray and the incident ray are on the opposite sides of the normal.

The reflection that occurs at an interface of two optical media obeys the same reflection law so long as the incident angle does not exceed the total reflection angle (see Section 1.3).

Figure 1.21 illustrates the reflection law using three differently shaped interfaces: planar, convex, and concave. The optical axis of an interface is its

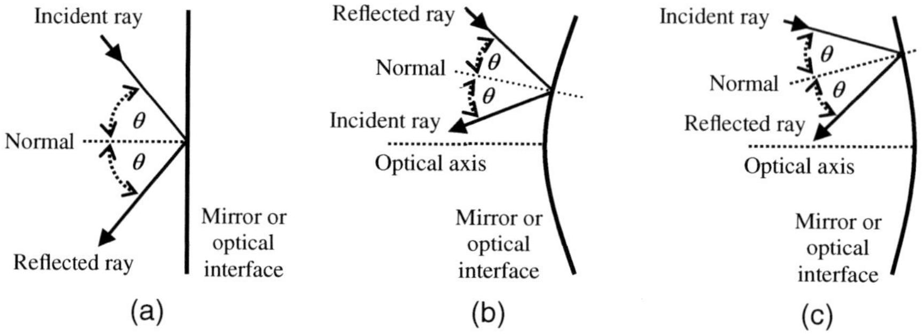

Figure 1.21 Reflection law using three differently shaped mirrors or optical interfaces: (a) planar, (b) convex, and (c) concave.

axis of symmetry. For a planar interface, the normal of the point at which the ray hits is the optical axis.

1.10.2 Mirror equation: lateral and axial magnifications

The thin lens equation [Eq. (1.15)] is also effective for mirrors and is rewritten here for convenience:

$$\frac{1}{o} + \frac{1}{i} = \frac{1}{f}, \tag{1.27}$$

where o is the object distance, i is the image distance, and f is the mirror focal length.

Only the rules of sign for mirrors are different from those for lenses:

1. A concave mirror is equivalent to a positive lens and has a positive focal length.
2. A convex mirror is equivalent to a negative lens and has a negative focal length.
3. The object is conventionally assumed at the left of the mirror with a positive object distance o.
4. The image can be either to the left or right side of the mirror with a positive or negative image distance i, respectively.
5. For spherical mirrors, their focal length always equals half their surface radius of curvature.

The lateral and axial magnifications for a mirror can be defined similar to those for a lens, only with different signs:

$$\begin{aligned} M_L &= -\frac{i}{o} \\ &= \frac{f}{f - o}, \end{aligned} \tag{1.28}$$

$$M_A = \frac{i^2}{o^2} = {M_L}^2. \tag{1.29}$$

M_A is always positive, which is an indication of the horizontal orientation of the image relative to the object, as will be shown in Fig. 1.22. It can be proven that for convex mirrors the image is always virtual.

1.11 Mirror Imaging

By tracing rays according to the reflection law and with Eqs. (1.27)–(1.29) in mind, the image of a mirror can be analyzed. Various rays can be traced from an object in front of a mirror to the mirror. Only two rays from the top of the object are usually traced. One ray is parallel to the optical axis of the mirror, and the other ray hits the cross-point of the mirror and the optical axis. The two rays are reflected by the mirror. Any other two rays can be selected to trace so long as the raytracing is convenient.

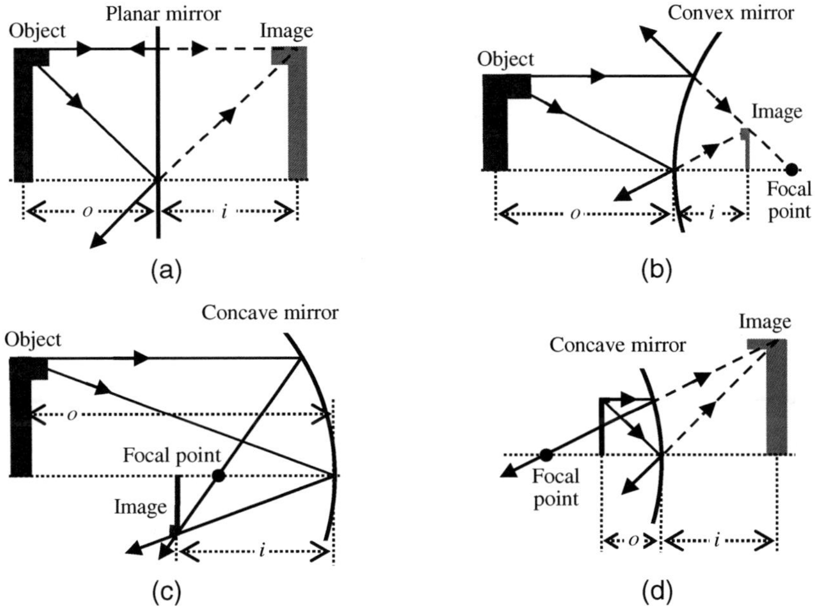

Figure 1.22 The raytracing technique is used to find the image of an object formed by three differently shaped mirrors. The solid lines represent rays. The dashed lines represent the backward extension of the rays. The spherical aberration is neglected. (a) Planar mirror. The image formed is positive, virtual, and the same size as the object. (b) Convex mirror. The image formed is positive, virtual, and smaller than the object. (c) Concave mirror with the object being placed outside the focal length. The image formed is negative, real, and smaller than the object. (d) Concave mirror with the object being placed inside the focal length. The image formed is positive, virtual, and larger than the object.

1.11.1 Planar mirror imaging

Figure 1.22 shows four examples. Figure 1.22(a) shows the raytracing for a planar mirror. Two rays are traced from the top of the object towards the mirror. The two reflected rays do not cross, but their back extensions do. The cross-point is the image of the top point of the object. Using the same raytracing technique, the image of any point on the object can be found, and thereby the whole image is found.

The image of a mirror can also be analyzed using Eqs. (1.27)–(1.29). For a planar mirror $f \to \infty$, Eqs. (1.27)–(1.29) lead to $i = -o < 0$, $M_L = 1$, and $M_A = 1$. The image is a virtual positive image at the right of the mirror. The horizontal orientation of the image shown here is $M_A > 0$.

1.11.2 Convex mirror imaging

Figure 1.22(b) shows the image formed by a convex lens using a raytracing technique with $o = -1.5f > 0$ (since for a convex lens $f < 0$ and o is always positive). Two rays are drawn from the top of the object toward the mirror. The two reflected rays do not cross. The backward extension of any ray parallel to the optical axis should pass through the focal point of the mirror if the spherical aberration of the mirror is neglected. The two backward extensions of these two rays cross and determine the position of the image of the object top. The whole image can be found using this raytracing technique and is positive, virtual, and smaller than the object.

The image can also be found analytically. Equations (1.27)–(1.29) lead to $i = 0.6f < 0$, the image is at the right side of the mirror, $M_L = -0.6f/(-1.5f) = 0.4f > 0$. The image is positive and 0.4 times the height of the object. $M_A = 0.16$, i.e., the image is much thinner than the object.

The image shown in Fig. 1.22(b) is approximately the right proportion, with M_L and M_A assumed to be constant to simplify the analysis. Both M_L and M_A are a function of o and can be treated as constant only when $\Delta o \ll f$. This situation is the same as the M_L and M_A for lenses.

1.11.3 Concave mirror imaging

Figure 1.22(c) shows the imaging of a concave mirror. The object is placed outside the focal length of the mirror with $o = 3f$. The raytracing is similar to those shown in Figs. 1.22(a) and (b), but the two reflected rays cross and determine that the image is real and negative.

Analytically, Eqs. (1.27)–(1.29) lead to $i = 1.5f$, $M_L = -0.5$, and $M_A = 0.25$. The image is thinner than the object. The drawing in Fig. 1.22(c) is approximately the right proportion.

Figure 1.22(d) shows the imaging of the concave mirror same as the mirror shown in Fig. 1.22(c). The difference is now the object is placed inside the focal length of the mirror with $o = 0.5f$. The two reflected rays do not cross, but their backward extensions cross and determine that the image is virtual and positive. From Eqs. (1.27)–(1.29), it can be found that $i = -f$, $M_L = 2$, and $M_A = 4$. The image is fatter than the object. The drawing in Fig. 1.22(d) is also approximately the right proportion.

1.12 Optical Aberrations

In most cases, an imaging lens cannot form a near-perfect image because of the presence of various aberrations in the lens. Aberrations can be significantly reduced by carefully designing the lens to use more-appropriate optical elements. However, more elements means higher cost, larger size, and heavier weight. In reality, the performance, cost, etc., of a lens must be balanced. In this section, several types of the most frequently seen optical aberrations are discussed.

Optical aberration is a complex subject. Optical books traditionally study aberrations in detail with a lot of mathematics, e.g., Born and Wolf,[7] which is necessary to manually locate the root of certain aberrations and reduce it. But with the help of optical design software, such exhaustive study is no longer necessary, particularly for those engineers and scientists who have limited time to study optical engineering. Therefore, this book includes only one section to briefly describe optical aberrations.

1.12.1 Spherical aberration

Three raytracing diagrams simulated by optical design software Zemax are plotted in Fig. 1.23(a)–(c), where the object point is on the optical axis and is infinitely away from a positive lens. The rays are parallel to the optical axis when they reach the lens and are focused by the lens to cross the optical axis. These rays are traced by Zemax with high accuracy, not just schematic drawings. All of the rays in Fig. 1.23 have a 0.55-μm wavelength.

The lens shown in Fig. 1.23(a) is an equi-convex lens made of Ohara S-FPL53 glass with an index of 1.44. Rays with different heights are focused at different locations, and a sharp and clean focused spot does not exist. The focused rays are said to have severe spherical aberration. The term "focal length" usually means the focal length for paraxial rays (rays with near-zero height), as shown in Fig. 1.23(a).

The magnitude of spherical aberration decreases as the lens surface radii of curvature increase. If Ohara S-LAH79 glass with a large refractive index of 2.00 is used to make a lens of the same size and focal length, the surface radii are longer, according to Eq. (1.10) and shown in Fig. 1.23(b), and the spherical aberration is less severe than the spherical aberration shown in Fig. 1.23(a).

If the small-index Ohara S-FPL53 glass is still used to make a lens of the same size but with a longer focal length of 50 mm, as shown in Fig. 1.23(c), the spherical aberration is barely noticeable within the resolution of the diagram since the lens surfaces are even more flat.

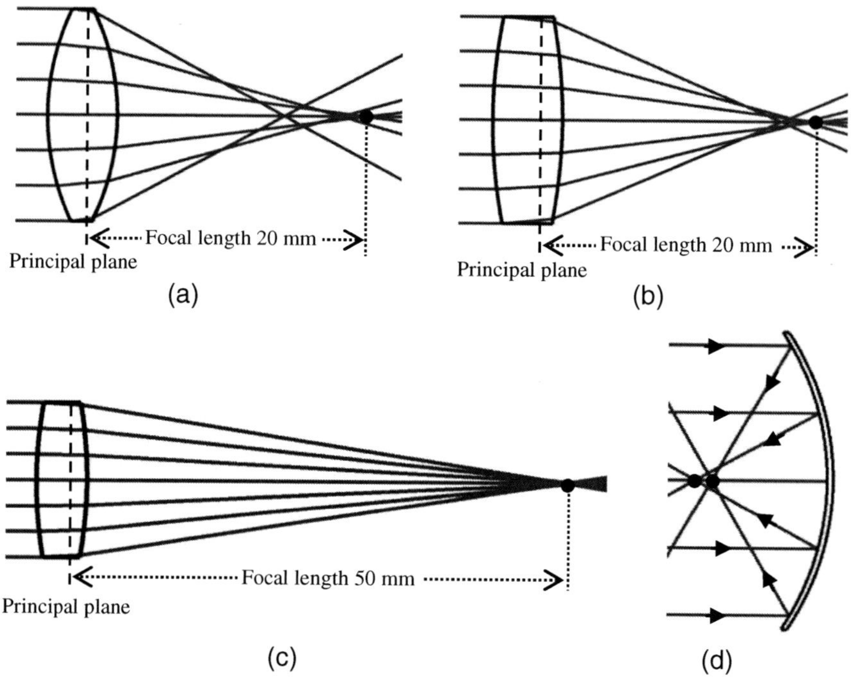

Figure 1.23 Zemax-generated raytracing diagrams to illustrate spherical aberration for (a)–(c) three lenses and (d) one mirror. The object point is on the optical axis and is infinitely distant. The rays are parallel when they reach the lens or mirror. (a) The lens has a 15-mm diameter, a 5-mm central thickness, a 20-mm focal length, and is made of Ohara S-FPL53 glass with a relatively small refractive index of 1.44. The spherical aberration is very severe. Higher rays are focused at a point closer to the lens. (b) The lens has the same size and focal length as those of the lens in (a) but is made of Ohara S-LAH79 glass with a relatively large refractive index of 2.00. The lens surfaces are flatter, and the spherical aberration is less severe. (c) The lens has a size and material the same as those of the lens in (a) except the focal length is 50 mm. The lens surfaces are flatter, and the spherical aberration is not apparent. (d) Spherical aberration also exits in the rays reflected by a mirror. The higher rays are focused at a point closer to the mirror.

The following conclusions can be drawn:

1. For a given lens size, a longer focal length will result in a smaller spherical aberration.
2. For a given lens size and focal length, a larger-refractive-index glass will result in a smaller spherical aberration. However, larger-index glasses usually have a smaller Abbe number and will lead to larger color aberration (discussed in Section 1.12.4).
3. For a given glass type and focal length, a smaller lens size will result in a smaller spherical aberration because only those rays with a smaller height can pass through the smaller lens.

Mirror reflection also has spherical aberrations; Fig. 1.23(d) shows an example. The rays are traced accurately by Zemax. Higher rays are focused by the concave spherical mirror at a point closer to the mirror, similar to lens spherical aberration. Spherical aberration is the primary aberration that degrades the image quality and must be reduced during the optical design process.

1.12.2 Coma

The spherical aberration discussed previously applies to incident rays that are parallel to the optical axis. The focused spot is on the axis. When the ray bundle incident on the lens has an angle to the optical axis, the focused spot is not on the axis, and the spherical aberration becomes "coma." Figure 1.24(a) shows the same lens as that in Fig. 1.23(a), but the ray incident angle is 20° rather than 0°. The rays are poorly focused. The focused spot has a coma-like intensity pattern. Again, only one wavelength of 0.55 μm is used to avoid color aberration.

Because coma is a type of spherical aberration, it exits the images formed by a mirror as well. Figure 1.24(b) shows a mirror the same as that in Fig. 1.23(d), but the ray incident angle is 10° instead of 0°. The focused spot has a coma-like intensity pattern too.

1.12.3 Astigmatism

The raytracing diagram shown in Fig. 1.24 is not symmetric about the optical axis. Only the rays in the plane of the page or the tangential plane are plotted.

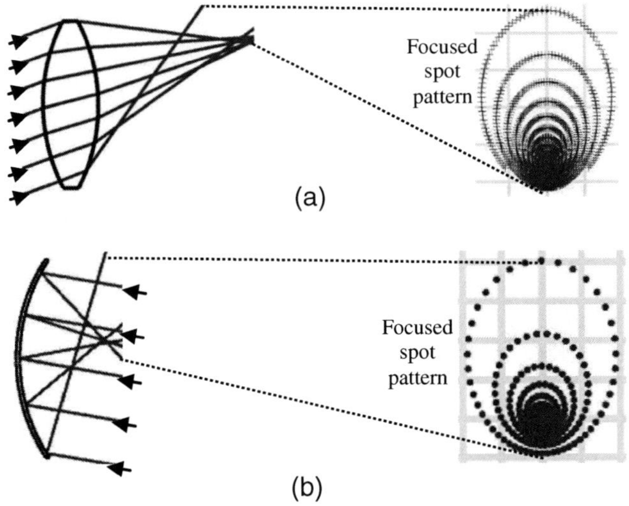

Figure 1.24 (a) The lens is the same as that in Fig. 1.23(a), but the incident ray bundle has an angle of 20° to the optical axis. (b) The mirror is the same as that in Fig. 1.23(d), but the ray bundle has an angle of 10° to the optical axis.

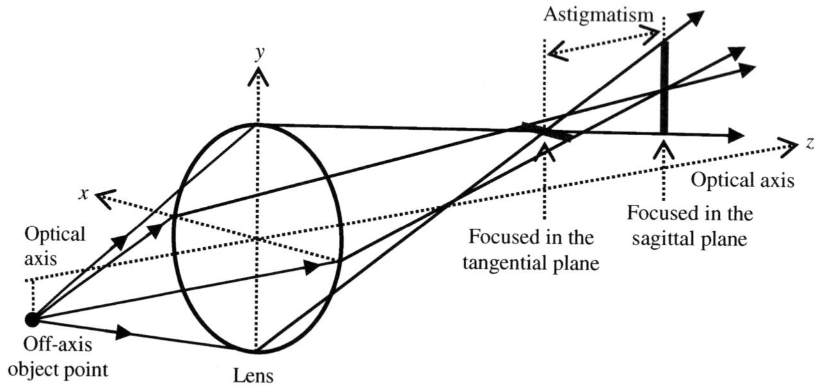

Figure 1.25 Illustration of astigmatism. When the light source is off axis in the *y-z* plane, the focusing points in the *x-z* and *y-z* planes are not at the same location. The axial distance between the two focusing points is the magnitude of astigmatism.

The behavior of the rays in the plane perpendicular to the page plane, particularly in the sagittal plane, is unknown. Figure 1.25 illustrates the situation with a 3D sketch.

The optical axis of the lens is the z axis, and the object point is at the y-z plane, which is the tangential plane. The x-z plane is the sagittal plane. As shown in Fig. 1.25, the focused spots in the tangential and sagittal planes are not at the same point. This phenomenon is common in optics and is called "astigmatism." In most cases, the focusing point in the tangential plane is closer to the lens. The axial distance between the two focusing points is the magnitude of the astigmatism. Just like spherical aberration, astigmatism will degrade the image quality and must be reduced during the optical design process.

1.12.4 Color aberrations

The refractive indices of optical materials are not constants; they vary as the wavelength varies, and a longer wavelength usually produces a smaller index: $dn/d\lambda < 0$, where n is the refractive index, and λ is the wavelength. This phenomenon is called "color dispersion." The magnitude of color dispersion is related to a parameter called the "Abbe number" (see Section 4.1.2 for details). A larger Abbe number means weaker color dispersion, which is usually desired.

Equation (1.10) shows that the focal length of a lens is proportional to the inverse of the refractive index of the lens material. So, the focal length of a lens is not a constant. A longer wavelength leads to a smaller index and longer focal length, and the results are called "color aberrations." There are two types of color aberration: longitudinal color and lateral color.

The lens used in Fig. 1.23(a) is re-plotted in Fig. 1.26(a) as an example. This lens uses Ohara S-FPL53 glass with a small refractive index of 1.44 and a large Abbe number of 94.9. This lens has a 15-mm diameter, a 5-mm central

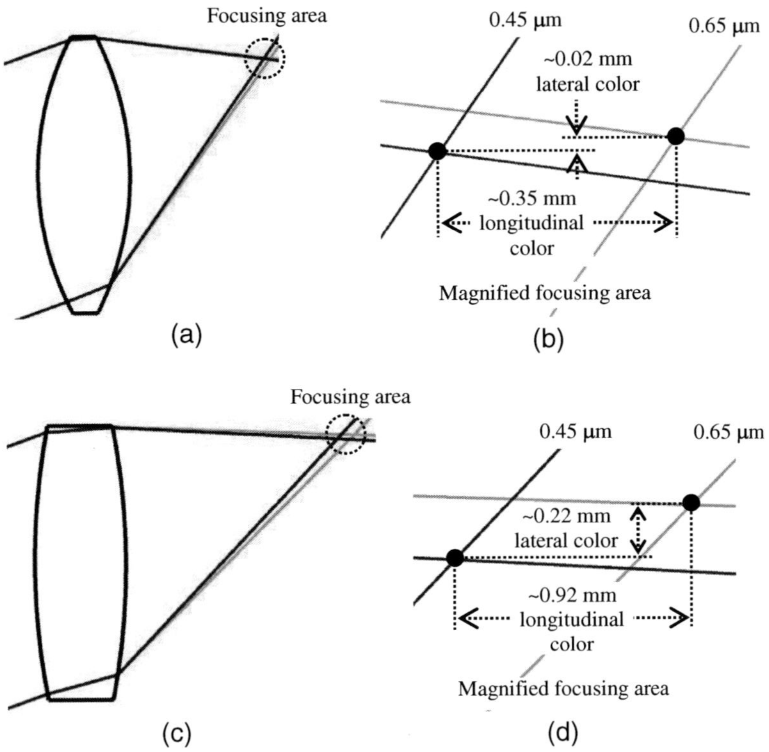

Figure 1.26 Two Zemax-generated raytracing diagrams illustrate color aberrations. Two wavelengths of 0.45 μm and 0.65 μm are used to plot. (a) A 20-mm-focal-length lens made of Ohara S-FPL53 glass focuses a ray bundle with an angle of 20° to the optical axis. The glass has a small refractive index of 1.44 and a relatively large Abbe number of 94.7. (b) The magnified focusing area marked by the dotted-line circle in (a). The longitudinal- and lateral-color-aberration magnitudes are noted. (c) A 20-mm-focal-length lens made of Ohara S-LAH79 glass focuses a ray bundle with an angle of 20° to the optical axis. The glass has a large refractive index of 2.00 and a relatively small Abbe number of 28.3. (d) The magnified focusing area marked by the dotted-line circle in (c). The longitudinal- and lateral-color-aberration magnitudes are much larger than the color aberrations shown in (b) because the smaller Abbe number results in large color aberrations.

thickness, and a 20-mm focal length. The incident rays shown in Fig. 1.26(a) have a 20° angle to the optical axis. In Fig. 1.23(a), only one wavelength of 0.55 μm is used, whereas Fig. 1.26(a) uses a blue-color wavelength of 0.45 μm and a red-color wavelength of 0.65 μm to show the color aberration. Figure 1.26(b) shows the magnified view of the focusing area marked by the dotted-line circle in Fig. 1.26(a).

Figure 1.26(b) shows that the two wavelengths are focused at different locations. The 0.45-μm wavelength is focused at a point closer to the lens than the 0.65-μm wavelength because of the glass dispersion. The horizontal distance between the two focused points is ~0.35 mm; this is the "longitudinal color aberration." Longitudinal color, also called "axial color," is approximately proportional to the focal length of the lens. The vertical distance between the two focused points is ~0.02 mm; this is the "lateral color aberration," which is approximately proportional to the incident angle of the rays. Both the longitudinal and horizontal color are approximately proportional to the inverse of the Abbe number value. The Abbe number value of S-FPL53 glass is 94.9, which is very large, and thus the glass has very weak color dispersion.

The lens in Fig. 1.23(b) is re-plotted in Fig. 1.26(c). It uses Ohara S-LAH79 glass with a large refractive index of 2.00 and a small Abbe number of 28.3. This lens has a 15-mm diameter, a 5-mm central thickness, and a 20-mm focal length. The incident rays shown in Fig. 1.26(c) have a 20° angle to the optical axis. In Fig. 1.23(b), only one wavelength of 0.55 μm is used, whereas in Fig. 1.26(c) a blue-color wavelength of 0.45 μm and a red-color wavelength of 0.65 μm are used to show the color aberration. Figure 1.26(c) shows that the two wavelengths are focused at different locations. Figure 1.26(d) shows the magnified view of the focusing area marked by the dotted-line circle in Fig. 1.26(c). The color aberrations in Fig. 1.26(d) is larger than those in Fig. 1.26(b) because the Abbe number of glass S-LAH79 is much smaller than the Abbe number of glass S-FPL53. The magnitude of the longitudinal and lateral color aberrations is ~0.92 mm and ~0.22 mm, respectively.

1.12.5 Field curvatures

In Fig. 1.27 a Zemax-generated raytracing diagram for a single lens is used to explain field curvature. This single lens is made of N-BK7 glass with 3-mm central thickness, 11-mm diameter, and 15-mm focal length. The three incident angles of rays are 0°, 10°, and 20°. The wavelength of the rays is 0.55 μm. This lens has severe coma. To avoid unnecessary complexity and to simplify the illustration, only two meridional rays are traced for every field angle. The real chief ray is not traced; instead, a conceived "coma-free" chief ray is manually drawn to help explain the situation. The three focused spots for the three field angles are marked by dark solid dots.

Every lens has a basic curved image surface, called a Petzval surface, as shown in Fig. 1.27. For a simple thin lens, the axial distance between the Petzval surface and the ideal flat image surface is given by $h^2/(2nf)$, where h is the image height, n is the refractive index of the lens material, and f is the lens focal length. The axial position of the focused spot moves toward the lens as the field angle increases. The tangential image surface is the dashed curve that links these three focused spots. Positive lenses introduce inward field curvatures. Negative lenses introduce outward field curvatures. This phenomenon is called field curvature. The image surface of a single lens is more severely curved than the Petzval surface, which is also shown here.

The raytracing and the field curvature shown in Fig. 1.27 are in the tangential plane. In the sagittal plane, a simple positive lens will also introduce

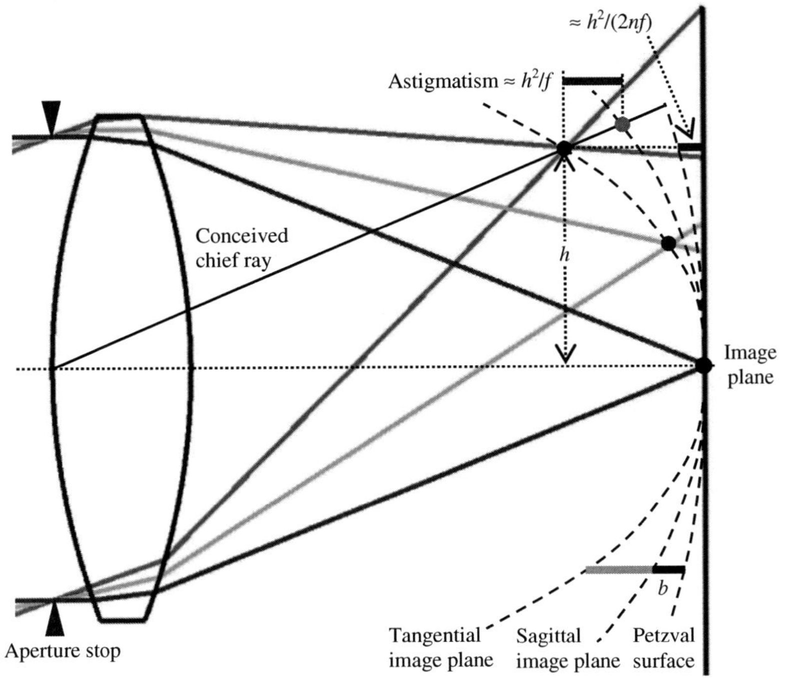

Figure 1.27 Zemax-generated raytracing diagram for a single lens shows a severe, inwardly curved image surface in the tangential plane. The sagittal image surface, which is perpendicular to the book page, is less curved than the tangential image surface but more curved than the Petzval surface.

an inward field curvature with a magnitude smaller than the field curvature magnitude in the tangential plane, as illustrated in Fig. 1.27. The axial distance between the tangential and sagittal image surfaces is about twice the axial distance between the sagittal image surface and the Petzval surface, as indicated by the two thick grey lines in Fig. 1.27.

The axial distance between the two focused points on the tangential and sagittal image surfaces along the same chief ray is the astigmatism, as marked in Fig. 1.27. The magnitude of astigmatism is about h^2/f, a parameter that is special in determining the magnitude of field curvatures.

For a multi-element lens, the field curves in both tangential and sagittal planes can have complex profiles. Since all of the sensors used to sense the image produced by a lens are planar, field curvature will cause defocusing and reduce the image sharpness, particularly at large field angles. Field curvature can often be mostly corrected by combining several positive and negative lens elements with properly selected surface curvatures and glasses or/and the use of some aspheric lenses. A well-designed lens should have a nearly flat image plane in both tangential and sagittal planes.

1.12.6 Wavefront errors and optical path difference

All optical aberrations are some type of wavefront error. For example, when a lens focuses rays to a point, the ideal wavefront is a convergent sphere. The center of the sphere is the focal point. All rays will travel along the radii of the wavefront sphere and be focused at the center. However, because of the presence of various types of aberrations, the real wavefront deviates from the ideal sphere, and the rays will not be focused at the same spot. If a real wavefront is "toroidal," there is astigmatism.

The concept of optical path difference (OPD) quantifies the deviation of a real wavefront from a spherical reference wavefront. Two types of OPD are frequently used: peak-to-valley OPD_{PV} and root-mean-square OPD_{RMS}. Figure 1.28 shows two examples. OPD_{PV} is the largest OPD between the wavefront and a reference sphere over the entire wavefront. OPD_{RMS} is calculated by the following equation using the data of a series of sampling points on the wavefront and the reference sphere:

$$OPD_{RMS} = \sqrt{\frac{\sum_{i=1}^{m} OPD_i^2}{m}}, \tag{1.30}$$

where m is the number of sampling points used to calculate the OPD_{RMS}. There is no rule about how many sampling points should be used. Common sense of mathematics applies here.

OPD_{RMS} contains more complete information about the wavefront and is widely used to evaluate the quality of a wavefront. For example, the wavefront error shown in Fig. 1.28(a) is more severe than the wavefront error shown in Fig. 1.28(b), where the two OPD_{PV} are the same but the OPD_{RMS} shown in Fig. 1.28(a) is larger. Since a real wavefront can have various complex shapes,

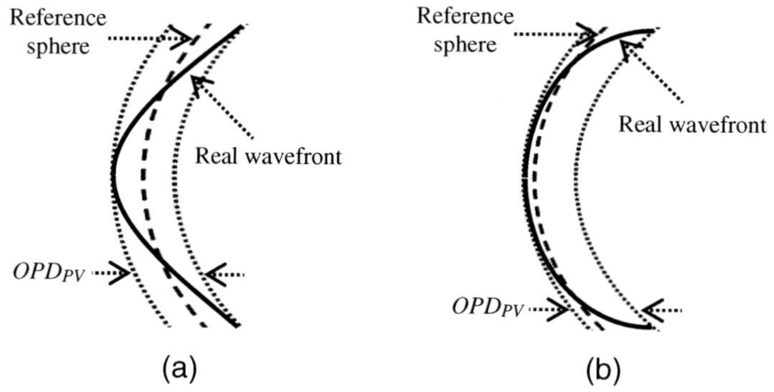

Figure 1.28 The solid curves depict two real wavefronts. The dashed curves are the two reference spheres selected to compare with the real wavefront. The distance between the real wavefront and the reference sphere is the OPD. The dotted curves define the peak-to-valley OPD_{PV}.

there is no simple relation between the OPD_{PV} and OPD_{RMS} because the wavefront can have various shapes. For a mixture of low-order aberrations, $OPD_{PV} \approx 4.5OPD_{RMS}$. It is usually assumed that $OPD_{PV} \approx 5OPD_{RMS}$.

1.12.7 How to read an OPD diagram

Optical software can plot the OPD diagram for a lens to evaluate the lens quality. Figure 1.29 shows a Zemax-generated OPD diagram for the double

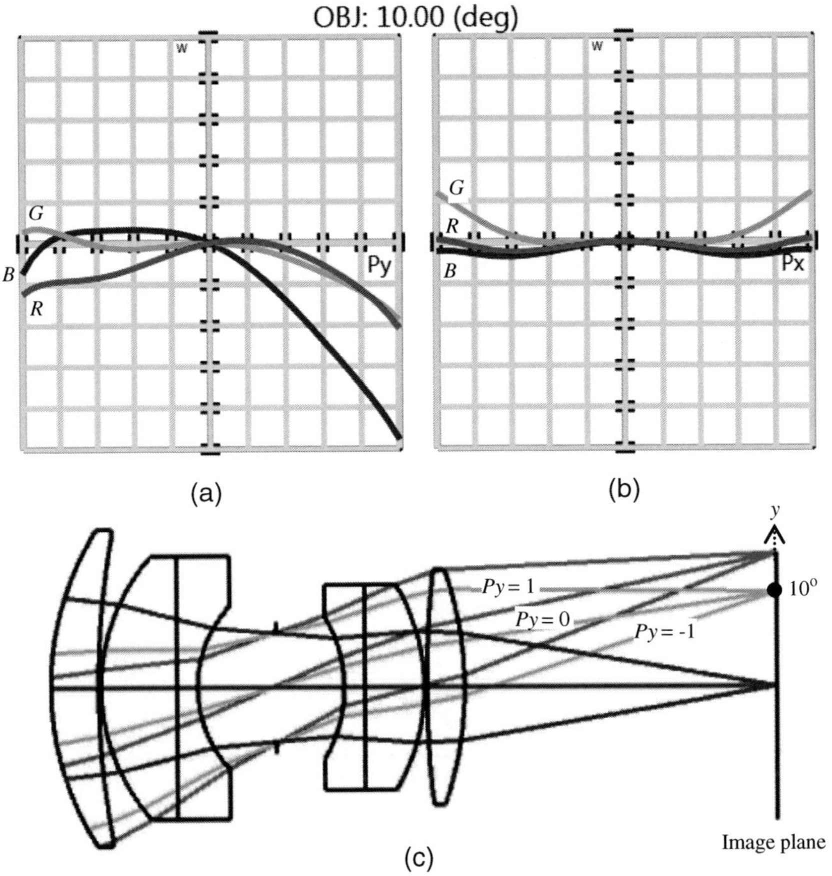

Figure 1.29 Zemax-generated OPD diagram at the image plane for the double Gauss lens shown on Fig. 1.10 for (a) a field angle of 10° in the sagittal direction (perpendicular to the page) and (b) 10° in the tangential direction (page plane), and RGB three colors. In (a) and (b), the vertical axis is the OPD with a unit of "wave" (the wavelength of He-Ne laser). (a) OPD vs. P_y for $P_y = -1$ to $P_y = 1$. P_y is marked in Fig. 1.29(c). The R and G colors have an OPD up to 2 waves. The B color has the largest OPD up to 5 waves. (b) OPD vs. P_x for $P_x = -1$ to $P_x = 1$, which is in the sagittal plane perpendicular to the page and is not marked in Fig. 1.29(c). The OPD is symmetric about $P_x = 0$ and is smaller than the OPD shown in (a). (c) Re-plot of the raytracing diagram for the double Gauss lens to illustrate P_y. The y direction is along the page plane and is the tangential direction. The 10° field and the $P_y = -1$, 0, and 1 are marked.

Gauss lens shown in Fig. 1.10, which is re-plotted here in Fig. 1.29(c) for convenience.

For any given field angle, $P_y = 0$ is the chief ray, $P_y = -1$ is the low ray, and $P_y = 1$ is the high ray, as marked in Fig. 1.29(c) for the field angle of 10°. All of the rays for any given field angle in the tangential plane are in the range of $-1 \leq P_y \leq 1$.

The OPD in the tangential direction is shown in Fig. 1.29(a) and is larger than the OPD in the sagittal direction shown in Fig. 1.29(b). This phenomenon is commonly seen and can also be seen in the CTF diagram in Fig. 1.33.

OPD is closely related to image quality. The Rayleigh criterion states that the $OPD_{PV} < 0.25$ wave or $OPD_{RMS} < 0.07$ wave can be translated to diffraction-limited quality.

1.13 Evaluation of Image Quality

Many lenses are used to generate an image of an object on a sensor for display. The image quality evaluation is an important subject. Image resolution and image distortion are the two most important parameters used to describe the image quality.

Several criteria are used to describe image resolution. Among these, the contrast transfer function is the most commonly used. The Rayleigh criterion (another criterion from the same Rayleigh mentioned earlier) is also frequently used. These two criteria are related.

1.13.1 Image resolution: Rayleigh criterion

Image resolution involves the diffraction of optical waves, which will be discussed later in Chapter 2. The results are presented here in advance.

The image of an infinitely small object point formed by a perfect lens has a certain size because of diffraction. The intensity profile of the image spot contains a central lobe and several side lobes with gradually decreasing intensity, as shown in Fig. 2.12(a). The spatial resolution limit of this lens is defined by Rayleigh as when the maximum of one image spot falls on the first minimum of the next image spot, as shown by the solid curves in Fig. 1.30(a). In such a case, the intensity profile of the two-image-spot combination has a central dip of ~74% of the maximum intensity of the two-image-spot combination, as shown by the dashed curve in Fig. 1.30(a), and the two image spots are said to be just resolvable.

The distance R between the two spots is the resolution limit of the lens. Any two image spots with a distance between them smaller than R is not resolvable (see Fig. 1.30(b)). Any two image spots with a distance between them larger than R is resolvable (see Fig. 1.30(c)).

This criterion is the Rayleigh criterion and is widely used to define the resolution limit of lenses. The image spots can either be diffraction limited

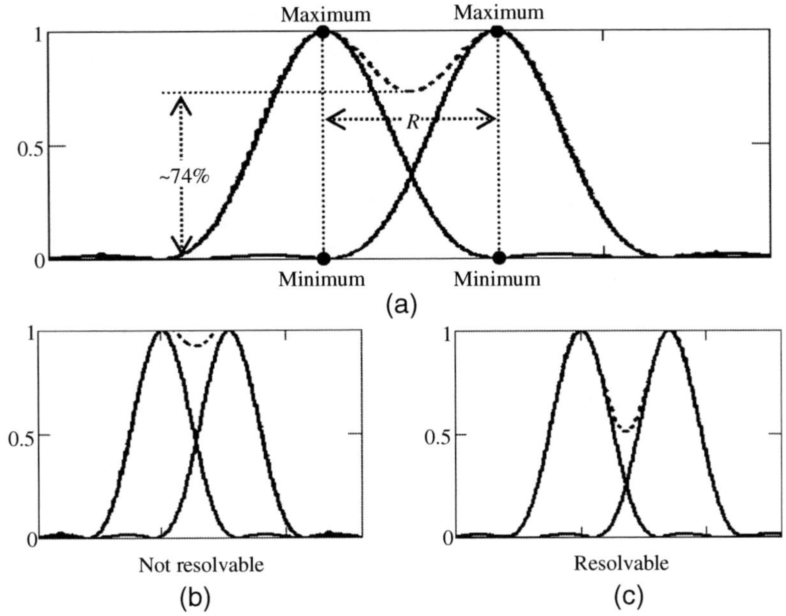

Figure 1.30 Rayleigh criterion of resolvable images. The solid curves are the intensity profiles of image spots. The dashed curves are the intensity profile of the two-image-spot combination. (a) The two image spots are just resolvable according to the widely used Rayleigh criterion. Distance R between the centers of the two spots is the image resolution of the lens used to generate these image spots. (b) The two image spots are too close to be resolvable. (c) The two image spots are clearly resolvable.

(aberration free) for a very-high-quality lens or have severe aberration for a low-quality lens. A lens that can form smaller focused spots has higher image resolution.

1.13.2 US Air Force resolution test chart

The US Air Force (USAF) resolution test chart, as shown in Fig. 1.31, is widely used as an object to measure the resolution of an image lens. The lens under test generates an image of the chart. The resolution of the lens can be found by comparing the chart and the image of the chart.

A resolution test chart contains many black-and-white bars. All of these bars are divided into several groups. Each group contains several bar elements with different widths or spatial frequencies. The chart is designed so that the spatial frequency R of a bar can be calculated by

$$R = 2^{group+\frac{element-1}{6}}, \tag{1.31}$$

where *group* and *element* are the group number and element number, respectively, as marked in Fig. 1.31.

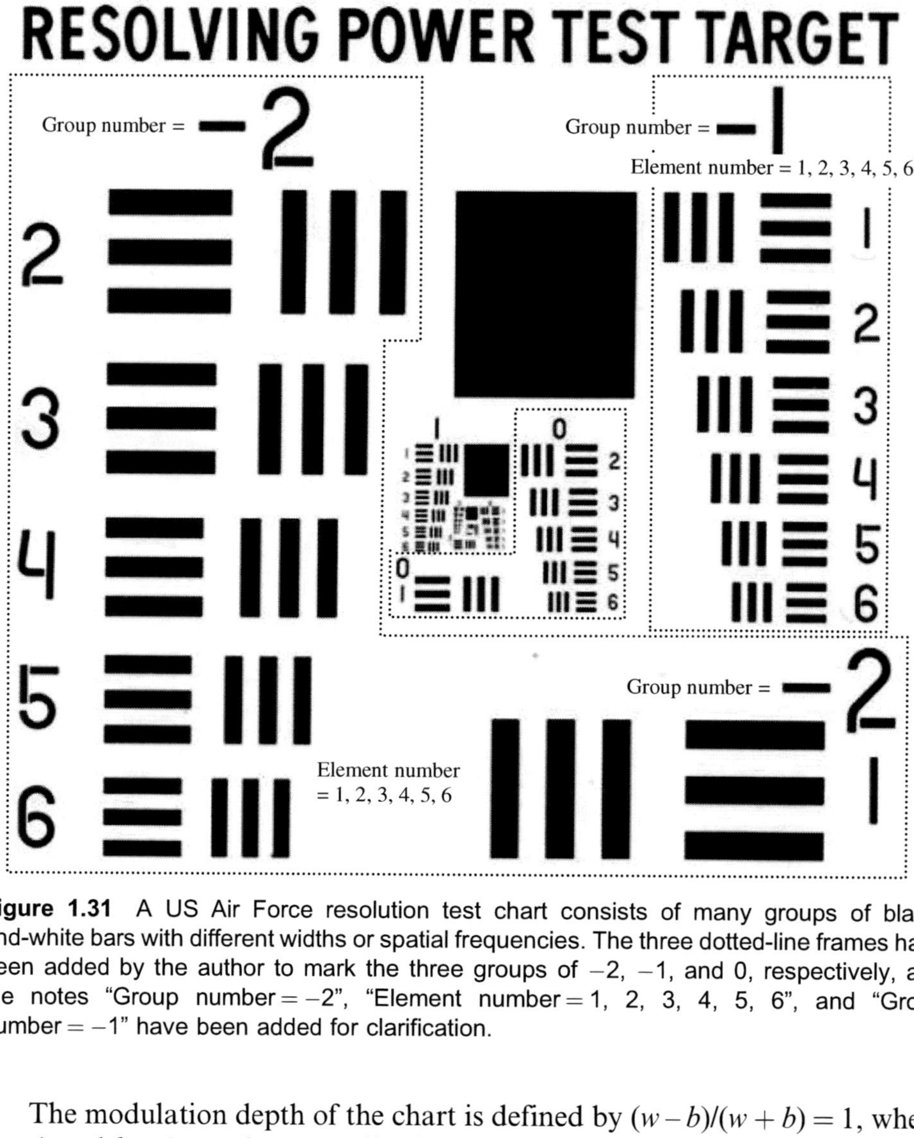

Figure 1.31 A US Air Force resolution test chart consists of many groups of black-and-white bars with different widths or spatial frequencies. The three dotted-line frames have been added by the author to mark the three groups of −2, −1, and 0, respectively, and the notes "Group number = −2", "Element number = 1, 2, 3, 4, 5, 6", and "Group number = −1" have been added for clarification.

The modulation depth of the chart is defined by $(w - b)/(w + b) = 1$, where $w = 1$ and $b = 0$ are the normalized optical intensities of the black-and-white bars, respectively. Detailed information about the USAF resolution test chart can be found in the literature.[8]

1.13.3 Image resolution: contrast transfer function

The spatial resolution of an imaging lens can be quantized by the contrast transfer function (CTF) or modulation transfer function (MTF). These two quantities are related, and the CTF is the more commonly used of the two.

The CTF of an imaging lens can be measured using a USAF resolution test chart or any similar chart, such as a Siemens star chart. The value of the

CTF is defined as the modulation depth of the image of a target with a modulation depth of 1. The mathematical form of the CTF is

$$CTF = \frac{I_w - I_b}{I_w + I_b}, \tag{1.32}$$

where I_b and I_w are the intensities of the images of the black/white bar, respectively, and the intensities of the target are normalized to $b = 0$ and $w = 1$.

Because of the diffraction, the image of a black/white vertical edge formed by a perfect lens is a gradient grey area, which is called the "line spread function." For an imperfect lens, the presence of various types of aberrations increases the width of this gradient grey area, as shown in Fig. 1.32(b), where the grey area width $2s$ is exaggerated. If the black-and-white bar width P is larger than s, the portions of the image outside the grey area still have

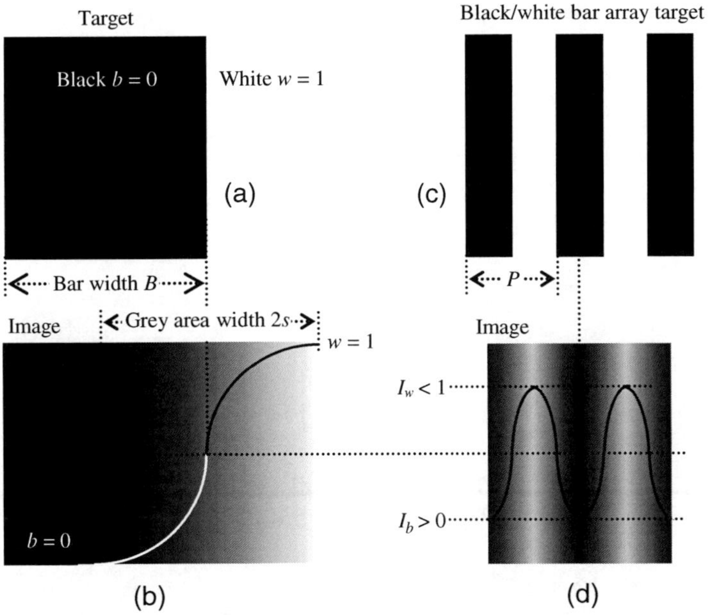

Figure 1.32 Illustration the relation between black-and-white targets and their images. (a) A black-and-white edge target with normalized intensity of $b = 0$ and $w = 1$. (b) Image of the black-and-white edge target formed by a lens is a gradient grey area with width $2s$ determined by the diffraction and aberrations of the lens. The curve is the intensity level of the grey area. Inside the grey area, the modulation depth of the image is $CTF = (I_w - I_b)/(I_w + I_b) < 1$. Outside the grey area, the modulation depth of the image is $CTF = (w - b)/(w + b) = 1$. (c) A black-and-white bar array target with spatial period P and normalized intensities of $b = 0$ and $w = 1$. (d) Image of the black-and-white bar array. When the bar width is $< s$, the modulation depth of the image is always < 1. As the bar width decreases, the image modulation depth decreases. When the bar width is much smaller than s, the modulation depth of the image approaches zero.

normalized intensities of 1 and 0, respectively, as shown in Fig. 1.32(b), and the image *CTF* is still 1.

If the black-and-white bar width *P*/2 is smaller than *s*, the image only consists of periodic grey areas, as shown in Fig. 1.32(d). The dark portion of the image is no longer black and has a normalized intensity $I_b < 1$. The light portion of the image is no longer white and has a normalized intensity $I_w > 0$. The image *CTF* is < 1, and the *CTF* value decreases as the ratio *P*/*s* decreases. The lens spatial resolution limit is widely defined as $CTF = 0.3$. The test bar width of the corresponding image bar width for $CTF = 0.3$ is the resolution limit of the lens under test. For $CTF = 0.3$, higher-quality lenses have a smaller *s* value and can resolve a smaller *P* (narrower test bar). The CTF will eventually decrease to zero for any lenses with a very small *P*/*s*.

When a test chart is used to measure the resolution limit of an imaging lens, the distance between the test chart and the lens must be specified because the same bars will look smaller if the distance increases. More frequently, people use the image bar width instead of the test chart bar width to specify the resolution limit of a lens. There is a one-to-one relation between the widths of the test bar and the image for a given lens focal length and the distance between the test chart and the lens.

The widths of the test chart bar and image bar are often expressed in terms of cycle/mm or line pair/mm. One cycle or one line pair means one black-and-white bar pair.

1.13.4 How to read a CTF diagram

Figure 1.33 shows a CTF diagram produced by Zemax for the double Gauss lens shown in Figs. 1.10 and 1.29(c):

1. The vertical axis "Square Wave MTF" means the CTF.
2. The horizontal axis "Spatial frequency in cycles per mm" is the image bar width. For example, 100 cycles/mm means that there are 100 black-and-white bar pairs in one millimeter in the image, and each bar has a 5-μm width.
3. For a given lens, the CTF varies as the angle of incident ray varies. A larger incident ray angle usually results in a smaller CTF. The CTFs for 0°, 10°, and 14° are plotted in Fig. 1.33.
4. For any angles of ray incidence, the CTF in the tangential and sagittal directions are represented by the dotted and solid curves in Fig. 1.33, respectively, and are often different, which means the lens has astigmatism. The black-and-white bars shown at the left of Fig. 1.33 illustrate this phenomenon. For a 0° incident angle, the rays are symmetric about the optical axis. The concepts of tangential and sagittal are invalid.
5. The top black curve in the diagram is the "diffraction-limited" CTF curve that is the theoretical best possible CTF curve that can be

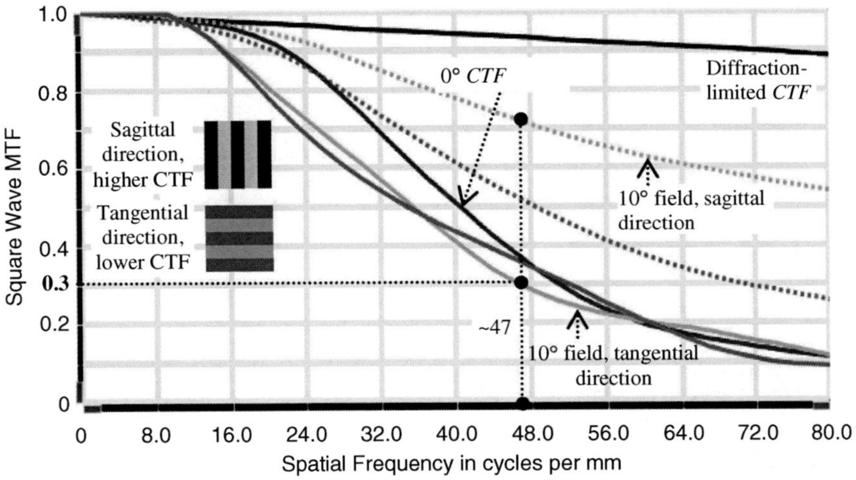

Figure 1.33 A CTF diagram generated by Zemax for the double Gauss lens shown in Figs. 1.10 and 1.29. The CTF value is a function of bar spatial frequency. The CTF curves for three field angles of 0°, 10°, and 14° are plotted. The CTF in the tangential and sagittal directions are plotted by dotted and solid curves, respectively; these two curves are usually different. At field center (0°), the CTFs in the tangential and sagittal directions coincide. The spatial resolution limit ($CTF \geq 0.3$) is ~47 cycle/mm in the tangential direction for a 10° field, as marked.

obtained only by a perfect lens of the same *F*-number. Different lenses have different diffraction-limited CTF curves, which are always smaller than 1.

6. The values of all CTF curves start as 1 at 0 cycles/mm, which means that the black-and-white bar width is infinitely large and decreases as the spatial frequency increases.
7. All CTF curves at a given spatial frequency are much smaller than the diffraction-limited CTF curve. This phenomenon indicates that this lens has large aberrations.
8. Spatial frequency of ~47 cycle/mm is about the resolution limit of this lens. Below this spatial frequency, all CTF curves >0.3. 47 cycle/mm translates to a ~10-μm bar width in the image.
9. The CTF for a 10° field in the tangential direction is much smaller than the CTF in the sagittal direction. This phenomenon can also be seen in the OPD diagram in Fig. 1.29.

1.13.5 Image resolution: modulation transfer function

Traditionally, a test chart with a sinusoidal periodic intensity pattern is also used. The normalized intensity of such a chart gradually varies between 0 and 1. The image modulation depth of such a test chart is the MTF. Because the sinusoidal intensity pattern does not have a sharp black-and-white edge, the value of the MTF is always smaller than the value of the CTF for the same spatial frequency and the same lens. The equations linking the CTF to MTF are[9]

$$MTF(\nu) = \frac{\pi}{4}\left[CTF(\nu) + \frac{CTF(3\nu)}{3} - \frac{CTF(5\nu)}{5} + \frac{CTF(7\nu)}{7} \dots\right] \quad (1.33)$$

or

$$CTF(\nu) = \frac{\pi}{4}\left[MTF(\nu) - \frac{MTF(3\nu)}{3} + \frac{MTF(5\nu)}{5} - \frac{MTF(7\nu)}{7} \dots\right], \quad (1.34)$$

where ν is the spatial frequency. Nowadays, the USAF resolution test chart is the most widely used test chart, and therefore the CTF is more widely used than the MTF.

All of the optical aberrations described in Section 1.12 can reduce the CTF or MTF of a lens. Increasing either value requires an overall improvement of the lens and can be achieved by carefully designing the lens using more appropriate optical elements.

1.13.6 Effect of sensor pixel size: Nyquist sampling theorem

The image formed by a lens is often detected by a 2D sensor array and sent to a display device. The sensor pixel size will have an impact on the resolution of the image sent to the display device.

The Nyquist sampling theorem[10] states that when sampling a periodic signal, at least two samplings per period are needed to recover the signal. In the case here, one period of signal consists of one black-and-white bar pair, and one sampling means one pixel. The resolution limit of the lens determines the smallest image of the bars that the lens can produce. When the pixel size of the sensor is larger than the resolution limit of the lens, the detector cannot fully resolve the image generated by the lens, so the quality of the lens is partially wasted, and vice versa. For example, to fully utilize the resolution of the image in Fig. 1.33, the sensor pixel must be smaller than 10 μm.

In the visible range, CCD arrays are widely used as the sensor. The pixel size of CCD arrays can be as small as a couple of microns now. Most lenses in the visible range have an image resolution lower than the sensor resolution and limit the resolution of the lens/sensor combination.

1.13.7 Image distortion

Any optical image is more or less distorted. The image distortion of a lens can be analyzed by imaging a grid target. There are two typical types of image distortion: barrel distortion and pincushion distortion, as shown in Fig. 1.34.

Barrel distortion shrinks the image, whereas pincushion distortion stretches the image. Some optical lenses produce images with a mixture of

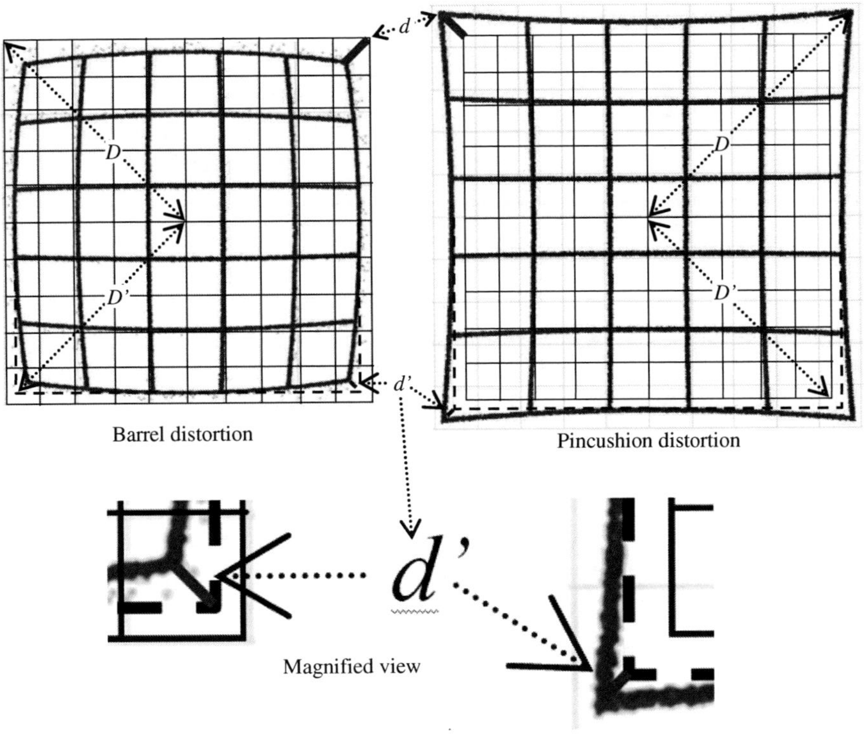

Figure 1.34 Two typical types of image distortion: barrel and pincushion. The thin line grid is the image of a grid target without distortion plotted here for comparison. The thick curved grids are Zemax-generated distorted images of the grid target generated by a lens under test. The very thick and short grey lines at the top right and left corners mark the absolute distortion value *d*. The relative distortion *d*/*D* is what is really sought, where *D* is the half-diagonal of the grid. Another way to define the relative distortion is *d*′/*D*′, where *d*′ is shown in the magnified view. Distortion *d*′/*D*′ ignores the image size change caused by distortion.

these two types of distortion. Figure 1.34 depicts pure barrel and pincushion distortions. The magnitude of the relative image distortion at a certain location in the image is the distorted amount divided by the radial position of this location. The maximum magnitude of the relative image distortion often appears at the image corner and is marked by the two thick, short grey lines at the top corner in Fig. 1.34. The maximum relative distortion is the length of these thick, short grey lines divided by the half-diagonal of the grid image. The maximum relative distortions in both images in Fig. 1.34 are ~8%.

The image size compression or stretching caused by distortion can often be compensated by intentionally designing a lens with a "too large" or "too small" image, respectively. Lens users are annoyed by the curved images of straight lines. Based on this argument, the image compression or stretching can be excluded from the distortion, and the maximum relative distortion can be defined by d'/D', as shown in Fig. 1.34.

Image distortion can be minimized by carefully designing the lens using more appropriate optical elements, which is an issue when balancing the performance and cost. For a camera lens, image distortion with a couple of percent magnitude is acceptable. For an inspection lens, the required distortion can be below 0.1%.

1.14 Illumination Optics versus Imaging Optics

Thus far, only imaging optics has been discussed, i.e., where an image of an object is formed. Every image point has an exclusive one-to-one relation with an object point, as illustrated in Fig. 1.35(a).

Illumination optics is different from imaging optics and is also a part of geometrical optics. In illumination applications, certain illumination intensity patterns (usually a uniform intensity pattern or a flat-top pattern) are desired. Many light sources, such as a tungsten bulb, have a certain source structure. If imaging optics is used to handle the illumination light, the light source structure will be imaged in the illumination pattern. Such a result is not desired. Therefore, illumination optics intentionally avoids forming any images, as illustrated in Fig. 1.35(b). Light from many source points can reach the same point on the working plane, or light from one source point can reach many points on the working plane. That is why illumination must be discussed in a separate section. Generally speaking, illumination optics is simpler than imaging optics, in terms of the number of variables involved and the complexity of the merit function, because the requirements for an illumination pattern are often not specified to a high accuracy. Illumination optics is also less widely used than imaging optics.

Since illumination optics does not focus the light from an object to form an image, all of the aberrations discussed in Section 1.12 are no longer relevant. Aberrations are sometimes utilized to generate the desired illumination pattern. The source wavelength can still be an issue. Besides the capability of generating

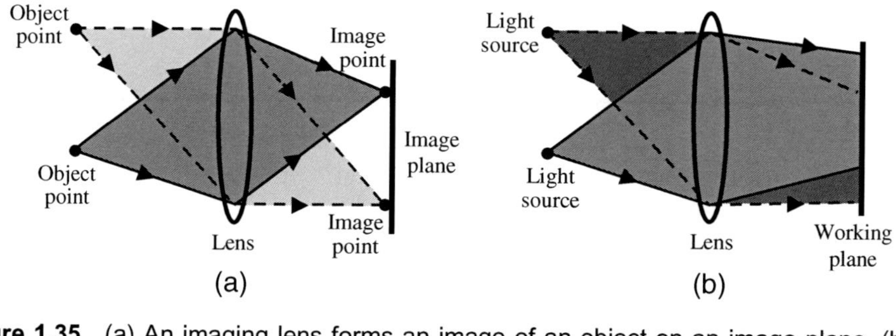

Figure 1.35 (a) An imaging lens forms an image of an object on an image plane. (b) An illumination lens forms the light from light source points to illumination patterns on a working plane. The combination of these illumination patterns is the desired illumination pattern. There is no image formed.

a certain illumination pattern, the light-collecting power is another important issue for image optics.

Optical design software for imaging optics are often capable of designing illumination optics as well, but their capability to design illumination optics is inferior. The designer sets the desired illumination pattern as the target in the merit function in the software, and the software will try to find a certain optical structure to meet the target. The raytracing technique used to design illumination optics is called "non-sequential raytracing," whereas the raytracing technique used when designing imaging optics is called "sequential raytracing."

Illumination optics deal not only with optical components to handle the light but also the light sources, including various types and shapes of light bulbs, lasers, and light-emitting diodes (LEDs). These light sources can have dramatically different characteristics and illuminations. Lenses used to work with different light sources can also be very different.

Section 12.4 will discuss the modeling of illumination lenses and light sources.

1.15 Radiometry

Radiometry is the science about handling light intensity. Among the units and terminologies used, some are unique and may be confusing. In this book, only the International System (SI) of units is used to avoid unnecessary complexity. The two widely used terminologies in radiometry are "radiance" and "irradiance." The former means the power per unit solid angle per unit area, and the latter means the power per unit area.

Many optical engineers and scientists are more experienced handling imaging optics than handing radiometry, although radiometry contains less substance than imaging optics. See Palmer and Grant[11] and McCluney[12] for more about radiometry.

1.15.1 Lambert's cosine law

One basic law in radiometry is Lambert's cosine law. The radiance characteristics of most diffusive surfaces can be approximately described by[13]

$$J(\theta) = J_0 \cos(\theta), \tag{1.35}$$

where J_0 is the maximum radiance that appears in the normal direction of the radiation source surface, θ is the angle between the direction of interest and the radiation normal, and $J(\theta)$ is the radiance in the θ direction.

One interesting characteristic of Lambertian sources is that the apparent radiance viewed in any direction is the same. In a direction θ off the source normal, the radiance falls from J_0 to $J_0\cos(\theta)$, and at the same time the source area being viewed is increased by a factor of $1/\cos(\theta)$, as illustrated in Fig. 1.36; these two factors compensate each other.

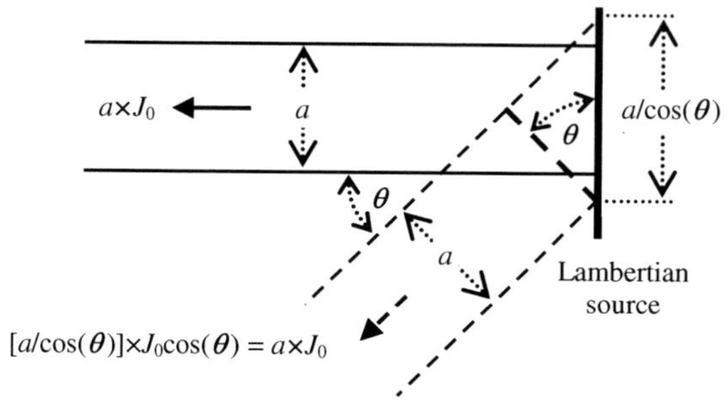

Figure 1.36 A Lambertian source has the same radiance in any direction.

1.15.2 Light-collecting power of a lens

One key specification of an illumination lens is its light-collecting power. Consider the case shown in Fig. 1.37, where the irradiance power P collected by this lens is given by

$$\begin{aligned} P &= \int AJ_0\cos(\theta)d\Omega \\ &= \int_0^R AJ_0\cos(\theta)\frac{2\pi r\cos(\theta)dr}{\left[\frac{o}{\cos(\theta)}\right]^2} \\ &= \frac{2\pi AJ_0}{o^2}\int_0^R r\cos^4(\theta)dr \\ &= 2\pi AJ_0\int_0^{\theta_M}\sin(\theta)\cos(\theta)d\theta \\ &= \pi AJ_0\frac{R^2}{o^2+R^2}, \end{aligned} \tag{1.36}$$

where A is the area size of the light source; o is the distance between the light source and the lens; i is the distance between the lens and the illuminated area with size A'; J_0 and J_0' are the peak radiance and irradiance of the light source and illuminated area, respectively; R is the radius of the lens aperture; θ and θ' are the angles between the point of interest on the lens and the optical axis, respectively; the largest value of θ is $\theta_M = \sin^{-1}[R/(o^2+R^2)^{0.5}]$; $r = o \times \tan(\theta)$ is the radial variable on the lens, $dr = o \times d\theta/\cos(\theta)^2$; $d\Omega = 2\pi\cos(\theta)rdr/[o/\cos(\theta)]^2$ is the incremental solid angle marked by the grey color ring on the lens imposes on the light source; and $d\Omega$ is integrated over the entire lens aperture.

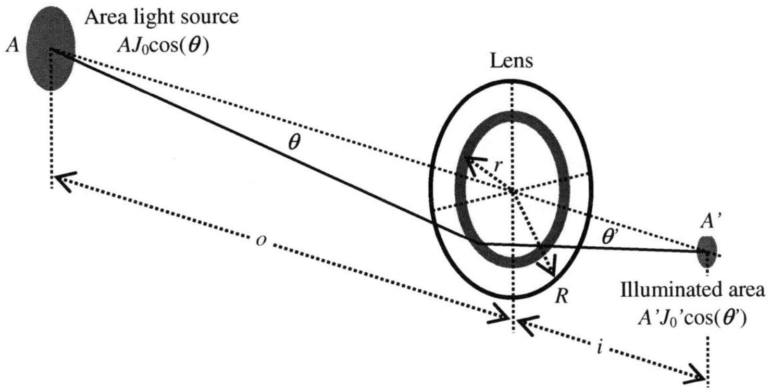

Figure 1.37 A lens with aperture radius *R* collects light power emitted by an area light source.

When $o \ll R$, Eq. (1.36) leads to $P = \pi A J_0$, which is the total power of a Lambertian source emitted to a half-sphere.

1.15.3 Inverse square law

For $o \gg R$, Eq. (1.36) reduces to $P = \pi A J_0 R^2/o^2 \sim 1/o^2$. This means that the light power collected by a lens is inversely proportional to the square of the distance between the lens and the light source, which is the inverse square law. When $R/o \to 0$, Eq. (1.36) leads to $P \to 0$, and the lens is too small or placed too far from the source, or both.

1.15.4 A point illuminated by a circular Lambertian source

Consider a point that is illuminated by a circular Lambertian source. The point is at the optical axis of the light source and a distance *o* away, as shown in Fig. 1.38. The source has a radius *R* and radiance $J_0\cos(\theta)$; the irradiance power the point receives can be calculated by

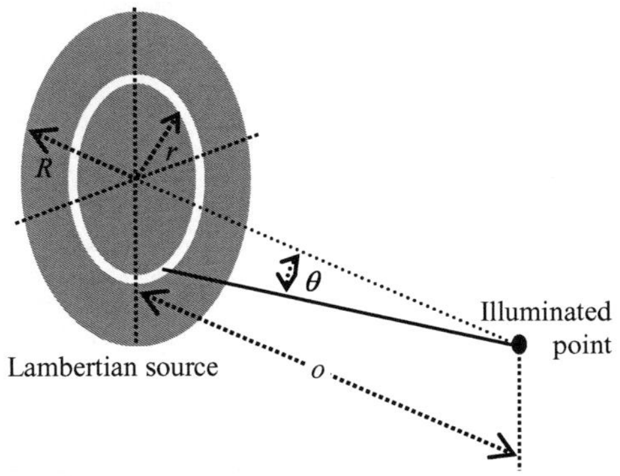

Figure 1.38 A circular Lambertian light source with radius *R* illuminates a point at the optical axis of the source and a distance *o* away.

$$
\begin{aligned}
P &= \int_0^R J_0 \cos(\theta) \frac{2\pi r \cos(\theta) dr}{\left[\frac{o}{\cos(\theta)}\right]^2} \\
&= 2\pi J_0 \int_0^{\theta_M} \sin(\theta)\cos(\theta) d\theta \\
&= \pi J_0 \sin^2(\theta_M) \\
&= \pi J_0 \frac{R^2}{o^2 + R^2},
\end{aligned}
\tag{1.37}
$$

where $2\pi\cos(\theta)rdr/[o/\cos(\theta)]^2$ is the incremental solid angle marked by the white ring on the source, $r = o \times \tan(\theta)$ is the radial variable on the source, $dr = o \times d\theta/\cos(\theta)^2$, and the largest value of θ is $\theta_M = \sin^{-1}[R/(o^2 + R^2)^{0.5}]$.

Note that the derivation of Eq. (1.37) is similar to the derivation of Eq. (1.36), and the results of these two equations are similar. Again, for $o \gg R$, Eq. (1.37) reduces to $P = \pi J_0 R^2/o^2 \sim 1/o^2$, and for $o \ll R$, Eq. (1.37) reduces to $P = \pi J_0$.

1.15.5 A working plane illuminated by a point source: the $\cos^4(\theta)$ off-axis relation

Consider a Lambertian point light source that illuminates a working plane perpendicular to the normal of the source radiance and has a distance o to the light source, as shown in Fig. 1.39. This radiance normal direction is the optical axis. A unit size area a on the optical axis at the working plane obviously receives the peak irradiance. For an off-axis unit size area b on the

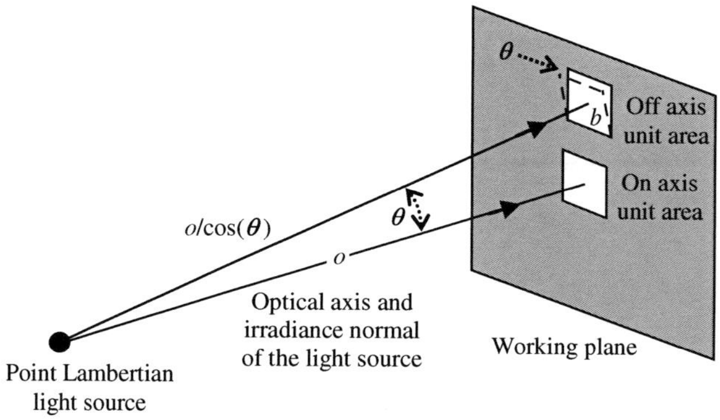

Figure 1.39 The radiance received by an off-axis unit area from a point Lambertian light source is $\cos^4(\theta)$ of the radiance received by the on-axis unit area, where θ is the off-axis angle.

working plane with an angle θ to the optical axis, the irradiance received is less than the irradiance received by a because of three factors:

1. The area projected by b in the direction of the light source is smaller than b by a factor of cos(θ), so the irradiance received by b is lowered by a factor of cos(θ).
2. The distance from the light source to area b is o/cos(θ). Thus, the irradiance received by b is lowered by a factor of $\cos^2(\theta)$, according to the inverse square law.
3. Finally, if the light source is Lambertian, the radiance in the direction of b is lower than the peak radiance in the normal direction by a factor of cos(θ).

The irradiance received by unit area b can be found by multiplying all three factors, i.e., $\cos^4(\theta)$ of the peak irradiance received by unit area a.

1.15.6 Etendue and radiance conservation

There are two conservation quantities in illumination optics: etendue and radiance. These two quantities are related, and familiarity with them can help one understand the performance of existing illumination lenses and design new illumination lenses.

Consider a lens with radius R that takes light from a light source a distance o away and transforms the light to illuminate an area a distance i away, as shown in Fig. 1.40, where h and h' are the half height of the light source and the illuminated area, respectively, and $\Omega = \pi R^2/o^2$ and $\Omega' = \pi R^2/i^2$ are the solid angle the lens imposes on the light source and the illuminated area, respectively. Note that the h' shown here is an ideal case. For a real lens, there are always some defects or aberrations; the real h' will be somewhat larger than the h' shown here. Thus, the relation $h/o \leq h'/i$ in the plane of the page can be established. A similar relation in the plane perpendicular to the page plane also holds. The relation

$$\frac{A}{o^2} \leq \frac{A'}{i^2} \tag{1.38}$$

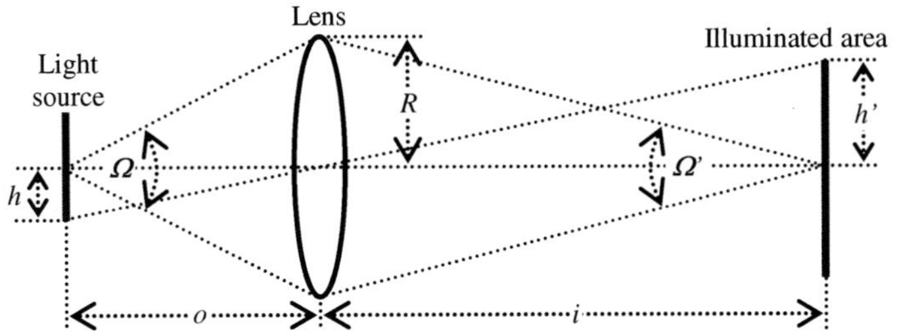

Figure 1.40 A lens with a radius R takes light from a light source a distance o away and illuminates an area a distance i away.

can then be established, where $A = h^2$ and $A' = h'^2$ are the size of the light source area and the illuminated area, respectively. Multiply both sides of Eq. (1.38) by πR^2 to produce

$$A\Omega \leq A'\Omega'. \tag{1.39}$$

$A\Omega$ is the etendue or throughput of the lens. Equation (1.39) is the etendue conservation law, which says that for a perfect lens the etendue remains a constant as the light propagates through the lens. If the lens is not perfect, etendue increases as the light propagates through it.

Etendue has applications in the design of illumination lens. For example, consider the use of a lens to couple all of the light power from a light source 100 mm away into a light guide 100 mm away. The light source size $A = 2$ cm^2, and the light-guide area size is $A' = 1$ cm^2. Then Eq. (1.39) is violated: only a portion of the light power can be coupled into the light guide, and the intended outcome will certainly fail.

When a light source and an illumination lens illuminate an area, the power that the illuminated area receives can be written as $T \times P$, where $T \leq 1$ is the transmission of the illumination lens, and P is the power collected by the lens from the light source. By taking an inverse of Eq. (1.39) and multiplying both sides by P, the result is

$$\frac{P}{A\Omega} \geq \frac{P}{A'\Omega'} \geq \frac{TP}{A'\Omega'} \geq \frac{RTP}{A'\Omega'}, \tag{1.40}$$

where $R \leq 1$ is the reflectivity of the area being illuminated. By definition, $P/(A\Omega) = J_0$ is the radiance of the light source, and $RTP/(A'\ \Omega') = J'_0$ can be considered as the radiance of the illuminated area. Then the radiance conservation law is

$$J_0 \geq J'_0. \tag{1.41}$$

Equation (1.41) states that the radiance of an area can never exceed the radiance of the light source that illuminates this area.

Etendue exists beyond geometrical optics; a similar relation also exists in wave optics and Gaussian beam optics, which will be discussed in Chapters 2 and 3.

1.15.7 Radiometry and photometry

Radiometry deals with the detection and measurement of electromagnetic radiation across the entire spectrum. The measured radiation is an absolute power. Photometry is a subfield of radiometry that scales the measured, absolute radiometric power by the spectral response (photopic curve) of human eyes. (The spectral response of the human eye is plotted later in Fig. 7.22.)

Radiometric quantities, particularly photometric quantities, may appear to be confusing or even odd. Table 1.1 summarizes the correspondence among the most commonly used radiometric quantities for SI units and photometric units.

By definition, 1 watt (W) of radiant power with a wavelength of 555 nm, which is the peak wavelength of the photopic curve, equals 683 lumens (lm). Then, from Fig. 7.22, 1 W radiant approximately equals 223 lm at 500 nm and approximately equals 429 lm at 600 nm.

There are other, less frequently used photometric quantities. For example, candela = lm/sr is also used. However, some of these quantities appear to be already or gradually becoming obsolete.

1.15.8 Blackbody

Any object radiates. The spectrum of the radiance is a function of the temperature described by Planck's law:[14]

$$I(\lambda,T) = \frac{2hc^2}{\lambda^5}\frac{1}{e^{hc/\lambda KT} - 1}, \tag{1.42}$$

where $I(\lambda, T)$ is the spectral radiance with units of $\mathrm{W/(Sr\cdot m^2 \cdot d\lambda)}$ (watt per solid angle per square meter and per meter wavelength), λ is the wavelength, T is the temperature in kelvin, $h = 6.62606957 \times 10^{-34}$ m^{2}kg/s is the Planck constant, $c = 3 \times 10^8$ m/s is the velocity of light in vacuum, and $K = 1.3806488 \times 10^{-23}$ m^{2}kg s^{-2}K^{-1} is the Boltzmann constant.

Equation (1.42) is plotted in Fig. 1.41 for six values of T from 250 K (−23 °C) to 500 K (227 °C) and in Fig. 1.42 for five values of T from 500 K to 4000 K. Note that the unit used in Figs. 1.41 and 1.42 is watt per solid angle per square meter area per nm spectrum.

Table 1.1 Correspondence among the most commonly used radiometric quantities of SI units and photometric quantities

Radiometric		Photometric	
Quantity	**Units**	**Quantity**	**Units**
Radiant power	W	Luminous flux	lm
Radiant intensity	W/sr	Luminous intensity	lm/sr
Irradiance	W/m^2	Illuminance	lm/m^2
Radiance	W/m^2·sr	Luminance	lm/m^2·sr

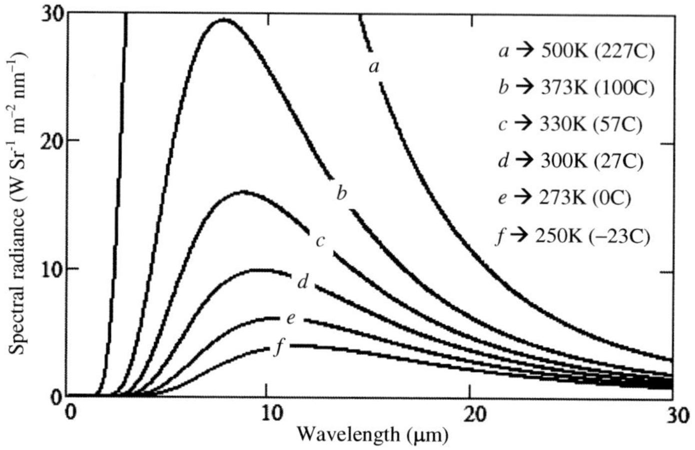

Figure 1.41 Plot of Eq. (1.42), the blackbody radiance spectra for six different temperatures from 250–500 K.

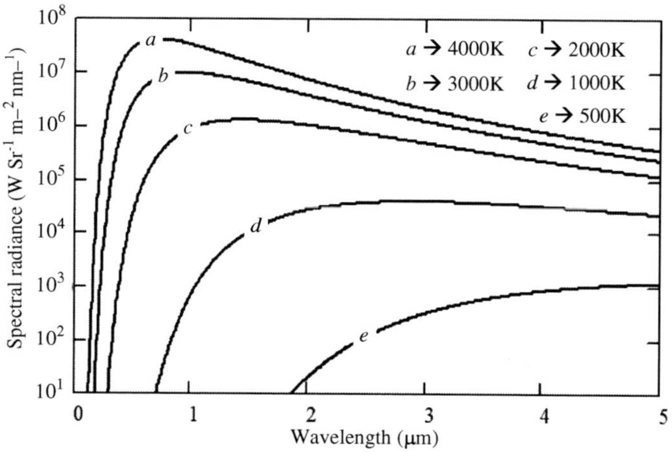

Figure 1.42 Plot of Eq. (1.41), the blackbody radiance spectra for five different temperatures from 500 K to 4000 K.

The fraction of power contained in a certain spectral range can be found by

$$F(T) = \frac{\int_a^b I(\lambda,T)dT}{\int_0^\infty I(\lambda,T)dT}, \tag{1.43}$$

where a and b are the lower and upper limits of the spectral range of interest, respectively. $F(T)$ is plotted in Fig. 1.43(a) for $a = 9$ μm and $b = 15$ μm. This spectral range is best for detecting objects with temperature around $T \approx 320$ K and has the highest fraction power of ~0.36.

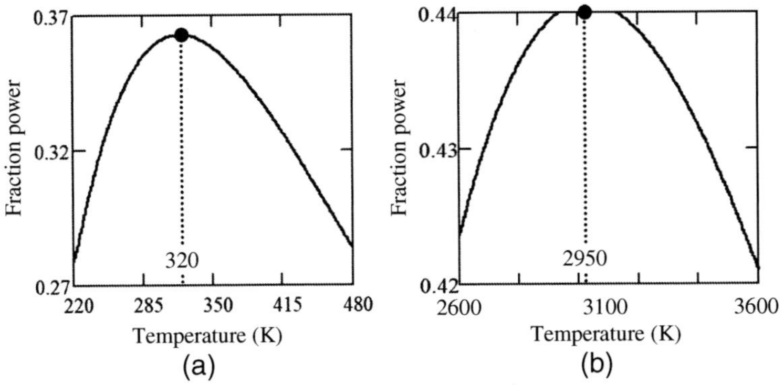

Figure 1.43 (a) The fractional power contained in the 9–15-μm range for temperatures from 220–480 K. (b) The fractional power contained in the 0.9–1.7-μm range for temperatures from 2600–3600 K.

For near-IR hyperspectral imaging applications, the spectral range of interest is about 0.9–1.7 μm. $F(T)$ is plotted in Fig. 1.43(b) for this spectral range. The highest fractional power of ~0.442 appears at $T \approx 2950$ K. Illumination lamps with a color temperature of ~2950 K have the best efficiency.

References

1. W. Smith, *Optical Engineering*, 4th ed., McGraw-Hill Education, New York (2007).
2. E. Hecht, *Optics*, 5th ed., Pearson Education, New York (2016).
3. Wikipedia, "Snell's law," https://en.wikipedia.org/wiki/Snell%27s_law
4. Wikipedia, "Total internal reflection," https://en.wikipedia.org/wiki/Total_internal_reflection
5. Wikipedia, "Paraxial approximation," https://en.wikipedia.org/wiki/Paraxial_approximation
6. Wikipedia, "Cardinal points (optics)," https://en.wikipedia.org/wiki/Cardinal_point_(optics)
7. M. Born and E. Wolf, "V Geometrical Theory of Aberrations," *Principles of Optics*, 7th ed., Cambridge University Press, Cambridge, England, pp. 228–260 (1999).
8. Wikipedia, "1951 USAF Resolution Test Chart," https://en.wikipedia.org/wiki/1951_USAF_resolution_test_chart
9. N. B. Nill, "Conversion Between Sine Wave and Square Wave Spatial Frequency Response of an Imaging System," Eqs. (10) and (11), http://citeseerx.ist.psu.edu/viewdoc/download?doi=10.1.1.914.2697&rep=rep1&type=pdf (2001)
10. Wikipedia, "Nyquist-Shannon Sampling Theorem," https://en.wikipedia.org/wiki/Nyquist%E2%80%93Shannon_sampling_theorem

11. J. Palmer and B. G. Grant, *The Art of Radiometry*, SPIE Press, Bellingham, WA (2009) [doi: 10.1117/3.798237].
12. R. McCluney, *Introduction to Radiometry and Photometry*, 2nd ed., Artech House Publishers, Norwood, MA (2014).
13. Wikipedia, "Lambert's Cosine's Law," https://en.wikipedia.org/wiki/Lambert%27s_cosine_law
14. B. E. A. Saleh and M. C. Teich, *Fundamentals of Photonics*, 2nd ed., Wiley & Sons, New York, Eqs. (9.2)–(9.18), p. 338 (2007).

Chapter 2
Wave Optics

A light wave is an electromagnetic wave, and the science of studying the wave properties of light is called wave optics. When the applications involve only spatial scales much larger than the light wavelength, the light wave can be approximately described by geometrical rays with good accuracy. When the applications involve spatial scales not much larger than the light wavelength, light must be treated as a wave. Such applications include light passing through a small opening, light being focused to a micron-size spot, and anything involving interference and diffraction.

2.1 Wave Equation

A light wave can be described by the wave equation

$$A(z,t) = A_0 \cos\left(2\pi n \frac{z+ct}{\lambda} + \phi\right), \tag{2.1}$$

where z is the propagation distance of the light from a reference point, t is the time, A_0 is the peak amplitude of light, n is the refractive index of the medium inside which the light is propagating, λ is the wavelength, c is the velocity of light in vacuum, and ϕ is the phase of the light at this reference point.

The frequency of light is so high (in the range of 10^{14} Hz for visible light) that no detector has a sufficiently fast response to catch the oscillation of light field. Therefore, the amplitude of light cannot be directly measured, but the time averaged intensity I of light can be directly measured by

$$I = \left[\overline{A(z,t)^2}\right] = \frac{A_0^2}{2}. \tag{2.2}$$

Note that the time average of $\cos^2[2\pi n(z+ct)/\lambda + \phi]$ is 1/2.

2.2 Polarization

2.2.1 Definition

The amplitude of an electromagnetic wave oscillates in a direction perpendicular to the propagation direction of the wave. This oscillation direction is the polarization direction. Polarization can be described by a vector, the direction of which is the polarization direction and the length of which is proportional to the amplitude of the wave, as illustrated in Fig. 2.1(a). There are three types of polarization: random, linear, and elliptical/circular.

2.2.2 Linear and random polarization

If there is only one polarization direction in a light wave and this direction remains unchanged, such a light wave is linearly polarized (see Fig. 2.1(a)).

Any polarization can always be considered as the combination of two orthogonal polarizations. Similarly, any two orthogonal polarizations are actually one polarization that is formed by the two orthogonal polarizations. Mathematically, a vector **P** describing a polarization can be decomposed to

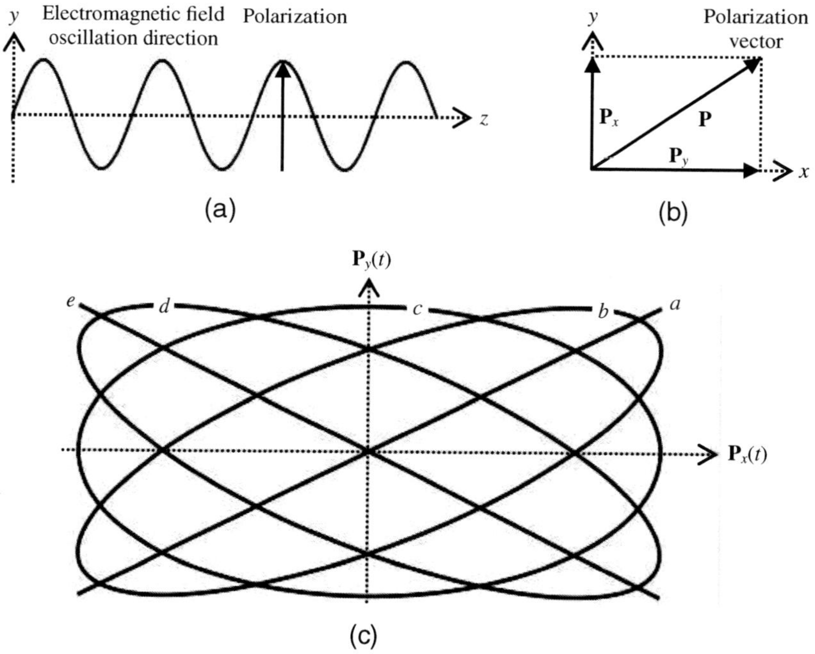

Figure 2.1 (a) Illustration of a light wave propagating in the *z* direction with linear polarization in the *y* direction. The light oscillation amplitude and direction can be denoted by a vector. (b) Any polarization vector **P** can be decomposed into two orthogonal polarization vectors $\mathbf{P}_x$ and $\mathbf{P}_y$, and vice versa. (c) Plot of the combination of two orthogonal polarization vectors $\mathbf{P}_x(t) = 2\cos[2\pi n(z + ct)/\lambda]$ and $\mathbf{P}_y(t) = \cos[2\pi n(z + ct)/\lambda + \phi]$. $\mathbf{P}_x(t)$ is twice as long as $\mathbf{P}_y(t)$ and has a phase ϕ behind $\mathbf{P}_y(t)$. The lines or curves marked by *a*, *b*, *c*, *d*, and *e* are for $\phi = 0$, $\pi/4$, $2\pi/4$, $3\pi/4$, and π, respectively.

two orthogonal vectors $\mathbf{P}_x$ and $\mathbf{P}_y$, respectively, and vice versa, as shown in Fig. 2.1(b). The way to decompose a vector or combine two vectors depends on the selection of the coordinate.

A light source usually consists of many micro sources independent of each other. Each micro source emits a light wave with a certain polarization direction. Further, the light wave emitted by one micro light source is not a long train but a series of short wave trains, each of which can have different polarization directions. As a result, at any moment, the combined wave of all of these waves has a linear polarization. But this linear polarization changes its strength and direction in a random manner. Such a type of polarization is called random polarization. Most light waves in nature have random polarization.

2.2.3 Elliptical and circular polarizations

In some cases, the linear polarization at an observing plane rotates its direction and changes its strength as time goes on. The locus of the tip of the polarization vector is an ellipse or a circle, which is a special case of ellipse. Such a polarization is called elliptical polarization or circular polarization. Elliptical polarization exists in nature and has unique applications.

An elliptical polarization can be considered as the combination of two orthogonal polarizations that have different lengths and a fixed phase difference ϕ between them. These two polarizations can be described by two vectors $\mathbf{P}_x$ and $\mathbf{P}_y$, given by

$$\begin{aligned} \mathbf{P}_x(t) &= 2\cos\left(2\pi\frac{z+ct}{\lambda}\right), \\ \mathbf{P}_y(t) &= \cos\left(2\pi\frac{z+ct}{\lambda}+\phi\right), \end{aligned} \tag{2.3}$$

where the length of $\mathbf{P}_x(t)$ is chosen to be twice longer than the length of $\mathbf{P}_y(t)$ for illustration, and at any observing plane, z is constant, and t is the only variable.

Figure 2.1(c) plots the tip locus of the combination of $\mathbf{P}_x(t)$ and $\mathbf{P}_y(t)$ using $\mathbf{P}_x(t)$ for the horizontal axis and $\mathbf{P}_y(t)$ for the vertical axis. Five different values of ϕ are used. The line marked by a is for $\phi=0$, this is a linear polarization, but can be considered as an elliptical polarization with the minor axis being equal to zero. The curve marked by b is for $\phi=\pi/4$, i.e., an elliptical polarization. The elliptically shaped curve marked by c is for $\phi=2\pi/4$ (also an elliptical polarization). The elliptically shaped curve marked by d is for $\phi=3\pi/4$ (another elliptical polarization). The line marked by e is for $\phi=\pi$, i.e., a linear polarization.

If $\mathbf{P}_x(t)$ and $\mathbf{P}_y(t)$ have the same length and $\phi=\pi/2$, the ellipse marked by c reduces to a circle, which is circular polarization. If the length of $\mathbf{P}_x(t)$ is more than twice longer than the length of $\mathbf{P}_y(t)$, the polarization ellipse will be longer that those shown in Fig. 2.1(c).

Note that because elliptical polarization rotates so fast as the time goes on, the rotation cannot be directly detected, but the polarization details can be found by using a polarizer or a retarder.

2.3 Reflectance and Transmittance of an Optical Interface

2.3.1 Reflectance and transmittance

When a light wave is incident on the interface of two optical media along the normal of the interface, the reflectance R of the interface is given by

$$R = \frac{(n_1 - n_2)^2}{(n_1 + n_2)^2}, \tag{2.4}$$

where n_1 and n_2 are the refractive indices of the two materials, respectively. The energy conservation law requires $R + T = 1$, where T is the transmittance of the interface. T can be found by

$$T = 1 - R = \frac{4n_1 n_2}{(n_1 + n_2)^2}. \tag{2.5}$$

Equations (2.4) and (2.5) are irrelevant to the polarization status of the light.

When a light wave is incident on an optical interface with an angle to the interface normal, as sketched in Fig. 2.2, the polarization status of the incident light affects the reflectance, and the situation becomes more complex: a portion of the wave is reflected, and another portion of the wave transmits through the interface.

For the convenience of analysis, two planes are defined. The tangent plane P is defined by the interface normal and the incident ray or, in this case, the

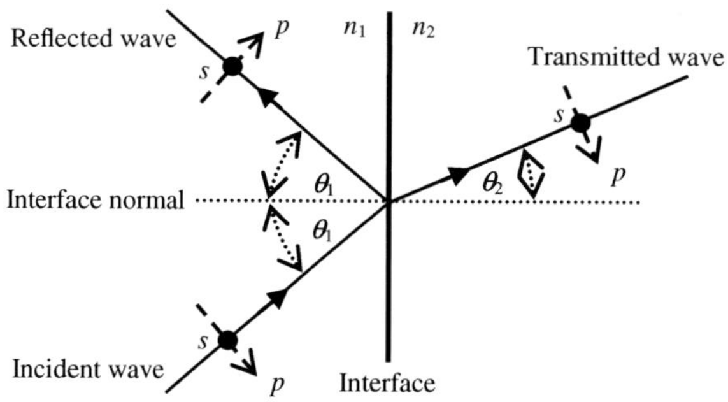

Figure 2.2 A light wave is incident on a planar interface of two optical materials. θ_1 and θ_2 are the incident (reflection) angle and the refraction angle of the waves, respectively. n_1 and n_2 are the two refractive indices of the two materials, respectively. Polarization in the three waves can be decomposed to polarization p in the plane of the page and polarization s perpendicular to the page.

plane of the page. Polarization in the P plane is called p-polarization, and polarization perpendicular to the P plane is called s-polarization, as marked in Fig. 2.2. Any polarization in the incident, reflected, and transmitted waves can be decomposed to p- and s-polarizations.

The p-polarization reflection coefficient (amplitude reflection) $r_p(\theta_1)$ of the interface is a function of the incident angle θ_1 and the two indices n_1 and n_2, given by[1]

$$r_p(\theta_1) = \frac{n_1\left[1 - \frac{n_1^2}{n_2^2}\sin(\theta_1)^2\right]^{0.5} - n_2\cos(\theta_1)}{n_1\left[1 - \frac{n_1^2}{n_2^2}\sin(\theta_1)^2\right]^{0.5} + n_2\cos(\theta_1)}. \tag{2.6}$$

The p-polarization reflectance (power reflection) of the interface is

$$R_p(\theta_1) = |r_p(\theta_1)|^2. \tag{2.7}$$

The p-polarization transmittance of the interface is $T_p(\theta_1) = 1 - R_p(\theta_1)$.

The s-polarization reflection coefficient (amplitude reflection) $r_s(\theta_1)$ is also a function of incident angle θ_1 and the two indices n_1 and n_2, given by[1]

$$r_s(\theta_1) = \frac{n_1\cos(\theta_1) - n_2\left[1 - \frac{n_1^2}{n_2^2}\sin(\theta_1)^2\right]^{0.5}}{n_1\cos(\theta_1) + n_2\left[1 - \frac{n_1^2}{n_2^2}\sin(\theta_1)^2\right]^{0.5}}. \tag{2.8}$$

The s-polarization reflectance (power reflection) of the interface is

$$R_s(\theta_1) = |r_s(\theta_1)|^2. \tag{2.9}$$

The s-polarization transmittance of the interface is $T_s(\theta_1) = 1 - R_s(\theta_1)$.

Note that for $\theta_1 = 0$, Eqs. (2.6) and (2.8) reduce to Eq. (2.4), $R_p(0) = R_s(0) = R$.

$R_p(\theta_1)$, $T_p(\theta_1)$, $R_s(\theta_1)$, and $T_s(\theta_1)$ are plotted in Fig. 2.3: (a) the left-side medium is air with $n_1 = 1$, and the right-side medium is N-BK7 glass with $n_2 = 1.52$; and (b) the left-side medium is N-BK7 glass with $n_1 = 1.52$, and the right-side medium is air with $n_2 = 1.52$.

2.3.2 Brewster's angle and total reflection

Figure 2.3(a) shows that around a special incident angle of $\theta_1 \approx 56.7°$, the p-polarization has zero reflectance. The reflected light has pure s-polarization. This special angle is called "Brewster's angle." As θ_1 approaches 90°, the reflectance of the p- and s-polarizations approaches 1.

Figure 2.3(b) shows that around a special angle of $\theta_1 \approx 33.3°$, the p-polarization has zero reflectance. The reflected light has pure s-polarization. This angle is also Brewster's angle. Above $\theta_1 \approx 41.2°$, both p- and

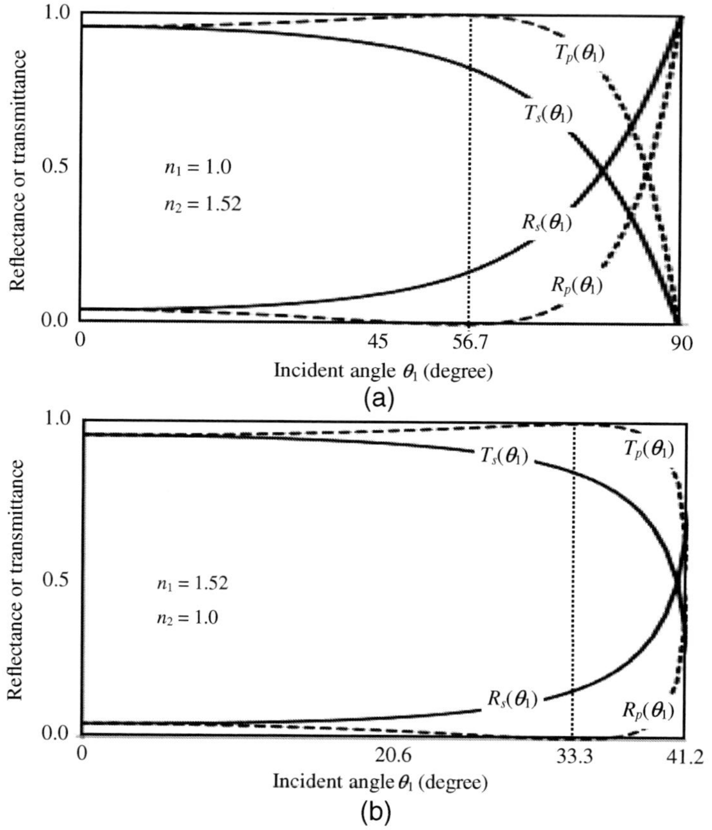

Figure 2.3 Reflectance and transmittance for *p*- and *s*-polarizations occurred at the interface of two optical materials with refractive indices n_1 and n_2, respectively. $R_s(\theta_1)$ and $R_p(\theta_1)$ are the reflectance of the *s*- and *p*-polarization, respectively. $T_s(\theta_1)$ and $T_p(\theta_1)$ are the transmittance of the *s*- and *p*-polarization, respectively. (a) $n_1 = 1$ and $n_2 = 1.52$. Around the incident angle of $\theta_1 \approx 56.7°$, the *p*-polarization has 0 reflectance, and the reflected light has pure *s*-polarization. (b) $n_1 = 1.52$ and $n_2 = 1$. Around the incident angle of $\theta_1 \approx 33.3°$, the *p*-polarization has 0 reflectance, and the reflected light has pure *s*-polarization. Above $\theta_1 \approx 41.2°$, both *p*- and *s*-polarizations are totally reflected. This phenomenon is called "total reflection."

s-polarizations are totally reflected. This phenomenon is called "total reflection." Total reflection only happens when the light wave propagates from a medium with a larger index to a medium with a smaller index. That is how an optical fiber can confine rays inside the fiber. Brewster's angle is utilized to make polarizer.

2.3.3 Phase change in reflections

Equation (2.6) demonstrates that for $n_1 < n_2$, $r_p(\theta_1)$ is always a real and negative number. Since the incident wave is assumed to be a unit wave that has a phase of 0, the real and negative $r_p(\theta_1)$ means it has a phase change of π

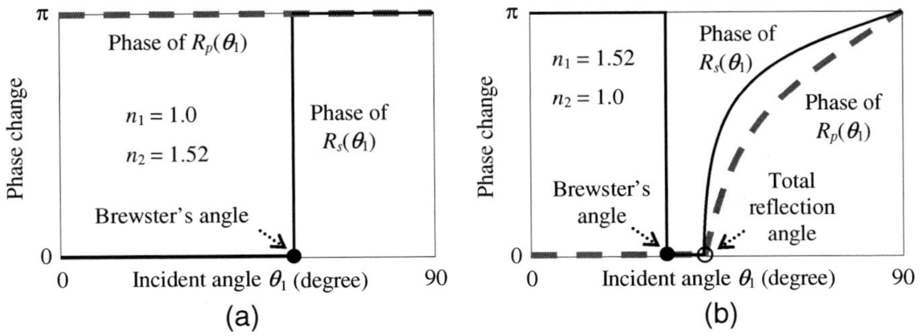

Figure 2.4 Phase change occurred at an optical interface as a function of the incident angle. Thin solid lines/curves are for *s*-polarized light. Thick lighter dash lines/curves are for *p*-polarized light. Phase change can be continuous or abruptly: (a) $n_1 = 1.0$ and $n_2 = 1.52$; (b) $n_1 = 1.52$ and $n_2 = 1.0$.

compared with the incident wave or the interface reflection causes a phase change of π to $r_p(\theta_1)$. The phase of $r_p(\theta_1)$ is plotted in Fig. 2.4(a).

For $n_1 > n_2$, $r_p(\theta_1)$ starts as a real and positive number, which means no phase change occurs during the reflection. Once θ_1 is above a certain value, $r_p(\theta_1)$ becomes a complex number, which means the phase of $r_p(\theta_1)$ starts changing, as shown in Fig. 2.4(b).

It can be seen from Eq. (2.8) that for $n_1 < n_2$, $r_s(\theta_1)$ starts as a real and positive number, which means no phase change occurs during the reflection. Once θ_1 is above a certain value, $r_s(\theta_1)$ becomes real and negative, which means the phase of $r_s(\theta_1)$ have a phase change of π, as shown in Fig. 2.4(a).

The situation for $n_1 > n_2$ is more complex. $r_s(\theta_1)$ starts as a real and negative number, then becomes real and positive once θ_1 is above a certain value. Once θ_1 is above another larger value, $r_s(\theta_1)$ becomes complex. The phase of $r_s(\theta_1)$ is plotted in Fig. 2.4(b).

2.4 Interference

Similar to mechanical waves, light waves can interfere with each other and result in interference. Two-wave interference and multiwave interference are commonly studied.

2.4.1 Two-wave interference

Figure 2.5 illustrates the case of combining two waves *a* and *b* of the same amplitude and frequency. If *a* and *b* are in phase, the amplitude of *a* and *b* add up, and the resultant wave *c* has an amplitude twice as large, as shown in Fig. 2.5(a). Such a phenomenon is called "constructive interference."

If *a* and *b* are out of phase, the amplitude of *a* and *b* offset each other, and the amplitude of the resultant wave *c* is zero, which means that wave *c* no long exists, as shown in Fig. 2.5(b). Such a phenomenon is called "destructive interference."

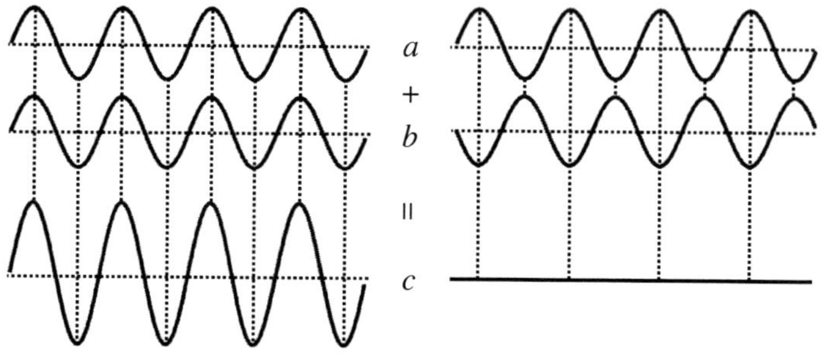

Figure 2.5 Illustration of two-wave interference. Waves *a* and *b* have the same amplitude. The combination of *a* and *b* leads to an interference wave *c*. (a) Waves *a* and *b* are in phase, and the amplitude of wave *c* is the sum of the amplitude of waves *a* and *b*. (b) Waves *a* and *b* are out of phase, the amplitude of waves *a* and *b* offset each other, and wave *c* either has zero amplitude or no longer exists.

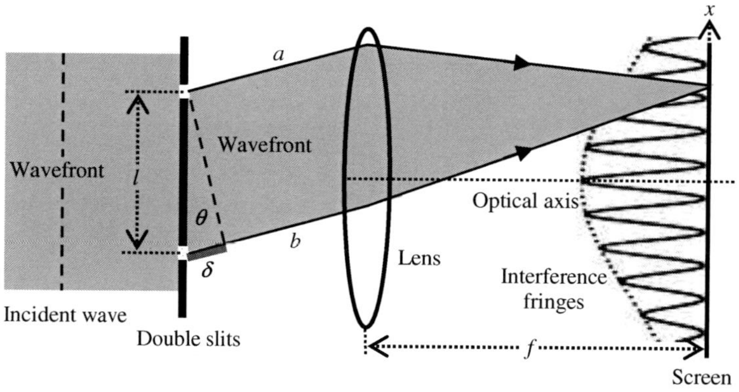

Figure 2.6 Schematic of a Young's double-slit experiment.

The famous Young's double-slit experiment, performed by Thomas Young in 1801,[2] was the first experiment to prove that light is a wave. Figure 2.6 shows the schematic of a Young's double-slit experiment. A light wave is incident on two slits that are carved on an opaque plate. The distance between the two slits is l. The two slits let two subwaves a and b pass through. A lens focuses waves a and b onto an observing screen. Waves a and b meet at the screen, and interference occurs between them.

For any wave propagation direction θ, The optical path difference between waves a and b is $\delta = l \times \sin(\theta)$, as shown in Fig. 2.6. As the observing point on the screen varies in the x direction, θ varies, and so does δ. Therefore, the phase difference between waves a and b varies alternately in the x direction from constructive to destructive. The intensity of light varies from maximum to zero on the screen in the x direction and forms intensity fringes. These fringes are called interference fringes. The fringe intensity is actually modulated by single-slit diffraction, shown by the dotted-line envelope in Fig. 2.6. Light diffraction is neglected in this section but will be discussed in Section 2.5.1.

Two-wave interference can be described mathematically. Let waves a and b be described by $A_0\sin(2\pi\nu t)$ and $A_0\sin(2\pi\nu t + 2\pi\delta/\lambda)$, respectively, where A_0 is the amplitude, ν is the frequency of the wave, and t is the time. At a point on the screen, the combined intensity of waves a and b is given by

$$\begin{aligned} I(\theta) &= \left\{A_0 \sin(2\pi\nu t) + A_0 \sin\left[2\pi\nu t + 2\pi\frac{\delta}{\lambda}\right]\right\}^2 \\ &= 4A_0{}^2 \sin^2\left[2\pi\nu t + \pi\frac{l\sin(\theta)}{\lambda}\right]\cos^2\left[\pi\frac{l\sin(\theta)}{\lambda}\right] \\ &= A_0{}^2\left\{1 + \cos\left[2\pi\frac{l\sin(\theta)}{\lambda}\right]\right\} \\ &\approx A_0{}^2\left\{1 + \cos\left[2\pi\frac{lx}{\lambda f}\right]\right\}. \end{aligned} \tag{2.10}$$

When deriving Eq. (2.10), the following facts are considered: the light intensity $I(\theta)$ is the square of the amplitude; the average of the very fast varying term $\sin^2[2\pi\nu t + \pi l\sin(\theta)/\lambda] = 0.5$; and $\sin(\theta) = x/(x^2 + f^2)^{0.5} \approx x/f$ since $f \gg x$. Equation (2.10) shows that $I(\theta)$ does vary nearly periodically as a function of x on the screen and forms interference fringes.

The process of analyzing interference is by first summing the amplitudes of the waves involved and then squaring the resultant amplitude to obtain the intensity. If the process is transposed, the square of the amplitude of the individual wave involved is first taken to obtain the intensity of every wave, and then the resultant intensities are summed; the final result will contain no interference fringes and is wrong.

2.4.2 Multiwave interference

Some optical devices, such as a diffraction grating and a Fabry–Pérot etalon (an optical etalon), create multiwaves and utilize multiwave interference. The multiwave interference to be studied in this section is created by an optical etalon with two parallel surfaces and thickness d, as shown in Fig. 2.7. The plate has a refractive index n_2 and is sandwiched in between two other materials with indices n_1 and n_3, respectively.

One light wave with unit amplitude is incident on the front surface of the etalon along the normal direction. In order to better illustrate the case, an incident angle > 0 and refraction angle $\theta > 0$ are drawn in Fig. 2.7 so that the multireflections and multitransmissions can be seen clearly.

A portion of the incident wave is reflected by the front surface of the etalon and is denoted by a_1. Another portion of the wave transmits through

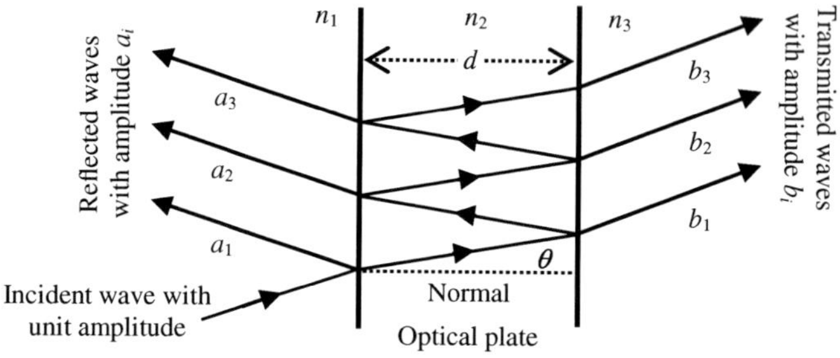

Figure 2.7 An optical wave is incident on an optical etalon with two parallel surfaces and thickness d. n_1, n_2, and n_3 are the refractive indices of the media before, inside, and after the etalon, respectively. The etalon can create multireflected waves a_i and multitransmitted waves b_i. Interference occurs among all a_i waves, as well as among all b_i waves.

the front surface of the etalon and is incident on the back surface of the etalon, where a portion of this wave is reflected, another transmits through the back surface of the etalon, and is denoted as b_1. The wave reflected by the back surface is then incident on the front surface of the etalon, a portion of this wave is reflected, another portion of this wave transmits through the front surface, and it is denoted as a_2. This process is on-going. Multireflected waves a_i and multitransmitted waves b_i are thereby created.

Each transmission lets a portion of the wave energy escape the etalon, and the amplitude of the wave remaining inside the etalon gradually falls. Eventually, the amplitude of the wave inside the etalon will decrease to zero. Figure 2.7 shows only the first three reflected waves, whose amplitudes are denoted by a_{1-3}, respectively, and the first three transmitted waves, whose amplitudes are denoted by b_{1-3}, respectively.

Let $R_1^{0.5}$ and $R_2^{0.5}$ be the amplitude reflection coefficients (R_1 and R_2 are the reflectance) and $T_1^{0.5}$ and $T_2^{0.5}$ be the amplitude transmission coefficients (T_1 and T_2 are the transmittances) of the front and the back surfaces of the etalon, respectively; the following equations can be written:

Equation	**Note**
$b_1 = T_1^{0.5} T_2^{0.5}$	The phase of b_1 is taken as the reference phase.
$b_2 = T_1^{0.5} T_2^{0.5} R_1^{0.5} R_2^{0.5}$	
$b_3 = T_1^{0.5} T_2^{0.5} R_1 R_2$	For p-polarization, normal reflection inside a glass does not cause any phase change, as shown in Fig. 2.4(b).
. . .	
$b_p = T_1^{0.5} T_2^{0.5} R_1^{0.5(p-1)} R_2^{0.5(p-1)}$	For s-polarization, twice the normal reflections inside a glass cause a 2π phase change, as shown in Fig. 2.4(b). Thus the phase change caused by reflections need not be considered here.

The phase difference δ between the two successive transmitted waves is given by

$$\delta = 2\pi \frac{2n_2 d}{\lambda \cos(\theta)}. \quad (2.11)$$

Note that for normal incidence, the refraction angle θ is 0. The amplitude of all of the transmitted rays combined is given by

$$\begin{aligned} E_T(\delta) &= \sum_{p=1}^{\infty} b_p e^{-i\delta} \\ &= T_1^{0.5} T_2^{0.5} \sum_{p=1}^{\infty} \left[R_1^{0.5} R_2^{0.5} e^{-i\delta}\right]^{p-1} \\ &= \frac{T_1^{0.5} T_2^{0.5}}{1 - R_1^{0.5} R_2^{0.5} e^{-i\delta}}. \end{aligned} \quad (2.12)$$

When deriving Eq. (2.12), the fact that $R_1^{0.5} R_2^{0.5} \cos(\delta) < 1$ is considered. The total transmitted power $I_T(\delta)$ is given by

$$\begin{aligned} I_T(\delta) &= |E_T(\delta)|^2 \\ &= \frac{(1 - R_1)(1 - R_2)}{1 + R_1 R_2 - 2{R_1}^{0.5} {R_2}^{0.5} \cos(\delta)}, \end{aligned} \quad (2.13)$$

where the relations $R_1 + T_1 = 1$ and $R_2 + T_2 = 1$ are used.

The total reflected power can be obtained by

$$\begin{aligned} I_R(\delta) &= 1 - I_T(\delta) \\ &= \frac{R_1 + R_2 - 2{R_1}^{0.5} {R_2}^{0.5} \cos(\delta)}{1 + R_1 R_2 - 2{R_1}^{0.5} {R_2}^{0.5} \cos(\delta)}. \end{aligned} \quad (2.14)$$

For a given optical etalon, the plate thickness d and index n are fixed, and the phase δ change can only be caused by a wavelength change.

Equations (2.13) and (2.14) are plotted in Fig. 2.8 against wavelength λ for a 2-mm-thick N-BK7 glass plate with a refractive index $n_2 = 1.52$ and a normal incidence $\theta = 0$. Note that once the wave energy enters the glass etalon, a larger R_1 and R_2 value means that less energy escapes the glass for each reflection, more reflections occur inside the plate, and more subwaves are generated. The interference effect joined by more waves is stronger. In Fig. 2.8, the transmitting fringes are much narrower for $R_1 = R_2 = 0.9$ than for $R_1 = R_2 = 0.5$. It is common to coat the two surfaces of a Fabry–Pérot etalon to $R_1 = R_2 = 0.99$ for extremely narrow fringes.

The curves plotted in Fig. 2.8 show that an optical etalon or an optical layer can be used as a spectral bandpass filter or spectral band reflector.

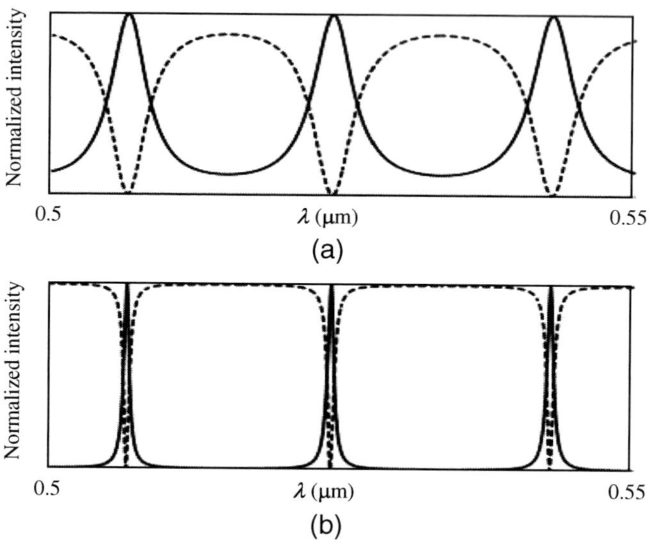

Figure 2.8 Multiwave interference patterns for a 2-mm-thick N-BK7 glass plate with normal incidence. The solid curve is the intensity of the total transmitted waves described by Eq. (2.13). The dashed curve is the intensity of the total reflected waves described by Eq. (2.14). The two surfaces of the glass plate are coated for (a) $R_1 = R_2 = 0.5$ and (b) $R_1 = R_2 = 0.9$.

All dielectric antireflection coatings and high-reflection coatings utilize such multiwave interferences.

2.5 Diffraction

Diffraction is a wave phenomenon and is somewhat similar to interference. Two types of diffraction are often encountered: slit and circular.

2.5.1 Single-slit diffraction

Figure 2.9 shows the schematics of a single-slit diffraction experiment. A planar wave is incident on a slit of width d carved on an opaque plate. The two edges of the slit clip the wave and spread the wave passing through the slit. A lens is placed behind the single slit to focus the wave passing through the slit onto a screen. A downward vertical coordinate y with it origin at the up edge of the slit is established for analysis purposes.

Consider a wave in a randomly selected direction θ to the normal of the screen. This wave can be considered as consisting of many subwaves or rays. Let the ray from $y = 0$ be the reference ray with zero phase and be described by its amplitude A_0. Then, a ray at y has a phase of $2\pi y\sin(\theta)/\lambda$, as shown in Fig. 2.9, and can be described by $A_0\sin[2\pi y\sin(\theta)/\lambda]$. The amplitude of the whole wave in the θ direction can be obtained by integrating all of the rays over the slit from $y = 0$ to $y = d$. The result is

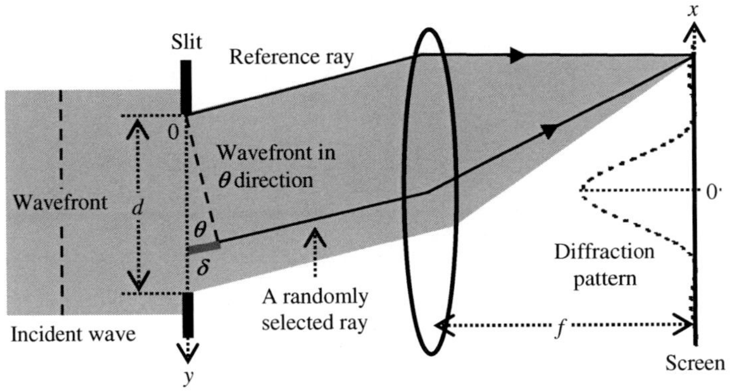

Figure 2.9 Schematics of a slit diffraction experiment. A slit of width *d* is carved on an opaque plate. A planar wave is incident on the slit. A lens with focal length *f* focuses the wave passing through the slit onto a screen, where the wave forms the diffraction intensity pattern.

$$A(\theta) = A_0\left[\int_0^d \sin\left[\frac{2\pi y \sin(\theta)}{\lambda}\right] dy\right]. \tag{2.15}$$

The intensity $I(\theta)$ of the whole wave in the θ direction can be obtained by taking the square of $A(\theta)$. The result is

$$\begin{aligned} I(\theta) &= A(\theta)^2 \\ &= A_0{}^2\left[\int_0^d \sin\left[\frac{2\pi y \sin(\theta)}{\lambda}\right] dy\right]^2 \\ &= I_0 \frac{\sin\left[\frac{\pi d \sin(\theta)}{\lambda}\right]^2}{\left[\frac{\pi d \sin(\theta)}{\lambda}\right]^2}, \end{aligned} \tag{2.16}$$

where $I_0 = A_0{}^2$. $I(\theta)$ is the single-slit diffraction intensity pattern.

The normalized $I(\theta)$ ($I_0 = 1$) as a function of θ is plotted in Fig. 2.10 for $d = 0.01$ mm and $\lambda = 0.5$ μm. A single-slit diffraction intensity pattern contains a central lobe and several weaker side lobes, though only two side lobes are plotted in Fig. 2.10. There are intensity maxima, and the intensity minima are zero.

In the intensity maxima directions, the constructive interference dominates. In the intensity minima directions, destructive interference occurs. The angular radius θ_a of the central lobe can be found by letting $\sin[\pi d\sin(\theta)/\lambda] = 0$ or $\pi d\sin(\theta)/\lambda = \pi$. Solving for θ, $\theta_a = \sin^{-1}(\lambda/d)$ is obtained. A smaller λ/d value results in a large θ_a for a given wave, which means a stronger diffraction effect.

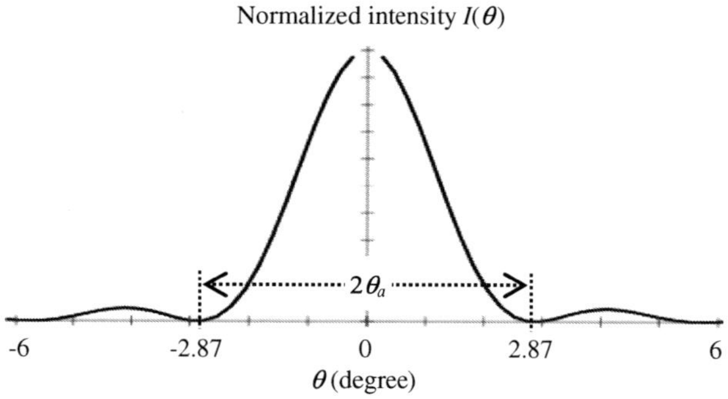

Figure 2.10 Single-slit normalized diffraction intensity pattern as a function of the diffraction angle θ for a slit width $d = 0.01$ mm and wavelength $\lambda = 0.5$ μm. The angular radius of the central intensity lobe is $\theta_a = 2.87°$.

In Eq. (2.16), sin(θ) can be replaced by x, which is the coordinate on the screen, as $\sin(\theta) = x/(f^2 + x^2)^{0.5} \approx x/f$, where f is the focal length of the lens, and in most cases $x \ll f$. The radius of the central lobe is given by $a \approx f\sin(\theta_a) \approx f\lambda/d$.

When studying double-slit interference, the single-slit diffraction effect is not considered. An accurate double-slit interference intensity pattern should be the multiplication of Eqs. (2.10) and (2.16), and the intensity of the interference fringes should be modulated by the diffraction pattern, as illustrated in Fig. 2.6.

2.5.2 Circular diffraction

The most commonly seen diffraction is circular diffraction. For example, a lens, which usually has a circular aperture, is used to manipulate a light wave. The lens aperture will cause circular diffraction. The principle behind circular diffraction is the same as the principle behind single-slit diffraction, but former is a 2D problem; the mathematics involved is more complex than that for the latter and is not addressed in this book.

Figure 2.11 shows the schematic of a circular diffraction experiment. A lens with a focal length f focuses a planar wave incident on the lens. The lens aperture of diameter D clips the wave and causes circular diffraction. The intensity pattern $I(\theta)$ of a circularly diffracted wave on the focal plane can be found by using the Kirchhoff diffraction integral:[3,4]

$$I(\theta) = \left\{ 2\frac{J_1\left[\frac{\pi D}{\lambda}\sin(\theta)\right]}{\frac{\pi D}{\lambda}\sin(\theta)} \right\}^2, \tag{2.17}$$

where θ is the angular variable, J_1 is the first-order Bessel function of the first kind, and λ is the wavelength. sin(θ) in Eq. (2.17) can be replaced by

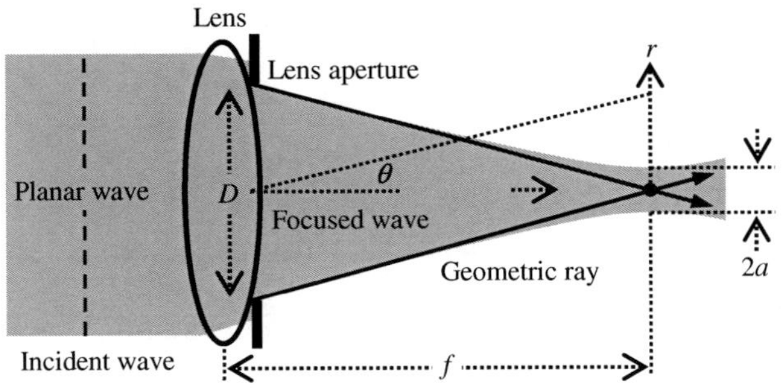

Figure 2.11 Schematics of a lens with a circular aperture of diameter *D* and focal length *f* focuses a planar wave. The lens aperture clips the wave and causes circular diffraction. θ is an angular variable. Each θ value corresponds to a value of the radial variable *r* at the lens focal plane. Even though the lens is aberration free and perfect, the real focused spot will have the smallest possible size of 2*a*, which is the diameter of a Airy disk (exaggerated here for illustration). The geometric rays depicted by the solid lines are only an approximation and leads to an erroneous, infinitely small focused spot depicted by a solid dot.

$\sin(\theta) = r/(f^2 + r^2)^{0.5}$, where r is the radial variable at the lens focal plane. Then $I(\theta)$ in Eq. (2.17) should be written as $I(r)$.

Figure 2.12(a) plots $I(r)$. The circular diffraction intensity pattern is similar to the intensity pattern of single-slit diffraction and contains a central lobe and several side lobes. The central lobe radius a can be found by letting $J_1[(\pi D/\lambda)r/(f^2 + r^2)^{0.5}] = 0$ and solving for r:

$$a = \frac{1.22f\lambda}{D}. \tag{2.18}$$

a is proportional to f/D and λ.

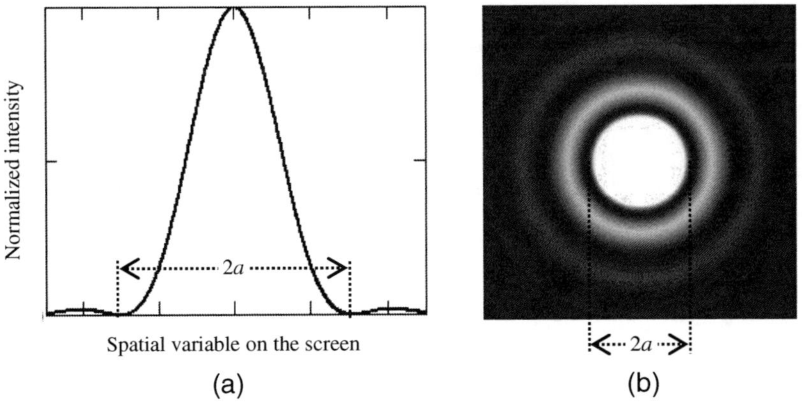

Figure 2.12 (a) A circular diffraction intensity pattern. (b) A simulated 2D greyscale diagram of a circular diffraction intensity pattern. The side rings are enhanced for demonstration.

Equation (2.18) is often written as $a = 1.22\lambda(F/\#)$ or $a = 0.61\lambda/(\mathrm{NA})$, where $F/\# = f/D$ is the F-number, and $NA = D/(2f)$ is the numerical aperture of the focusing lens.

Consider a numerical example where $D = 20$ mm, $\lambda = 0.5$ μm, and $f = 100$ mm; the central lobe radius is $a = 1.22\lambda f/D = 3.05$ μm.

A simulated 2D greyscale diagram of a circular diffraction intensity pattern is plotted in Fig. 2.12(b). The central intensity lobe is the famous "Airy disk" and defines the smallest possible focused spot for a given lens and wavelength, or it defines the lens resolution limit. The Airy disk contains about 84% of the total energy contained in the wave. The remaining 16% energy spreads in the side lobes. As a comparison, geometrical optics, which is only an approximation, says that if the lens is free of aberrations, the focused spot size is infinitely small, as depicted by the solid lines and the solid dot in Fig. 2.11.

If an Airy disk a few microns in size is not an issue in the application, geometrical optics is a good approximation and can be used to simplify the analysis. However, if the Airy disk size is an issue (e.g., a sensor with 2-μm pixels is used to detect the focused spot), then such an Airy disk will reduce the sensor resolution and cannot be neglected.

Equation (2.18) can be written as $a(D/f) = 1.22\lambda$ = constant for a given wavelength. $a(D/f)$ is the etendue for wave optics, where D/f is the solid angle that the lens aperture imposes on the focal point. It is impossible to reduce the lens aperture size D while maintaining the Airy disk size a because the etendue, if changed, can only be increased. It is possible to increase the lens aperture size D while keeping the Airy disk a unchanged by introducing aberration to the lens.

2.5.3 Comparison with a Gaussian pattern

Figure 2.13 plots the intensity profiles of a circular diffraction, a slit diffraction, and a Gaussian fitting. The parameters of these three curves are adjusted so that their central lobes almost overlap.

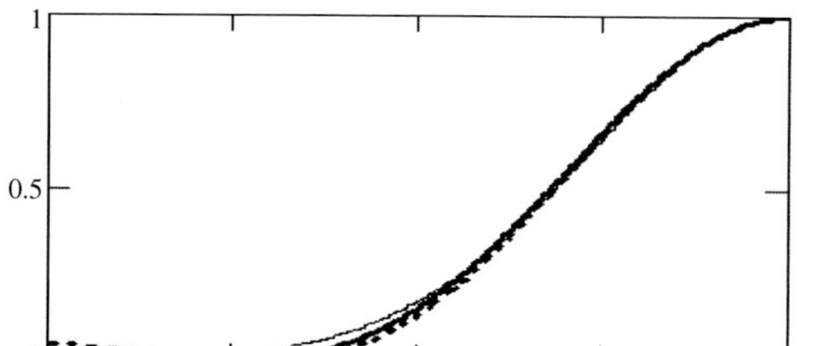

Figure 2.13 Comparison among the intensity profiles of circular diffraction (thick solid curve), slit diffraction (thick dashed curve), and Gaussian (thin solid curve). Only half of the patterns are plotted in order to show more details.

The three central lobes are fairly similar, the Gaussian profile has a long tail and no side lobe, and the slit diffraction pattern has side lobes larger than the side lobes of the circular diffraction pattern. Figure 2.13 reveals that for most applications, a Gaussian fit to the diffraction pattern will be sufficiently accurate. The mathematics for a Gaussian pattern is much simpler, and some important results obtained for Gaussian beams can be readily used. For example, a Gaussian beam will remain Gaussian as it travels. If the accuracy requirement is not high, it can be assumed that the intensity pattern of both the slit and circular diffractions remain Gaussian at any observation distances.

References

1. B. E. A. Saleh and M. C. Teich, "Reflection and Refraction," *Fundamentals of Photonics*, 2nd ed., Wiley & Sons, Inc., New York, pp. 209–214 (2007).
2. Wikipedia, "Young's Interference Experiment," https://en.wikipedia.org/wiki/Young%27s_interference_experiment
3. M. Born and E. Wolf, "Kirchhoff's Diffraction Theory," *Principles of Optics*, Cambridge University Press, Cambridge, England, pp. 417–429 (2001).
4. W. Smith, *Modern Optical Engineering*, 3rd ed., McGraw-Hill, New York, Eqs. (6.18) and (6.19), p. 159 (2000).

Chapter 3
Gaussian Beam Optics

3.1 Gaussian Beam Equations

Neither geometrical optics nor wave optics can accurately describe the characteristics of a laser beam. The propagation characteristics of a laser beam must be described by a set of three Gaussian equations:[1]

$$\begin{aligned} w(z) &= w_0\left[1+\left(\frac{M^2\lambda z}{\pi w_0^2}\right)^2\right]^{1/2} \\ &= w_0\left[1+\left(\frac{z}{z_R}\right)^2\right]^{1/2}, \end{aligned} \tag{3.1}$$

$$\begin{aligned} R(z) &= z\left[1+\left(\frac{\pi w_0^2}{M^2\lambda z}\right)^2\right] \\ &= z\left[1+\left(\frac{z_R}{z}\right)^2\right], \end{aligned} \tag{3.2}$$

$$I(r,z) = I_0(z)e^{-2r^2/w(z)^2}, \tag{3.3}$$

where $w(z)$ is the $1/e^2$ intensity radius of the beam at z; z is the axial distance from the point of interest to the waist of the laser beam; w_0 is the $1/e^2$ intensity radius of the beam waist;

$$z_R = \frac{\pi w_0^2}{M^2\lambda} \tag{3.4}$$

is an important parameter called the "Rayleigh range;" $M^2 \geq 1$ is the M-squared factor that describes the deviation of this beam from a perfect laser beam with $M^2 = 1$; λ is the wavelength; $R(z)$ is the beam wavefront curvature radius at z; $I(r,z)$ is the beam intensity radial distribution at a cross-section plane at z; r is the radial variable at a cross-section plane; and $I_0(z)$ is the beam peak intensity at a cross-section plane at z.

The frequently used terms "near field" and "far field" are qualitatively defined as $z \sim z_R$ and $z \gg z_R$, respectively. The z value between the near field and far field is the intermediate field. It can be found from Eqs. (3.1) and (3.4) that at $z = z_R$, $w(z) = 2^{0.5} w_0$.

Equations (3.1) and (3.2) are paraxial solutions of the wave propagation equation, the solving of which is not covered here (see Siegman[2] for more details).

3.2 Beam Size Characteristics

Equation (3.1) shows that the minimum beam size $w(z)$ appears at $z = 0$, where $w(0) = w_0$, that is why w_0 is called a beam waist. The beam waist is located somewhere inside the laser cavity when the laser beam is created, and the beam propagation distance is measured from the beam waist. Since it is not convenient to locate the beam waist inside a laser cavity, the beam waist is often assumed to be located at the output window of the laser.

At the near field $z \sim z_R$, $w(z)$ is not a linear function of z and must be treated by Gaussian beam optics. According to Eq. (3.1) at the far field $z/z_R \gg 1$, $w(z)$ reduces to

$$\begin{aligned} w(z) &= \frac{w_0 z}{z_R} \\ &\approx \frac{M^2 \lambda z}{\pi w_0}. \end{aligned} \tag{3.5}$$

Equation (3.5) contains two important results. First, the far-field beam size is proportional to the propagation distance z, which can significantly simplify the analysis of far-field beam behaviors. Second, the far-field beam size is inversely proportional to the beam waist radius w_0. In many applications, people want to deliver a laser beam to a target far away with a size as small as possible for high energy density. To achieve this goal, the beam waist w_0 must be made as large as practically possible. Sun[3] provides a detailed discussion on this topic.

According to Eq. (3.5), the far-field $1/e^2$ intensity half-divergence θ can be found by

$$\begin{aligned} \theta &= \frac{w(z)}{z} \\ &\approx \frac{M^2 \lambda}{\pi w_0}. \end{aligned} \tag{3.6}$$

Equation (3.6) can be written as $\theta w_0 = M^2\lambda/\pi$ = constant for a given wavelength. That is the etendue of Gaussian beam. It's impossible to decrease the value of w_0, while maintain the value of θ unchanged. Since etendue, if changed, can only increase. It is possible to increase the value of w_0 and maintain the value of θ by increasing the value of the M^2 factor (i.e., disrupt the beam).

For a Gaussian beam with an intensity described by $I(x) = \exp(-2x^2/w_0^2)$, its full width at half magnitude (FWHM) radius w_{FWHM} can be found by letting $I(x) = 0.5$ and solving for x:

$$\begin{aligned} w_{\mathrm{FWHM}} &= w_0 \left[\frac{\ln(0.5)}{-2}\right]^{0.5} \\ &\approx 0.589 w_0, \end{aligned} \tag{3.7}$$

or, in reverse, $w_0 \approx 1.7 w_{\mathrm{FWHM}}$. The laser industry tends to use w_{FWHM} to specify the laser beam size, whereas the optics community uses w_0 to specify the width of a Gaussian pattern. When discussing laser beam size, the beam size must be clearly defined to avoid confusion.

3.3 Beam Wavefront Characteristics

At the waist, the wavefront of a beam is planar, the wavefront radius of curvature is infinite, and, as given by Eq. (3.2), $R(0)$ is infinite. At the near field $z \sim z_R$, $R(z)$ is not a linear function of z and must be treated by Gaussian beam optics. As the beam propagates, the value of $R(z)$ initially drops. The value of z at which $R(z)$ is minimum, as well as the minimum value of $R(z)$, can be found by solving $dR(z)/dz = 0$ for z and $R(z)$. The results are $z = z_R$ and $R(z_R) = 2z_R$.

At the far field $z/z_R \gg 1$, $R(z)$ reduces to $R(z) \approx z$ according to Eq. (3.2). Such a wavefront is spherical and can be treated by geometrical optics.

3.4 Intensity and Power

The total power P_0 of a laser beam can be found by

$$P_0 = 2\pi I_0 \int_0^\infty e^{-2r^2/w_0^2}\, r dr = \frac{I_0}{2}\pi w_0^2. \tag{3.8}$$

Equation (3.8) can be written as

$$I_0 = \frac{2P_0}{\pi w_0^2}. \tag{3.9}$$

It is easy to measure the total power P_0 of a laser beam but difficult to measure the beam's on-axis intensity I_0. Equation (3.9) connects these parameters, and Fig. 3.1(a) provides further illustration.

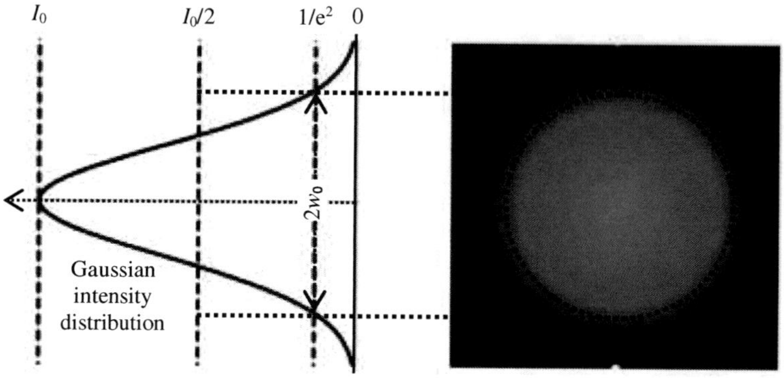

Figure 3.1 The total power of a Gaussian beam P_0 is related to the on-axis intensity I_0 by $P_0 = (I_0/2)\pi w_0^2$.

At distance z, the peak on-axis intensity $I_0(z)$ can be found by replacing w_0 by $w(z)$ in Eq. (3.8). The result is

$$I_0(z) = \frac{2P_0}{\pi w(z)^2}. \tag{3.10}$$

The power of a laser beam passing through a hole with radius a and a distance z from the beam waist is

$$\begin{aligned} P(z,a) &= 2\pi P_0 \int_0^a e^{-2r^2/w(z)^2} r dr \\ &= p_0\left(1 - e^{-2a^2/w(z)^2}\right). \end{aligned} \tag{3.11}$$

$a = w(z)$ leads to

$$\frac{P(z,a)}{P_0} = (1 - e^{-2}) \approx 0.865, \tag{3.12}$$

i.e., the fraction power encircled inside the $1/e^2$ intensity level is about 0.865.

3.5 Graphic Explanations of Laser-Beam-Propagation Characteristics

Figure 3.2 plots the propagation characteristics of two laser beams. Note the following three points:

1. At the near field, both $w(z)$ and $R(z)$ are nonlinear functions of the propagation distance z; Gaussian beam optics [Eqs. (3.1) and (3.2)] must be used to analyze the beams.

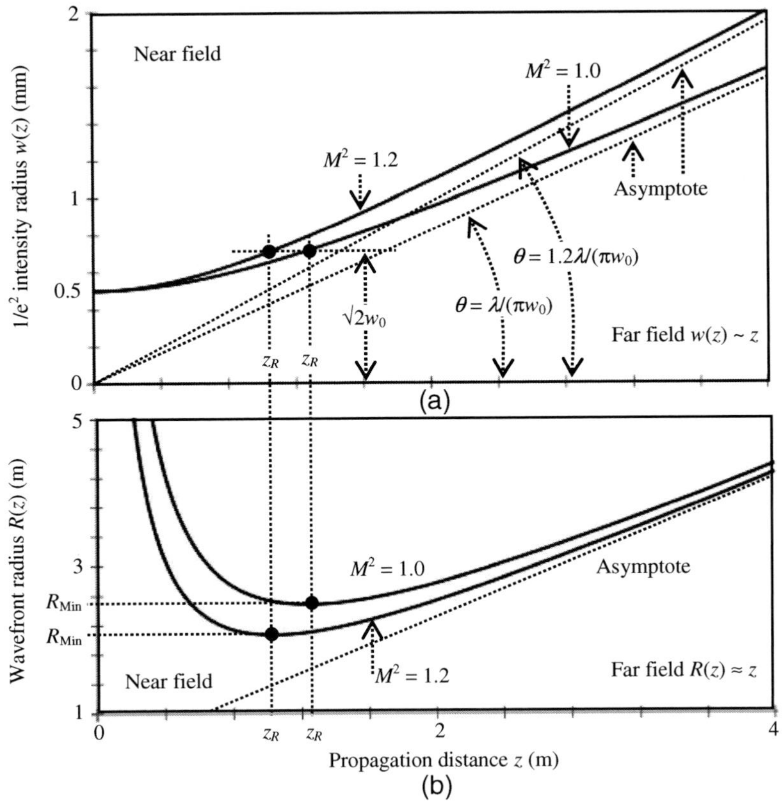

Figure 3.2 The propagation characteristics of two laser beams with $\lambda = 0.6328$ μm, $w_0 = 0.5$ mm, and $M^2 = 1$ and 1.2, respectively. The beam waist is positioned at $z = 0$. The two Rayleigh ranges z_R are marked. (a) The solid curves are the $1/e^2$ intensity radius $w(z)$ as a function of the propagation distance z. In the far field, $w(z)$ is proportional to z, and the beam divergence θ is constant, as shown by the two asymptotes. (b) The solid curves are the beam wavefront radius $R(z)$ as a function of the beam propagation distance z. The minimum $R(z)$ appears at $z = z_R$. In the far field, $R(z)$ is proportional to z, and the wavefront is spherical.

2. Before $z = 4$ m, both beams reach their far field. This can be verified by the fact that both $w(z)$ and $R(z)$ can be approximately represented by two asymptote lines that originate at the center of the beam waist. At the far field, the two beams reduce to two spherical waves with $R(z) \approx z$, and the two waves have constant divergent angles. Such waves can be treated as rays emitted from a conceived source point at the center of the beam waist, geometrical optical can be used to analyze the beams, and the mathematical process involved can be simplified.
3. The far-field divergence is proportional to the M^2 value. For a given beam waist size and wavelength, the best-quality beam has the smallest value of $M^2 = 1$ and the smallest possible far-field divergent angle.

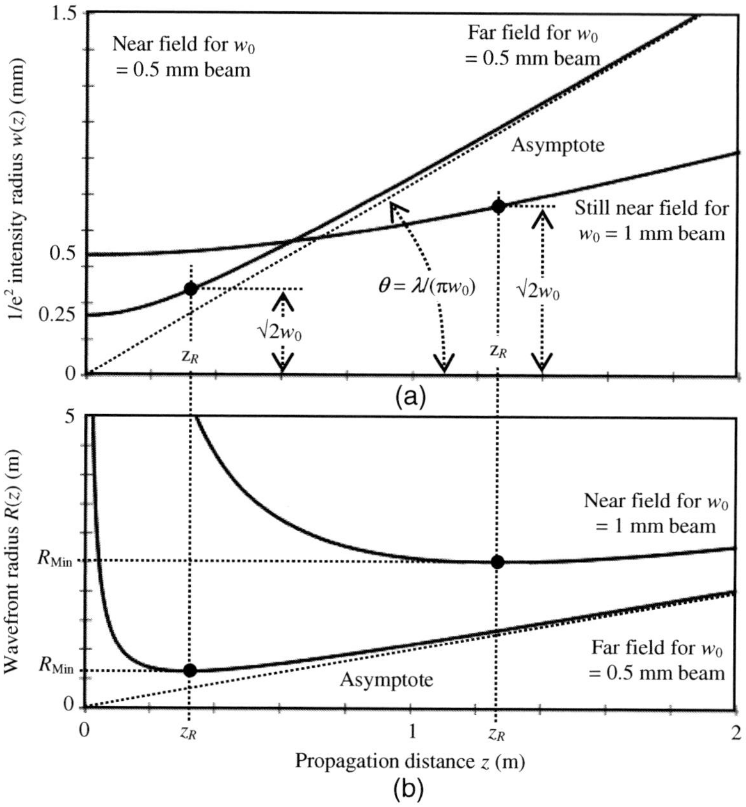

Figure 3.3 The propagation characteristics of two laser beams with $\lambda = 0.6328$ μm, $M^2 = 1$, and $w_0 = 0.5$ mm and 1 mm, respectively. The beam waist is positioned at $z = 0$. The two Rayleigh ranges z_R are marked. (a) The solid curves are the $1/e^2$ intensity radius $w(z)$ as a function of the propagation distance z. (b) The solid curves are the beam wavefront radius $R(z)$ as a function of the beam propagation distance z. The minimum $R(z)$ appears at $z = z_R$. Before $z = 2$ m, the $w_0 = 0.5$ mm beam reaches the far field, both $w(z)$ and $R(z)$ can be represented by asymptote lines. While the $w_0 = 1$ mm beam is still in the near field, both $w(z)$ and $R(z)$ are still curves.

Figure 3.3 plots the propagation characteristics of two other laser beams for further discussion. Before $z = 2$ m, the beam with $w_0 = 0.5$ mm already reaches the far field, both $w(z)$ and $R(z)$ can be represented by two asymptote lines. Conversely, the beam with $w_0 = 1$ mm is still in the near field around $z = 2$ m, where both $w(z)$ and $R(z)$ are still a little curved.

3.6 Thin Lens Equation for a Real Laser Beam

The thin lens equation [Eq. (1.15)] is initially derived to analyze geometrical rays passing through a lens but later modified to analyze a real laser beam propagating through a lens.[1] The modified Gaussian-beam-optics thin lens equation takes the following form:

$$\frac{i}{f} = \frac{\frac{o}{f}\left(\frac{o}{f} - 1\right) + \left(\frac{z_R}{f}\right)^2}{\left(\frac{o}{f} - 1\right)^2 + \left(\frac{z_R}{f}\right)^2}, \tag{3.13}$$

where the object distance o and the image distance i are defined as the distances between the lens and the waist of the input and output beam, respectively; z_R is the input-beam Rayleigh range; and f is the lens focal length. The object is the input beam waist, and the image is the output beam waist. The magnification m of the lens then becomes the ratio of the output beam waist size w_0' to the input beam waist size w_0 given by[1]

$$\begin{aligned} m &= \frac{w_0'}{w_0} \\ &= \frac{1}{\left[\left(\frac{o}{f} - 1\right)^2 + \left(\frac{z_R}{f}\right)^2\right]^{1/2}}. \end{aligned} \tag{3.14}$$

The derivation of Eq. (3.14) is complex and thus not covered here (see Sun[1]).

3.7 Collimating and Focusing: Maximum and Minimum Focusing Distance

There are four interesting results from Eqs. (3.13) and (3.14):

1. For $z_R/f = 0$, the beam waist size is negligible, the beam reduces to a point source, and geometrical optics applies. Equation (3.13) reduces to Eq. (1.15), and Eq. (3.14) reduces to Eq. (1.16), which is the lens magnification for geometrical optics.
2. When $o = f$, Eq. (3.13) results in $i = f$, which means that the waists of the input and output beams are located at the two focal points of the lens, respectively. This result is very different from geometrical optics, i.e., when the object is at the focal plane, the image is at infinity.
3. For $o = f$, the magnification reduces to $m = f/z_R$ according to Eq. (3.14). The situation is a laser beam that is being collimated, focused, or relayed, depending on the ratio of f/z_R. Figure 3.4 illustrates this situation with more details.
4. There is a maximum image distance i_{max} and a minimum image distance i_{min}. i_{max} and i_{min} can be found by differentiating i and o in Eq. (3.13) and letting $\Delta i/\Delta o = 0$. The results are

$$i_{max} = f\frac{\frac{2z_R}{f} + 1}{\frac{2z_R}{f}} \qquad \text{when } o = f + z_R, \tag{3.15}$$

and

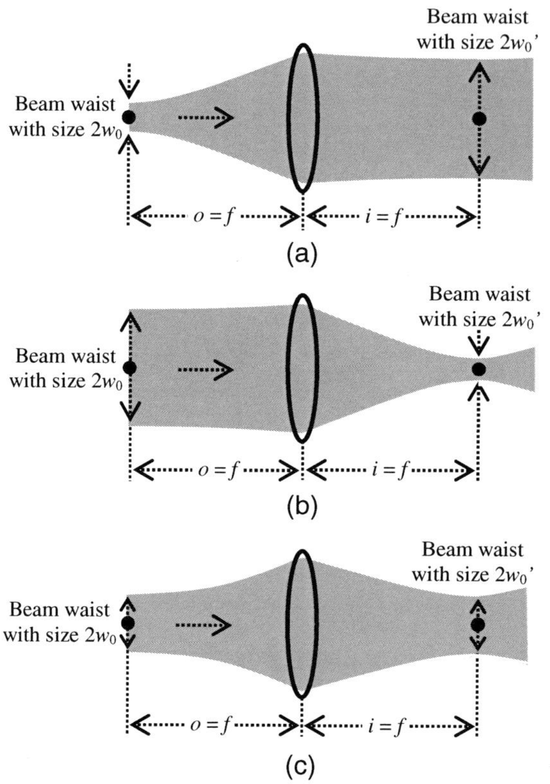

Figure 3.4 The waist of a laser beam is placed at the focal point of a positive lens, and the waist of the beam propagated through the lens appears at the other focal point of the lens. f is the focal length of the lens; the focal points of the lenses are marked by the solid dots; and the waist size of the input and output beams is marked by $2w_0$ and $2w_0'$, respectively. (a) $f > z_R$, $m = w_0'/w_0 > 1$, i.e., the beam is collimated. (b) $f < z_R$, $m = w_0'/w_0 < 1$, i.e., the beam is focused. (c) $f = z_R$, $m = w_0'/w_0 = 1$, i.e., the beam is relayed.

$$i_{\min} = f \frac{\frac{2z_R}{f} - 1}{\frac{2z_R}{f}} \qquad \text{when } o = f - z_R. \tag{3.16}$$

Any attempt to focus a laser beam at infinity will not succeed—another result of Gaussian beam optics that is very different from geometrical optics.

To present an overall view, Fig. 3.5 plots the Gaussian-beam-optics thin lens equation [Eq. (3.13)] and the Gaussian-beam-optics lens magnification equation [Eq. (3.14)].

3.8 Two Examples of Focusing or Collimating Laser Beams

3.8.1 Focusing a He-Ne laser beam

Assume the laser beam has a wavelength of $\lambda = 0.633$ μm and a $1/e^2$ intensity waist radius $w_0 = 0.5$ mm, and the focusing lens has a focal length $f = 10$ mm. According to Eq. (3.4), this beam has $z_R = 1.24$ m $\gg f = 10$ mm.

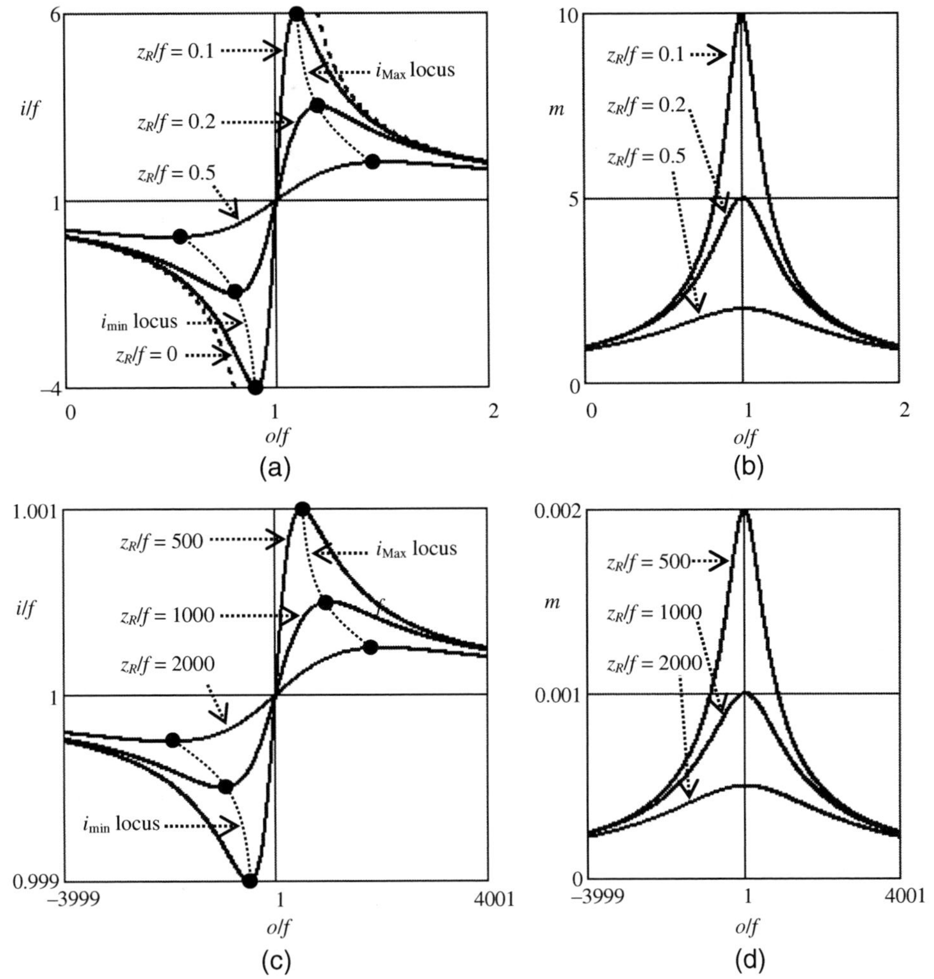

Figure 3.5 (a) and (c) Plots of the Gaussian-beam-optics thin lens equation Eq. (3.13) i/f vs. o/f for various z_R/f values. (b) and (d) Plots of Gaussian-beam-optics lens magnification Eq. (3.14) m vs. o/f for various z_R/f values. For $z_R/f > 1$, the lens focuses the beam, $m < 1$. For $z_R/f < 1$, the lens collimates the beam, $m > 1$. For $z_R/f = 0$, the beam reduces to a geometrical point source, the i/f vs. o/f relation is plotted by the dashed curve in (a) for comparison. i_{max} and i_{min} are marked by the solid dots in (a) and (c).

The spherical aberration of the lens is not considered since the F-number is $f/(2w_0) = 10$, which is very large.

To focus the beam, the beam waist must be placed at the focal point of the lens, as shown in Fig. 3.4(b), with $o = f$. After combining Eqs. (3.4) and (3.14), the $1/e^2$ intensity waist radius of the focused beam is

$$w_0' = \frac{M^2 f \lambda}{\pi w_0}. \tag{3.17}$$

He-Ne lasers have a high beam quality, and $M^2 \approx 1$, so Eq. (3.17) results in $w_0' = 4\ \mu\text{m}$.

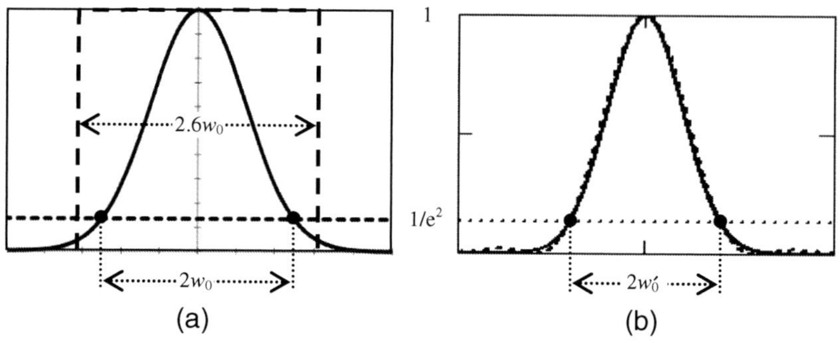

Figure 3.6 A comparison of focusing a Gaussian beam and a planar wave. (a) The solid curve is a Gaussian-beam-waist intensity profile with a $1/e^2$ intensity diameter of $2w_0$. The dashed square is a planar wave with a diameter of $2.6w_0$. (b) The solid curve is the intensity profile of the focused spot generated by the Gaussian beam shown in (a). The dashed line curve is the intensity profile of the focused spot generated by the planar wave shown in (a). These two focused spots have the same $1/e^2$ intensity diameters of $2w_0'$.

3.8.2 Collimating a laser diode beam

Most laser diode beams are elliptical and have a tiny waist size. For discussion purposes, the beam is assumed to be circular with $w_0 = 1$ μm. The other parameters are assumed to be $\lambda = 0.65$ μm, $f = 10$ mm, and $M^2 = 1.1$. According to Eq. (3.4), this beam has z_R = 0.0044 mm $\ll f = 10$ mm.

To collimate a laser beam, the beam waist must be placed at the focal point of the lens as well with $o = f$, as illustrated in Fig. 3.4(a). The $1/e^2$ intensity radius of the collimated beam can be found by combining Eqs. (3.4) and (3.14), or directly from Eq. (3.17). The result is $w_0' \approx 2.28$ mm. Note that the F-number of this collimating lens is 10 mm/(2 × 2.28 mm) ≈ 2.2. For such a small F-number, a single spherical lens will have severe spherical aberration that is neglected here, so an aspheric lens or spherical lens group must be used to minimize the spherical aberration.

3.8.3 Focusing a Gaussian beam versus focusing a planar wave

It is interesting to compare the focusing of a Gaussian beam and a planar wave. When a Gaussian beam waist has a $1/e^2$ intensity diameter of $2w_0$ and a planar wave has a diameter 1.3 times larger than that, as shown in Fig. 3.6(a), the two focused spots have the same $1/e^2$ intensity diameters, as shown in Fig. 3.6(b).

The result shown in Fig. 3.6 can be easily proven using Eqs. (2.18) and (3.17).

References

1. H. Sun, "Thin lens equation for a real laser beam with weak lens aperture truncation," *Op. Eng.* **37**, 2906–2913 (1998).

2. A. E. Siegman, "Wave Optics and Gaussian Beams" and "Physical Properties of Gaussian Beams," Chapters 16 and 17 in *Lasers*, University Science Books, South Orange, NJ, pp. 626–697 (1986).
3. H. Sun, "Analysis of delivering a laser diode beam with the smallest spot." *Opt. Eng.* **51**, 019701-1–019701-4 (2012).

Chapter 4
Optical Materials

4.1 Properties

Many types of materials are used to make optical components, which include hundreds of types of optical glasses; several types of optical polymers; crystals and infrared materials; and a few types of UV materials.

There are several important properties of optical materials, such as refractive index, Abbe number, transmission spectral range, partial dispersion or deviation of partial dispersion, refractive index temperature dependence, thermal expansion, chemical resistance, mechanical strength, softening and melting temperature, density, price, availability, etc. All of these properties are briefly discussed in this section.

Optical material vendors will provide data in their product catalogs for most of these material properties. Optical design software will also provide that information. Reference 1 is an excellent online information source that provides the refractive index, transmission range, and many other data for ~200 types of optical materials. References 2–4 are relatively new books on glasses and other optical materials; Musikan[5] is an older book on optical materials; Klocek[6] is about infrared materials; and Nikogosyan[7] is about optical crystals.

4.1.1 Refractive index

The most important parameter that describes these optical materials is the refractive index, which is not a constant. It varies as a function of wavelength. The varying index of optical materials causes color aberrations in optical systems.

This section includes several polynomial formulas that are equally capable of describing the refractive index of optical materials. Many coefficients are used in these formulas. Optical material manufacturers provide at least one set of these coefficients in their catalogs so that users have at least one formula to simulate the index. The symbol λ used in these formulas represents wavelength, and the unit of wavelength depends on the unit of the coefficients.

The Schott formula is

$$n(\lambda)^2 = a_0 + a_1\lambda^2 + a_2\lambda^{-2} + a_3\lambda^{-4} + a_4\lambda^{-6} + a_5\lambda^{-8}, \tag{4.1}$$

where a_0 to a_5 are coefficients.

The Sellmeier 5 formula is

$$n(\lambda)^2 - 1 = \frac{K_1\lambda^2}{\lambda^2 - L_1} + \frac{K_2\lambda^2}{\lambda^2 - L_2} + \frac{K_3\lambda^2}{\lambda^2 - L_3} + \frac{K_4\lambda^2}{\lambda^2 - L_4} + \frac{K_5\lambda^2}{\lambda^2 - L_5}, \tag{4.2}$$

where K_1 to K_5 and L_1 to L_5 are coefficients. The Sellmeier 4 formula uses the first four terms on the right side of Eq. (4.2), and the Sellmeier 3 formula uses the first three terms.

The Herzberger formula is

$$n(\lambda) = A + BL + CL^2 + D\lambda^2 + D\lambda^4 + E\lambda^6, \tag{4.3}$$

where $L = 1/(\lambda^2 - 0.028)$, and A to E are coefficients.

The extended formula

$$n(\lambda)^2 = a_0 + a_1\lambda^2 + a_2\lambda^{-2} + a_3\lambda^4 + a_4\lambda^{-4} + a_5\lambda^6 + a_6\lambda^{-6} + a_7\lambda^8 + a_8\lambda^{-8}, \tag{4.4}$$

where a_0 to a_8 are coefficients.

Figures 4.1(a) and 4.2(a) plot the refractive index $n(\lambda)$ of the most widely used optical glass, N-BK7, and UV-grade fused silica, respectively. The refractive index value drops as the wavelength increases, or $dn(\lambda)/d\lambda < 0$, and that even $dn(\lambda)/d\lambda$ is not a constant; in the short-wavelength range, the index value varies faster. All of these properties are common for almost all

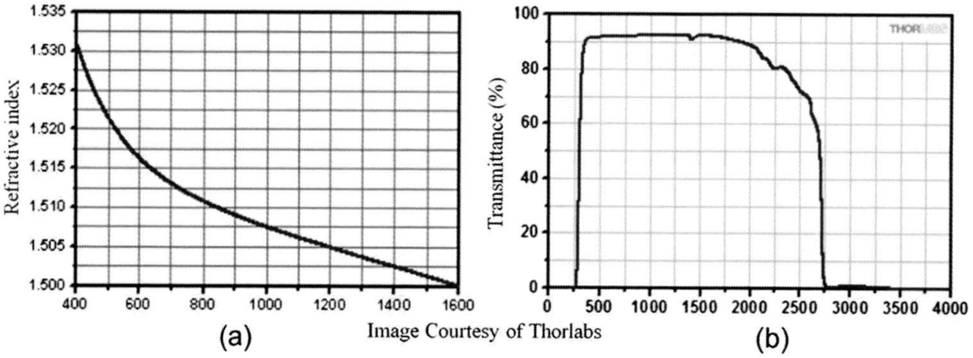

Figure 4.1 (a) Refractive index of N-BK7 glass as a function of wavelength. (b) Transmittance of a 10-mm-thick N-BK7 glass plate with uncoated surfaces. The peak transmittance is almost 100% if the two-surface reflectance of ~8.6% is excluded. The high transmittance range covers from visible to ~2 μm.

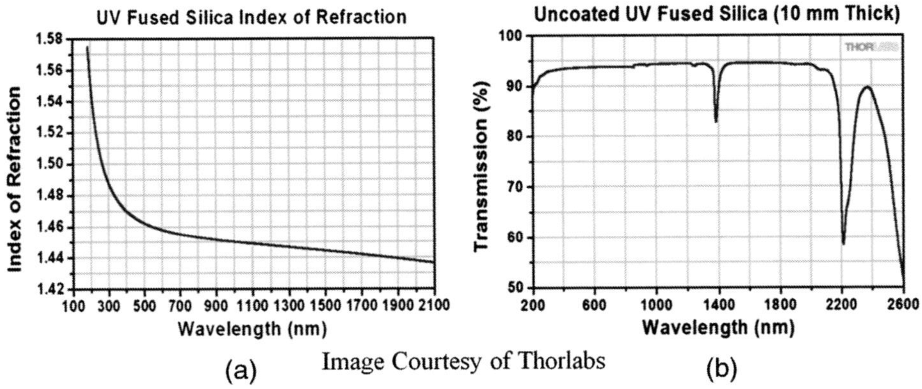

Figure 4.2 (a) Refractive index of UV-grade fused silica as a function of wavelength. (b) Transmittance of a 10-mm-thick UV-grade fused silica plate with uncoated surfaces. The peak transmittance is ~93% if the two-surface reflectance of ~6.8% is excluded. The high transmittance range covers from below 200 nm in the UV to ~2 μm, except for a dip around 1.4 μm.

optical materials. To simplify the situation, people usually use n_D, the index value at a wavelength of 589 nm, as the approximate index value of optical materials. Then, n_D is about 1.52 and 1.46 for N-BK7 glass and UV-grade fused silica, respectively. Optical glasses have an n_D in the range of about 1.4 to about 2.0, whereas optical polymers have an n_D around 1.55.

4.1.2 Abbe number

The Abbe number is a parameter that approximately describes how quickly the refractive index changes as the wavelength changes. The Abbe number V_D is defined as[8]

$$V_D = \frac{n_D - 1}{n_F - n_C}, \tag{4.5}$$

where n_F is the index value at a blue color of 0.486 μm, and n_C is the index at a red color of 0.656 μm. A larger value of V_D indicates that the values of n_F and n_C are closer to each other or the $n(\lambda)$ curve is relatively flatter, and this glass has relatively less color dispersion. The Abbe number is also called the V-number. Glass vendors routinely provide V-numbers in their product catalogs for all of their optical materials. Optical glasses have a V_D value between about 16 to 95, whereas optical polymers have a V_D value around either 30 or 55.

Figure 4.3 is a Schott glass map, also called an Abbe diagram, that lists many glasses made by Schott. Those with a large index tend to have a small Abbe number, and vice versa. The glass maps of other brands of glasses are similar.

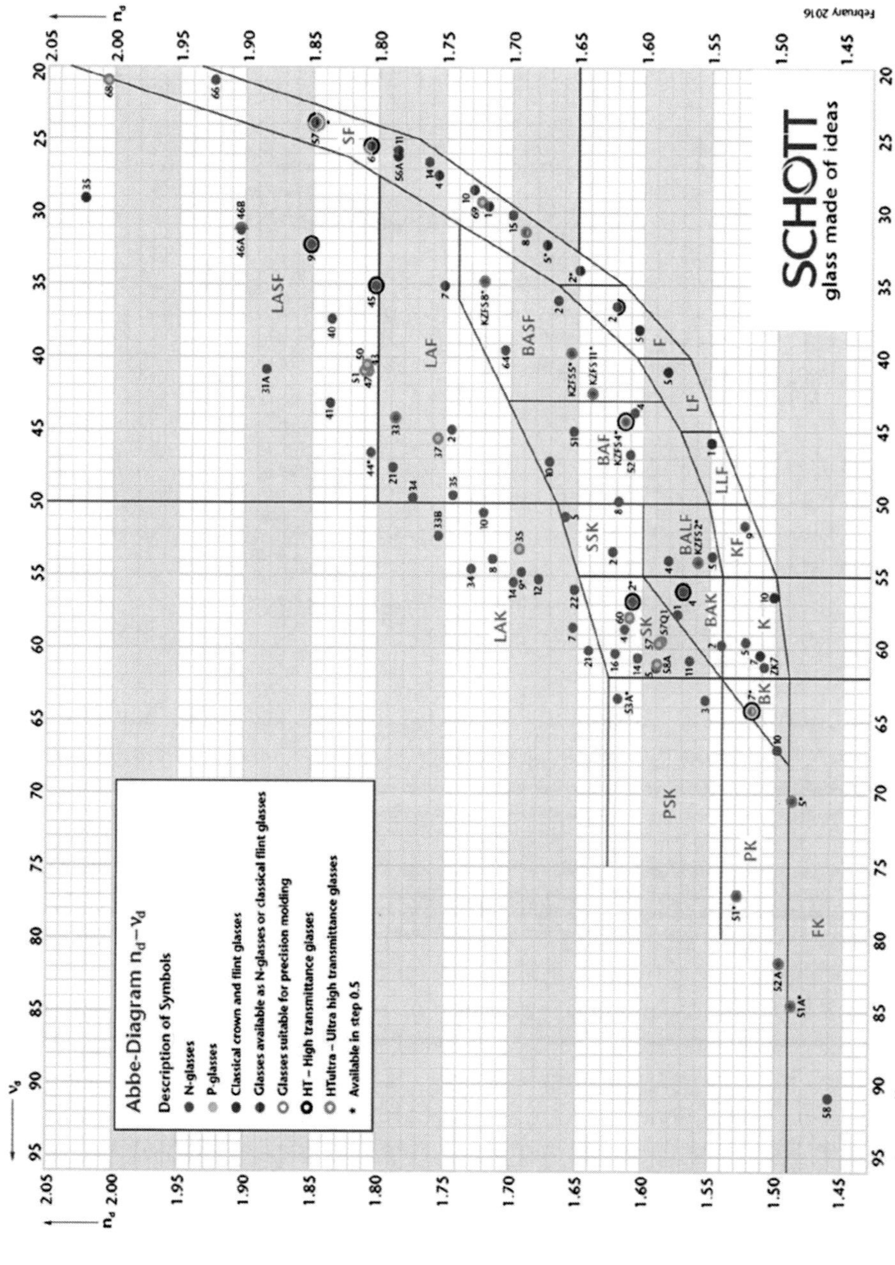

Figure 4.3 Schott glass map or Abbe diagram. The horizontal axis is the Abbe number, and the vertical axis is the refractive index. Reproduced with permission from Schott AG.

4.1.3 Transmission spectral range

Optical materials have a certain transmission spectral range. Inside this range the absorption is very low, outside this range the absorption is high and the optical materials can be opaque. Figures 4.1(b) and 4.2(b) plot the percentage transmittance of a 10-mm-thick N-BK7 glass and UV-grade fused silica plate with uncoated surfaces, respectively. Note that the two uncoated surfaces of the N-BK7 glass and UV-grade fused silica plates have a total reflection loss of about 8.6% and 6.8%, respectively. These reflection losses are included in Figs. 4.1(b) and 4.2(b).

Most optical glasses have a high transmission spectral range from the deep blue (below 400 nm) to the near IR (a couple of microns). Optical polymers have a high transmission spectral range from around 400 nm to 1 μm.

4.1.4 Deviation of partial dispersion

The deviation of partial dispersion (dPgF) is defined as[9]

$$P_{x,y} = \frac{n_x - n_y}{n_F - n_C}, \tag{4.6}$$

where x and y are two wavelengths of interest. Usually the Fraunhofer g line of 0.436 μm and F line of 0.486 μm are used. More than half of optical glasses have a dPgF > 0, and most optical polymers have dPgF = 0.

The dPgF supplements the Abbe number to further describe the refractive-index curve of optical materials. For approximate modeling, the value does not matter much, but for accurate modeling and design it does. Optical software contains the value of dPgF of all of the optical materials loaded in the software. The refractive index polynomials contain all of the information about the refractive index, including the Abbe number and the deviation of partial dispersion.

4.1.5 Refractive-index temperature dependence

The refractive index of optical materials varies as the temperature varies. Several different polynomials are used to describe the thermal index variation. These polynomials are equivalent. Equation (4.7) is one such polynomial:[10]

$$\Delta n_{\mathrm{Abs}}(\lambda,t) = \frac{n(\lambda,t)^2 - 1}{2n(\lambda,t)} \times \left[D_0(t-t_0) + D_1(t-t_0)^2 + D_2(t-t_0)^3 + \frac{E_0(t-t_0) + E_1(t-t_0)^2}{\lambda^2 - \lambda_{tk}^2} \right], \tag{4.7}$$

where $\Delta n_{\mathrm{Abs}}(\lambda,t)$ is the absolute thermal index variation; a nonlinear function of t; and D_0, D_1, D_2, E_0, E_1, and λ_{tk} are parameters. Equation (4.7) is plotted

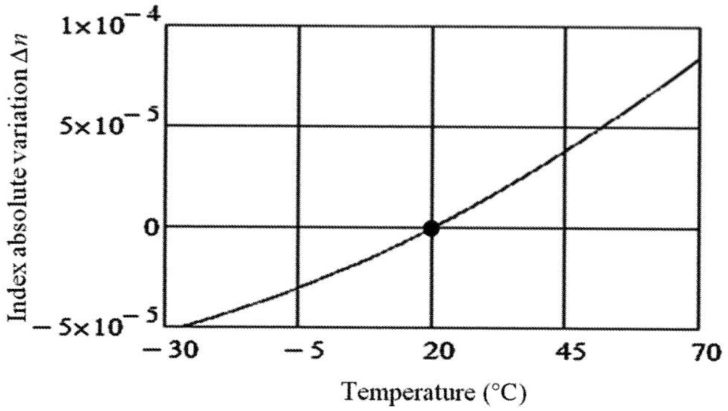

Figure 4.4 N-BK7 glass refractive-index absolute change as a function of temperature. The index at 20 °C is selected as the reference point.

in Fig. 4.4 for Schott N-BK7 glass using the data provided by Zemax with $\lambda = 0.589$ μm and $n(0.589$ μm, 20 °C$) = 1.5168$. Note that $\Delta n_{\mathrm{Abs}}(\lambda,t)/\Delta t \sim 10^{-6}$/°C > 0, which is the case for most glasses, but some glasses have $\Delta n_{\mathrm{Abs}}(\lambda,t)/\Delta t < 0$. For optical polymers, $\Delta n_{\mathrm{Abs}}(\lambda,t)/\Delta t$ can be either > 0 or < 0 with a magnitude ~10^{-4}/°C.

Since the index of air surrounding the lens also changes as the temperature changes, the relative-index temperature variation of a glass is the factor that affects the lens' thermal performance and is given by[11]

$$\left(\frac{\Delta n}{\Delta t}\right)_{\mathrm{Rel}} = \left(\frac{\Delta n}{\Delta t}\right)_{\mathrm{Abs}} - n_{\mathrm{Rel}}\left(\frac{\Delta n_{\mathrm{Air}}}{\Delta t}\right). \tag{4.8}$$

Note that on the right side of Eq. (4.8) the second term is much smaller than the first term.

Thermal index variation is the main cause of thermal defocusing. If an imaging lens will be used in a large temperature range, e.g., from −30 °C to +60 °C, thermal defocusing will be a severe issue. Athermal optical design must be applied; it properly matches the $\Delta n(t)/\Delta t$ of various optical materials and the thermal expansion of the optical system housing so that these effects will mostly offset each other.

Optical material vendors and optical software will provide thermal index change data for most the optical materials they offer or carry.

4.1.6 Thermal expansion

Optical glasses have a thermal expansion coefficient of ~10^{-6}/°C,[10] and optical polymers have a thermal expansion coefficient of ~10^{-5}/°C. The thermal expansion of optical materials will deform the shape of the optical components and affect the optical performance. Furthermore, if the temperature variance is large, the structure that holds the optical components must have thermal expansion close to the thermal expansion of the optical components or else the optical components will be either loosened or cracked when the temperature significantly varies. Optical material vendors and optical software provide thermal expansion data for most of the optical materials they offer or carry.

For imaging lenses, optical-material thermal expansion affects the image quality much less than the thermal index change does, and it is less of a concern. For a lens, the "temperature coefficient of optical path length," which includes the thermal index change as well as the thermal physical expansion of the material, is the factor to be considered, given by[12]

$$\frac{\Delta s}{\Delta t} = (n - 1)\alpha + \frac{\Delta n}{\Delta t}, \tag{4.9}$$

where s is the optical path, n is the glass index, and α is the linear thermal expansion coefficient of the glass.

4.1.7 Chemical stabilities

Generally speaking, optical glasses have good chemical stabilities. For most commercial applications, the chemical stability of glass is not a concern. For applications in hash environment, the chemical stability of the glasses used can be an issue.

Glass chemical stability has four main categories,[13] and each category has several grades (smaller grade numbers mean greater stability):

1. Climatic resistance (CR) with grades 1–4.
2. Stain resistance (FR) with grades 0–5.
3. Acid resistance (SR) with grades 1–53.
4. Alkali resistance (AR) and phosphate resistance (PR) with grades 1–4.

Table 4.1 lists the chemical stability grades for the widely used optical glasses N-BK7 and N-SF11. They are all very stable.

Glass vendors and optical software usually provide the chemical stability grades for many glasses they offer or carry. The chemical stabilities of some optical glasses (and most IR and UV materials) are not graded, although they can also be stable. The chemical stabilities of optical polymers are not graded either, but they are known to be unstable.

4.1.8 Density

Most optical glasses have a density from 2.2 to over 6 g/cc. Most optical polymers have a density slightly larger than 1 g/cc. For comparison, water has

Table 4.1 Chemical properties of N-BK7 and N-SF11.

Chemical properties	CR	FR	SR	AR	PR
N-BK7	1	0	1	1	1
N-SF11	1	0	1	1	1

a density of 1 g/cc. Optical material vendors and optical software provide the density data for all of the optical materials they offer or carry.

4.1.9 Mechanical and electrical properties

Mechanical properties include several aspects, such as Knoop hardness, abrasion factor, elastic properties, flexural strength, optical homogeneity, stress birefringence, and bubbles and inclusions.

Generally speaking, optical glasses have strong mechanical strength, and their strength is not a concern in most applications. Optical polymers are soft and easily scratched and deformed.

The electrical properties of a glass include the relative permittivity and volume resistivity. Glasses are good isolators, which can be a problem in some applications where electromagnetic shielding is required. This problem can be solved by applying an electrical conductive coating on the glass surface. Such a coating will cause certain transmission loss. This topic will be discussed in Section 5.8.

4.1.10 Melting temperature

Optical glasses have a melting temperature of several hundred degrees Celsius. Some optical glasses have relatively low melting temperature of a couple of hundred degrees Celsius and are suitable for molding. Optical polymers start softening above 100 °C or so. Glass vendors often specify which glasses of theirs are suitable for molding.

4.1.11 Price

The price of optical glasses is often expressed by an index. The price index of the most widely used optical glass N-BK7 is 1. The price index of most other optical glasses is a few times larger than that, and some are over 10. Like all other commercial products, the dollar price of optical glasses changes over time.

The final step of producing optical glasses is to mold the melting glasses into blanks with a few shapes and various sizes, such as a rod and disk, etc. Some blanks even have good surface quality and are semi-finished products. Lens vendors buy the right size glass blanks from glass vendors to grind lenses. For example, when a lens vendor needs to grind lenses with a 20-mm diameter, they will try to buy glass rods with a 25-mm diameter or so. A glass rod with a 75-mm diameter is clearly not a good choice because it takes additional labor to cut the 75-mm-diameter rod to smaller pieces, and a lot of glass material will be wasted.

There is no simple price number for optical glasses. The price depends on the shape and size of the glass blanks and on the glass quality. But $100 per kilogram of N-BK7 can be used as a very rough number. When developing optical systems, the optical glass price is often an issue but not a major issue. However, the optical polymer price is never an issue because

1. Optical polymer prices are usually less than one tenth the price of optical glasses.
2. An optical polymer of the same weight of has a much larger volume than glasses because of its low density and can be used to produce much more optical components.

4.1.12 Availability

Optical glass vendors usually classify their glasses into three categories: preferred, standard, and obsolete. As the names suggest, preferred glasses are frequently produced and have a relatively low price. Standard glasses are regularly produced but less than the preferred versions. There is a chance that some types of standard glasses are out of stock, in which case buyers may have to wait for next production run. Obsolete glasses are no longer produced, although some retailers may still have some in stock. The data for obsolete optical glasses is provided to help people model and analyze old optical designs that feature now-obsolete glasses.

For those glasses in production, the term "melting frequency" describes the "production frequency" for a certain type of glass and provides more detailed information about how frequently this type of glass is produced. There are usually five melt frequencies:

1. Highest,
2. High,
3. Medium,
4. Low,
5. Lowest, and
6. Discontinued (but some may remain in stock).

Glass vendor catalogs often specify the melting frequency of many types of glasses they produce, whereas optical software often does not have this information because the production frequency can change over time. Whenever possible, use glasses with at least a "high" melting frequency.

4.2 Optical Glasses

4.2.1 Glass types

Optical glasses can be divided into crown glasses and flint glasses. The former usually have relatively smaller refractive indices, which is often not desired, and relatively larger Abbe numbers, which is often desired. Conversely, the latter usually have larger indices and smaller Abbe numbers. There are also crown and flint mixed glasses, and other types of glasses. Glasses with a large index and a large Abbe number are highly desired, but unfortunately, no such glasses exist.

Older optical glasses often contain environmentally hazardous elements, such as lead and arsenide; however, many types of eco-friendly glasses have been developed to replace them. These new glasses have the same optical properties as their predecessors, but they contain no hazardous elements and have lighter densities.

Some types of optical glasses have a low melting temperature of a couple of hundred degrees Celsius. These glasses are suitable for molding. More than half of all available optical glasses can be processed by grinding, polishing, or diamond cutting.

Optical glasses have a refractive index value in the range of about 1.4 to 2.0, an Abbe number value in the range of about 16 to 95, and a high transmission spectral range from below 400 nm to a couple of microns. Some other properties of optical glasses have already been discussed previously in Section 4.1 and are not repeated here.

4.2.2 Glass brands

The most commonly used optical glass vendors are Ohara, Schott, CDMG, Hoya, and Corning. These brands offer tens to hundreds of different types of glasses. Many optical glasses from different brands have the same optical properties and are interchangeable.

4.3 Optical Polymers

Only about two dozen types of optical polymers are commonly used, much fewer than the types of optical glasses available: acrylic, PMMA (poly methyl methacrylate), polycarb (polycarbonate), polystyr (polystyrene), several Osaka Glass Chemical materials, several Zeonex materials, and a few Topas materials.

Optical polymers are soft, have a low melting temperature, and are chemically less stable than glasses. They cannot be ground or polished because of their viscosity; they can only be processed by machining, e.g., diamond turning or molding. Therefore, small-quantity polymer components are produced by machining and are more expensive than glass components that are made by the conventional grinding technique. However, large-quantity polymer components can be produced by molding and cost much less than glass components made by grinding because polymers are relatively easy to mold. Polymer lenses with strong aspheric surfaces and polymer components with various unconventional shapes can be made to achieve unique functions.

Optical polymers have a refractive index value around 1.55, an Abbe number value around either 30 or 55, and a high transmission spectral range from around 400 nm to 1 μm. The much smaller selection of optical polymers makes them less useful than optical glasses. Optical polymers are widely used to produce low-end optical components or devices for the visible range, such as cellphone camera lenses.

4.4 UV Optical Materials

Several type of materials are used in the UV range; see Table 4.2. All of the refractive indices listed in Table 4.2 can be found in Reference 1. Among these

Table 4.2 Commonly used UV materials and their properties. The same type of materials produced by different vendors can have apparently different properties. One type of material can have different grades, e.g., UV grade or IR grade, and have apparently different transmission ranges. The refractive indices defined at different wavelengths are different. Listed here are only some typical numbers. Transmission ranges defined at different percentages of transmission are also different. Some UV materials are also IR materials. An asterisk denotes a birefringent material.

Material Name	Transmission Range (nm)	Refractive Index
Calcite	~260–2700	~1.656 @ 0.633 μm ~1.485 @ 0.633 μm*
Calcium fluoride	~180–8000	~1.433 @ 0.633 μm
Fused silica	~180–2100	~1.457 @ 0.633 μm
Crystalline quartz	~170–3400	~1.542 @ 0.633 μm ~1.551 @ 0.633 μm*
Sapphire	~260–5000	~1.766 @ 0.633 μm ~1.758 @ 0.633 μm*

materials, fused silica is the most widely used. It is also widely used in the visible and short-IR ranges, as well as in mirror substrates, because fused silica has strong mechanical strength and low thermal expansion. The shortcoming of using fused silica is that the strong mechanical strength makes it difficult to fabricate, and the price of fused silica is over ten times higher than N-BK7 glass.

Because color dispersion $|dn(\lambda)/d\lambda|$ increases as the wavelength decreases for all optical materials, color dispersion is more severe in the UV range than in the visible range. Fortunately, most applications in the UV range deal with only one narrow spectral band. The strong color dispersion is not a concern. If color dispersion must be corrected in the UV range, there are no many materials to select.

4.5 IR Optical Materials

In the infrared range, the color dispersion $dn(\lambda)/d\lambda$ of all optical materials is relatively small. The main concern of material selection is not the color dispersion, but the transmission range.

Several types of material are commonly used in IR range and listed in Table 4.3. All of the refractive indices listed in Table 4.3 can be found in Reference 1.

The transmission range of some IR optical materials is above visible range, they look opaque to human eyes. The refractive index of some IR materials can be over 4, much larger than the index of optical glasses in the visible range. In air, the surface reflectance of a plate made of such high-index material can be ~40%. AR coating on component surfaces made of such materials is a must.

4.6 Mirror Substrates

Most mirrors are front surface mirrors. The mirror substrates are behind the reflection coating. Some optical qualities, such as color dispersion and

Table 4.3 Commonly used IR materials and their properties. The same type of materials produced by different vendors can have apparently different properties. One type of material can have different grades, e.g., UV grade or IR grade, and have apparently different transmission ranges. The refractive indices defined at different wavelengths are different. Listed here are only some typical numbers. Transmission ranges defined at different percentages of transmission are different. Some IR materials are also UV materials. AMTIR 1 to AMTIR 6 are six different types of IR materials made by Amorphous Materials, Inc. All of these materials are mixtures of several other IR materials. An asterisk indicates a birefringent material.

Material name	Transmission Range (μm)	Refractive Index
AMTIR 1 to AMTIR 6	~1–12	~2.6
Calcite	~0.26–2.70	~1.656 @ 0.633 μm
		~1.485 @ 0.633 μm*
Calcium fluoride	~0.18–8.00	~1.433 @ 0.633 μm
Fused silica	~0.18–2.10	~1.457 @ 0.633 μm
Crystalline quartz	~170–3400	~1.542 @ 0.633 μm
		~1.551 @ 0.633 μm*
Sapphire	~0.26–5.00	~1.766 @ 0.633 μm
		~1.758 @ 0.633 μm*
Silicon	~1–9	~3.90 @ 0.633 μm
Gallium arsenide (GaAs)	~2.5–14.0	~3.87 @ 0.633 μm
Germanium	~2–15	~5.58 @ 0.633 μm
Zinc selenide (ZnSe)	~0.6–16.0	~1.13 @ 0.633 μm
		~5.33 @ 1.65 μm

absorption, are not issues for a mirror substrate. The desired qualities for mirror-substrate materials are strong mechanical strength, low thermal expansion, and low price.

All low-cost glasses are good mirror-substrate materials. Metals, such as aluminum, are also used as mirror substrates. The three most commonly used materials for high-quality mirror substrates are fused silica (linear thermal expansion coefficient of 0.51×10^{-6}/°C), Zerodur (a glass–ceramic made by Schott with a linear thermal expansion coefficient of $\sim 0 \pm 0.1 \times 10^{-6}$/°C), and Pyrex (a borosilicate glass made by Corning with a linear thermal expansion coefficient of 3.25×10^{-6}/°C).

4.7 Crystals

Several types of crystals are used to fabricate lenses, wave plates, polarizers, etc., for applications from UV to IR or to make opto-electrical devices. This section briefly describes the properties of some commonly used crystals. A few other sections in this book also discuss crystals from different perspectives. Reference 14 provides data for over one hundred types of optical crystals.

4.7.1 Crystals for making optical components

Some crystals are anisotropic and have different refractive indices in two orthogonal directions. Such a crystal is considered to be "birefringent" or "birefractive." These two directions are called "ordinary" and "extraordinary" directions, respectively. Some other types of crystals are isotropic.

Crystals often used to make optical components include the following:

1. Quartz, which is an anisotropic crystal. Its refractive indices at 590 nm in the ordinary and extraordinary directions are 1.544 and 1.553, respectively. Quartz is mainly used to make wave plates, as described in Section 6.8.2.
2. Calcite, which is an anisotropic crystal. Its refractive index is from 1.9 at 190 nm to 1.5 at 1700 nm in the ordinary direction and from 1.6 at 190 nm to 1.4 at 1700 nm in the extraordinary direction. Calcite is mainly used to make polarizers, as described in Section 6.6.1.
3. Calcium fluoride (CaF_2), which is an isotropic crystal with broad transmission from 180 nm to 8 μm, low absorption, high damage threshold, high thermal expansion, and is mildly hydroscopic. CaF_2 is mainly used to make components for excimer laser applications.
4. Magnesium fluoride (MgF_2), which is an isotropic crystal with broad transmission from 160 nm to 7 μm and a very low index of 1.383 at 632.8 μm. MgF_2 is resistant to thermal and mechanical shock, has a higher energy damage threshold, and is extremely durable but mildly hydroscopic. MgF_2 is mainly used for applications involving laser pulses or for antireflection coatings.
5. Zinc selenide (ZnSe), which is an isotropic crystal with hexagonal and cubic two-crystal structures. ZnSe has an extremely broad transmission range from ~0.6 μm to 16 μm and is widely used to make IR optical components for CO_2 laser applications at the 10.6-μm wavelength. ZnSe is also used as a laser-diode active medium.

4.7.2 Crystals for opto-electrical devices and nonlinear applications

When an electrical field is applied to some types of crystal, the refractive indices of these crystals in the direction of the electrical field apparently vary. Birefringence is then created or enhanced in these crystals. When the electrical field applied to the crystals is modulated, the refractive index or birefringence of these crystals is modulated. The amplitude/intensity or phase or polarization state of beams passing through these crystals are also modulated. Several types of opto-electrical devices are developed based on this working principle. Section 9.8 provides a more information about opto-electrical devices.

The refractive index of some crystals has a nonlinear relation to the electrical field applied to the crystals. Such crystals are called "nonlinear optical crystals." The nonlinear optical phenomenon is more apparent as a laser beam, which has a very-high-intensity optical (electrical) field, passing through the crystal. Under certain conditions, nonlinear optical crystals can mix a few laser beams passing through the crystal and create a new laser beam whose frequency is the sum of other laser beams. Such a process is called "frequency doubling," "frequency tripling," etc.

Crystals often used for opto-electrical devices or nonlinear applications include the following:

1. Lithium triborate (LBO),[15] which is a nonlinear optical crystal. LBO is mainly used to double the frequency of a laser beam passing through it.
2. Lithium niobate ($LiNbO_3$),[16] which is a synthetic optical crystal mainly used to make various types of opto-electrical modulators.
3. Barium borate (BBO),[17] which is a popular nonlinear optical crystal used in many different applications.
4. Potassium titanyl phosphate (KTP),[18] which is a nonlinear optical crystal which is mainly used for frequency doubling of beams of laser diode pumped solid state lasers, and also used to make opto-electrical modulators.
5. Potassium dideuterium phosphate (KDP),[19] which is a nonlinear optical crystal widely used in the frequency doubling, tripling, and quadrupling of laser beams.

There are many more different types optical crystals currently available. Nikogosyan[7] provides a detailed discussion about optical crystals.

References

1. *Refractive Index Database*, https://refractiveindex.info/
2. P. Hartmann, *Optical Glass*, SPIE Press, Bellingham, WA (2014) [doi: 10.1117/3.1002595].
3. H. Bach and N. Neuroth, Eds., *The Properties of Optical Glass*, Schott Series on Glass and Glass Ceramics, Springer-Verlag, Berlin (1998).
4. J. Simmons, *Optical Materials*, Academic Press, Cambridge, MA (1999).
5. S. Musikan, *Optical Materials: An Introduction to Selection and Application*, CRC Press, Boca Raton, FL (1985).
6. P. Klocek, *Handbook of Infrared Optical Materials*, CRC Press, Boca Raton, FL (1991).
7. D. N. Nikogosyan, *Nonlinear Optical Crystals: A Complete Survey*, Springer, Berlin (2005).
8. "2.2 Dispersion and Abbe number," *Optical Glass*, Hoya Corporation USA Optics Division, www.hoyaoptics.com/pdf/OpticalGlass.pdf
9. "2.4 Relative partial dispersion and abnormal dispersion," *Optical Glass*, Hoya Corporation USA Optics Division. www.hoyaoptics.com/pdf/OpticalGlass.pdf
10. "Chapter 24 Thermal analysis", *Zemax Manual*.
11. "2.5 Temperature coefficient of refractive index ($\Delta n/\Delta T$)," *Optical Glass*, Eq. (7), Hoya Corporation USA Optics Division, www.hoyaoptics.com/pdf/OpticalGlass.pdf

12. "2.6 Temperature coefficient of optical path length ($\Delta s/\Delta T$)," *Optical Glass*, Eq. (8), Hoya Corporation USA Optics Division, www.hoyaoptics.com/pdf/OpticalGlass.pdf
13. *Chemical Properties*, Precision Micro-Optics, www.pmoptics.com/glass_chemical_property.html
14. "Optical and laser crystals," at website www.americanelements.com/optical-materials.html
15. Wikipedia, "Lithium triborate," en.wikipedia.org/wiki/Lithium_triborate
16. Wikipedia, "Lithium niobate," en.wikipedia.org/wiki/Lithium_niobate
17. Wikipedia, "Barium borate," en.wikipedia.org/wiki/Barium_borate
18. Wikipedia, "Potassium titanyl phosphate," en.wikipedia.org/wiki/Potassium_titanyl_phosphate
19. Wikipedia, "Potassium dideuterium phosphate," en.wikipedia.org/wiki/Potassium_dideuterium_phosphate

Chapter 5
Optical Coatings

Optical coatings can be discussed in terms of coating types or coating functions. There are two coating types: metallic coatings and dielectric coatings (also called thin film coatings). There are several coating functions: antireflection coatings, high-reflection coatings, beamsplitting coatings, polarization coatings, and spectral coatings. References 1–3 are relatively recent books about optical coatings.

5.1 Metallic Coatings

Metallic coatings are relatively simple and inexpensive. They can provide a broadband, high reflection that covers from the UV to the near-IR range. Commonly used coating materials include aluminum, silver, and gold. Figure 5.1 shows some typical reflectance curves of metallic coatings.

Such coatings are delicate, and they can be easily scratched, polluted, or oxidized. A thin layer of dielectric material is often applied on top of the metal coating for protection. The thin layer can also enhance the performance of metal coatings by either increasing the reflectance or expanding the high reflectance to the UV or mid-IR ranges. The very broadband reflection is the main advantage of metallic coatings.

5.2 Dielectric Coatings

Dielectric coatings are also called thin film coatings. There are single-layer and multilayer dielectric coatings.

Single-layer dielectric coatings are often used to reduce the surface reflectivity of glass. Such coatings are called antireflection (AR) coatings. The thickness of single-layer AR coatings is selected so that destructive interference occurs between the two waves reflected by the air/coating and coating/glass interfaces, respectively. Thereby, the intensity of the reflected wave is minimized. The performance of single-layer AR coatings can be optimized only for one wavelength and one incident angle. In fact, all dielectric coatings—single layer or multilayer, antireflection or high reflection—utilize interference.

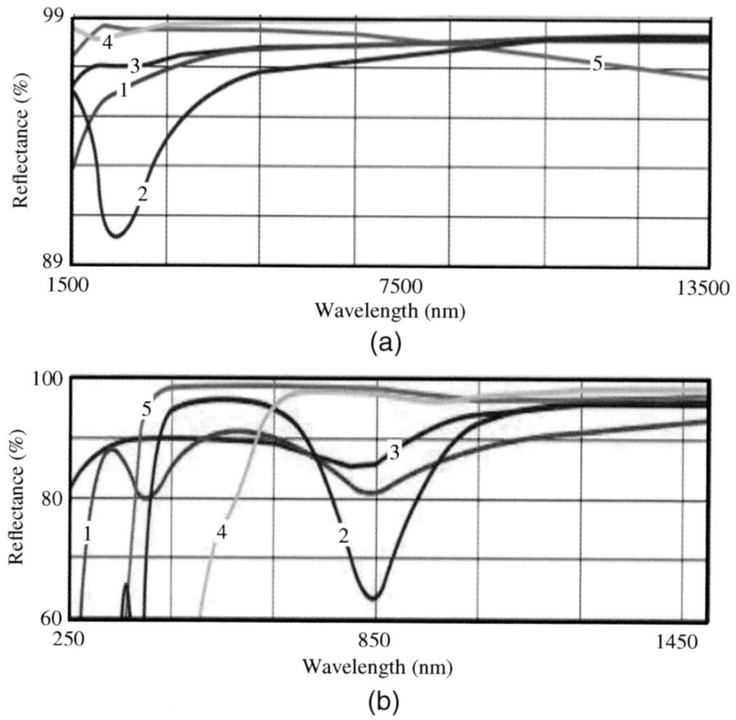

Figure 5.1 Typical reflectance curves of metallic coatings (a) from the near IR to mid-IR and (b) from the UV to near IR. 1: protected aluminum; 2: enhanced aluminum; 3: UV-enhanced aluminum; 4: protected gold; 5: protected silver. Image reprinted courtesy of Edmund Optics.

Multilayer dielectric coatings use several different types of dielectric materials alternatively with different layer thicknesses. The thickness of each coating layer and the index of each coating material are selected so that the very complex interference that occurs inside these layers will lead to various functions, such as antireflection, high reflection, narrow band, broadband, polarizing, etc. The number of layers in a multilayer coating varies from several to tens.

The design of a multilayer dielectric coating is a specialized task: only well-trained personnel using tailored software can perform such tasks. The fabrication process is also not easy, often requiring high-performance coating chambers and tight process control. The yield can be low, and the performance of each coating run can be slightly different.

5.3 Antireflection Coatings

The reflectance of an air and uncoated-glass interface is several percent. Each lens or other optical component has two surfaces, and one optical system may have many components in it. The sum of all of these surface reflection losses can be high. The energy reflected by the surfaces can bounce inside an optical system and produce noise background or ghost images. Therefore, AR coatings are now widely used on optical surfaces.

Figure 5.2 shows the residual reflectance of a single-layer magnesium fluoride antireflection coating on N-BK7 glass. The uncoated surface reflectance of N-BK7 glass is also plotted for comparison. Note that the residual reflectance varies as the ray incident angle varies. The curves shown in Fig. 5.2 are optimized for 0° incidence, which is the usual case, but they can be optimized for other incident angles, such as 45°. The reflectance of the bare surface of glass also varies as the ray incident angle varies. (This topic was discussed in Section 2.3.)

The residual reflectance shown in Fig. 5.2 is over 1% and is considered still too high for many applications. A multilayer dielectric AR coating can further reduce the residual reflectance. Figure 5.3 shows the residual reflectance of several typical coatings for visible and near infrared range. The residual reflectance at the intended wavelength or spectral band is well below 1%. There are many variations of multilayer dielectric coatings that perform differently than that shown in Fig. 5.3. For special requirements, consult optical coating vendors for solutions.

5.4 High-Reflection Coatings

Many applications require a very high reflectance in a limited spectral band. Any reflection outside the desired band may produce undesired noise and should be as low as possible, e.g., a laser line reflector. For these applications, the high-reflection metal coatings shown in Fig. 5.1 have excessive reflection bandwidth.

Specially designed multilayer dielectric coatings can provide very high reflectance only within the desired band. Figure 5.4 shows four typical

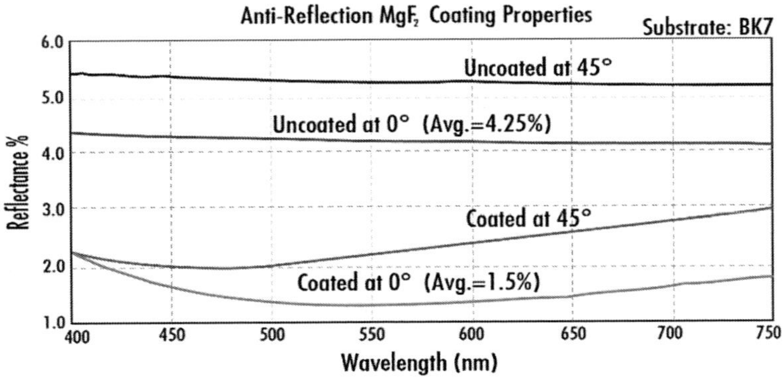

Image Courtesy of Edmund Optics

Figure 5.2 Reflectance of a single-layer magnesium fluoride AR coating. The uncoated surface reflectance of N-BK7 glass is also plotted for comparison.

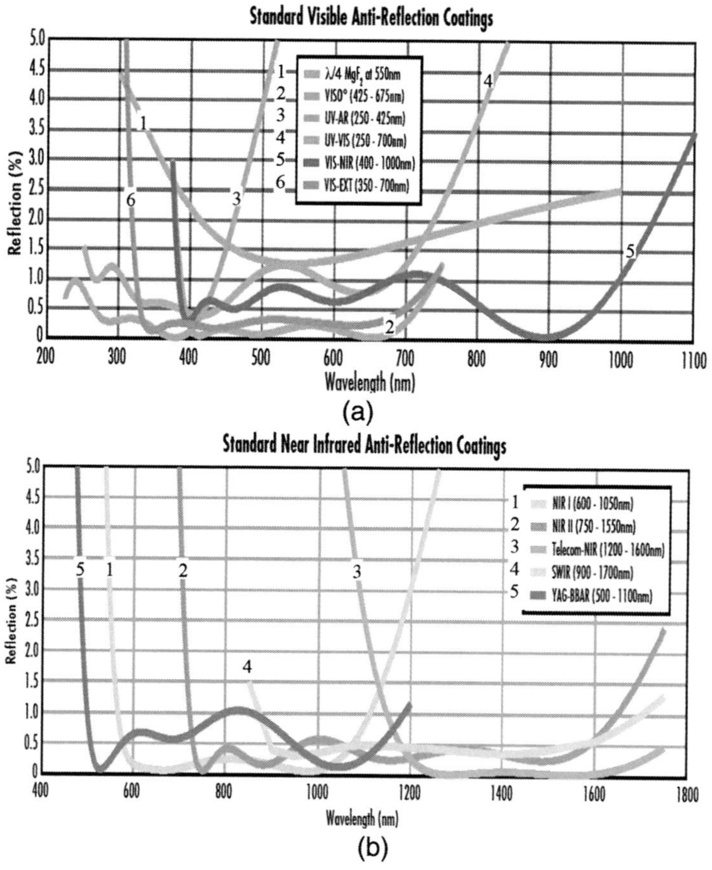

Figure 5.3 Residual reflectance of several multilayer AR coatings for the (a) visible and (b) near-IR range. The residual reflectance at the intended wavelength or band is well below 1%. Image reprinted courtesy of Edmund Optics.

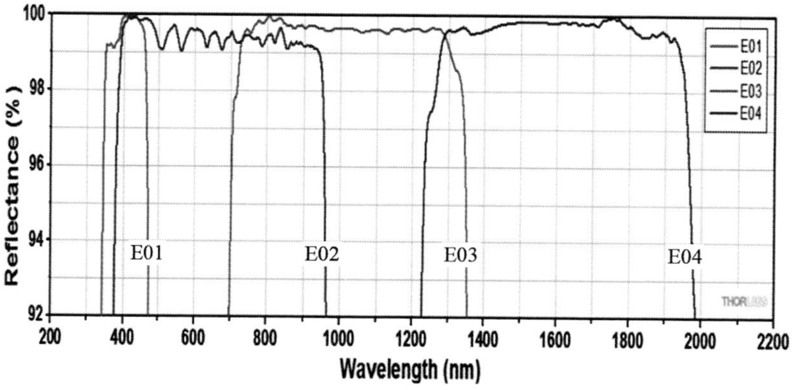

Figure 5.4 Four typical multilayer, dielectric high-reflection coatings offered by Thorlabs. Their reflectance is high inside the desired band and drops quickly outside the band. Image reprinted courtesy of Thorlabs.

high-reflectance multilayer dielectric coatings. Their reflectance is high inside the desired band and drops quickly outside the band. The E01 coating has a narrow band, intended for laser line reflection. The E02 coating has a band that covers the visible to near IR; it matches the working spectral range of CCD and CMOS sensors.

5.5 Beamsplitting Coatings

As the name suggests, a beamsplitting coating splits a beam incident on the coating into two. The coating can be a thin layer of metal if there is no special requirement for the splitting spectral range. If the splitting must happen within a certain spectral range, specially designed, multilayer dielectric coatings must be used.

A beamsplitting coating has a certain splitting ratio, such as 50:50, 70:30, 10:90, etc. Figure 5.5 shows a 30:70 beamsplitting coating offered by Thorlabs. Note that the splitting ratio actually varies as the wavelength varies, and the splitting ratio for *p*- and *s*-polarization is slightly different. These characteristics are common for beamsplitting coatings, so the specified splitting ratio is not an accurate number.

5.6 Polarizing Coatings

Some specially designed multilayer dielectric coatings can separate the *p*- and *s*-polarizing components of a beam incident on the coating. These coatings are called polarizing coatings. Polarizing coatings can also be considered as a type of beamsplitting coatings. A beam must be incident on the coating at a certain angle, usually 45°, so that both *p*- and *s*-polarization exist. If the beam is incident on the coating along the normal of the coating, the situation is symmetric about the normal, there are no *p*- and *s*-polarization, and the polarizing coating will not perform.

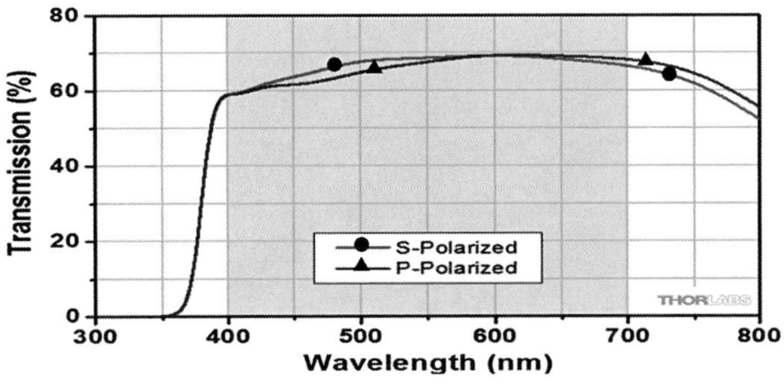

Figure 5.5 A 30:70 beamsplitting coating offered by Thorlabs. The actual splitting ratio varies as the wavelength varies, and the splitting ratio for *p*- and *s*-polarization is slightly different. Image reprinted courtesy of Thorlabs.

Polarizing coatings only work within certain spectral ranges. After a beam is split into a *p*-polarizing beam and an *s*-polarizing beam, the *p*-polarizing beam will still contain a small amount of *s*-polarization that leaks through, and vice versa. The leaky polarization is undesired and should be as low as possible. The extinction ratio specifies the ratio of the desired polarization power to the undesired polarization power. An extinction ratio greater than 100 is not difficult to achieve.

Figure 5.6 shows the transmission curve of a polarizing beamsplitting coating for 420–680 nm offered by Thorlabs. The *s*-polarization is reflected and has near-zero transmission. The *p*-polarization has high transmission. Therefore, the reflected and transmitted beams are polarized. The *p*-polarization transmission varies as the wavelength varies.

Note that some other optical components or structures can also split a beam based on polarization, such as a wired grid.

5.7 Spectral Coatings

There are other multilayer dielectric spectral coatings. Figure 5.7 shows four different types offered by Thorlabs: (a) shortpass filter, (b) longpass filter, (c) bandpass filter, and (d) notch (band-rejection) filter. Shortpass and longpass filters are specified by their peak transmission number and cutting-edge wavelength. Bandpass and notch filters are specified by their peak transmission number, central wavelength, and bandwidth.

The curves shown in Fig. 5.7 are just examples. All of these specifications can be adjusted by modifying the coatings. For any type of spectral filter, a high peak transmission, steep-cut edge or band edge, and narrow band are difficult to achieve.

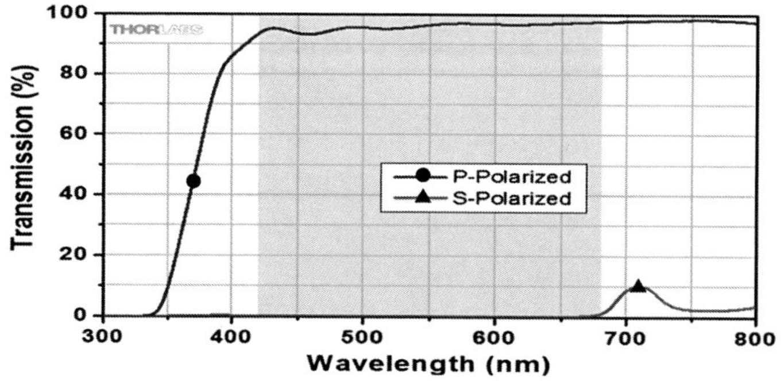

Figure 5.6 The transmission curve of a polarizing beamsplitting coating for 420 nm to 680 nm offered by Thorlabs. The *s*-polarization is reflected and has near zero transmission. The *p*-polarization has high transmission, which varies as the wavelength varies. Image reprinted courtesy of Thorlabs.

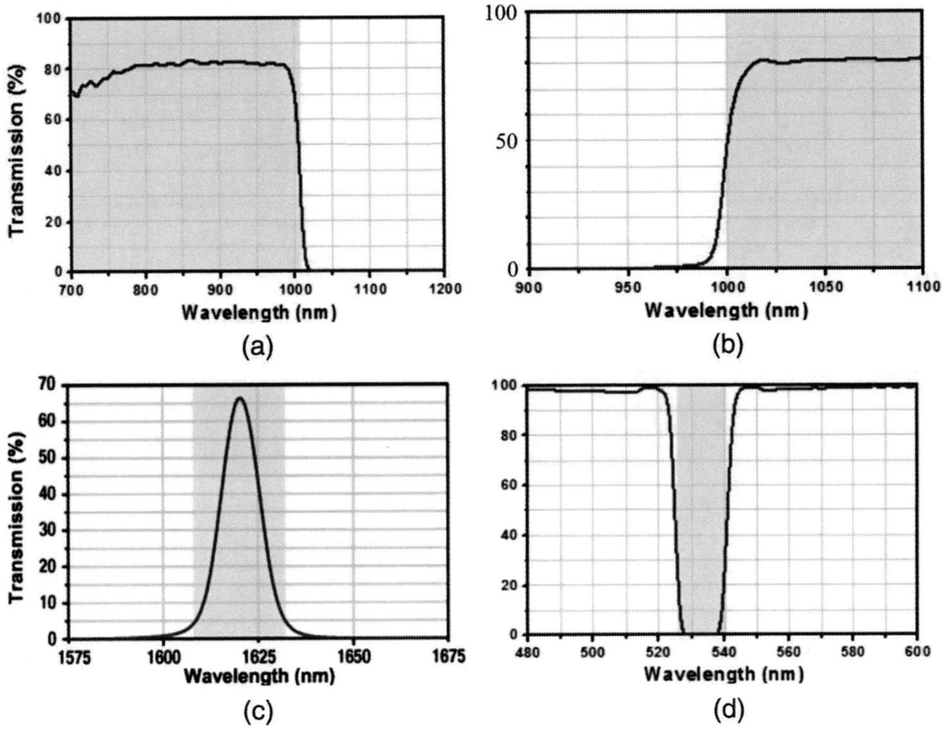

Figure 5.7 Four spectral coatings offered by Thorlabs: (a) shortpass filter, (b) longpass filter, (c) bandpass filter, and (d) notch (band-rejection) filter. Image reprinted courtesy of Thorlabs.

Note that the cutting-edge wavelength and central wavelength are specified for a certain incident angle, usually 0°. An increase in the incident angle will slightly reduce these specific wavelengths.

5.8 Electrical Conductive Coatings

In many applications, a lens views distant objects and forms an image onto a sensor array. Electronics are used to control the sensor and perform complex data processing. The electronics are mounted together with the lens inside a housing and must be shielded from outside electromagnetic interference, some of which may be intentional and hostile. A metal housing cannot completely block outside electromagnetic interference because a glass window is required in the metal house to let the lens look out. Therefore, the window must be coated by an optically transparent and electrically conductive coating to block the external electromagnetic field.

To define the electrical resistance of a coating, some special terminologies must be explained.[4] The resistance of a volume object is

$$R = \frac{\rho L}{A}, \tag{5.1}$$

where A is the cross-section of the volume, L is the length of the volume in the direction of the electrical current flow, and ρ is the resistivity in ohm-m. For a thin coating, ρ is the resistivity of the coating, and $A = W \times T$, where W and T are the width and thickness of the coating, respectively. Since T can be different for different types of coatings, the resistance of a thin film is often written as

$$R = \frac{\rho}{T} \cdot \frac{L}{W}. \tag{5.2}$$

where ρ/T is the sheet resistivity in ohms or "ohm per square," just to clarify that it is a sheet resistance.

The most widely used electrical conductive coating is the indium tin oxide (ITO) coating. Few companies can apply such a coating, much fewer than the number of companies that can apply dielectric and metal coatings. For ITO coating a glass surface, the ballpark numbers are ~50 ohm per square for sheet resistivity and 80–90% optical transmittance. A coating with smaller resistance can provide better shielding of electromagnetic interference, but it will be more difficult to coat and have lower optical transmittance. Usually only one surface of the window must be coated with ITO.

References

1. O. Stenzel, *Optical Coatings: Materials Aspects in Theory and Practice*, Springer-Verlag, New York (2016).
2. A. Piegari and F. Flory, *Optical Thin Films and Coatings: From Materials to Applications*, Woodhead Publishing, Cambridge, UK (2013).
3. P. W. Baumeister, *Optical Coating Technology*, SPIE Press, Bellingham, WA (2004) [doi: 10.1117/3.548071].
4. Wikipedia, "*Electrical Resistivity and Conductivity*," wikipedia.org/wiki/Electrical_resistivity_and_conductivity

Chapter 6
Optical Components

6.1 Lenses

Lenses are the most widely used optical component. There are many types of lenses, and each one has several parameters to describe its properties.

6.1.1 Lens specifications

The following parameters are used to specify a lens:

1. Material. Based on the type of material used to make the lens, the refractive index and Abbe number can be found, and the spectral range in which the lens is intended for use can sometimes be determined. The refractive index tolerance is often specified by $\pm\Delta$ without unit, where Δ is a certain number, e.g., ± 0.001. The Abbe number tolerance is often specified by $\pm\Delta$ with a unit of percentage, such as $\pm 0.8\%$.
2. Diameter and clear aperture. The edge of a lens is often polluted or scratched during the fabrication, handling, and mounting process. Therefore, the diameter of the usable part of a lens is usually about 80% of its physical diameter and is called a clear aperture. A clear aperture of 90% of the physical diameter is not rare. Lens diameter tolerance is often specified as $D+0/-\Delta$ with a unit of mm, where D is the nominal diameter of the lens, e.g., $10+0/-0.1$ mm. A lens with diameter tolerance of $D\pm\Delta$ will risk being too large to fit into an accurately made lens housing with a nominal inner diameter D.
3. Center thickness. This thickness is often specified with a tolerance of $\pm\Delta$ with a unit of mm, e.g., ± 0.05 mm.
4. Edge thickness. This thickness is sometimes specified for mounting purposes. A minimum edge thickness of 1 mm is often required for convenience of lens fabrication. For very small lens with a diameter of a couple of millimeters or less, the edge thickness can be thinner than 1 mm.
5. Focal length. This is an important parameter for lens users and is often specified with a percentage tolerance of $\pm\Delta$ % or $\pm\Delta$ mm. Since the focal length is a function of the two radii of surface curvature, the central thickness and the materials of the lens, once all of these other parameters

are specified with a tolerance range, there is no need to specify the focal length again. Therefore, focal length is usually not specified in lens drawings for lens vendors. As long as vendors make the lens material, lens central thickness, and the radii of the two surfaces of the lens correctly, the lens will naturally have the correct focal length.

6. Radii of surface curvature. For a given focal length, the orientation of a lens affects its performance. For example, if a planar-convex lens is used to focus a bundle of parallel rays, the planar surface of the lens needs to face the focused spot to minimize the spherical aberration, as shown in Fig. 6.1. Subsection 9.3.4 discusses this topic in more detail. So the radius of curvature of each surface must be specified. Also, for lens-mounting purposes, each surface shape must be known. The radius of surface curvature is often specified with a tolerance of $\pm\Delta$ % or ±mm, or fringe numbers, which will be discussed in Section 14.5.
7. *F*-number. The *F*-number is defined as the ratio of the lens focal length to its clear aperture. Once these two parameters are known, the *F*-number is known. The *F*-number is often abbreviated as *F*/#, e.g., *F*/1 or *F*/2. Figures 6.1(a) and (b) show a *F*/1.5 and a *F*/3 lens focusing three rays, respectively. A smaller *F*/# means faster focus; such a lens is also called a fast lens. On one hand, a faster lens has a theoretically smaller Airy spot or possibly a smaller focused spot, as shown by Eq. (2.18). On the other hand, faster focusing will result in severe spherical aberrations that will increase the focused spot size. What *F*/# is best depends on what lens is used. If the lens is aspheric with spherical aberration corrected, a smaller *F*/# will certainly result in smaller focused spot. If the lens is convex spherical-planar, like those shown in Fig. 6.1, the focused spot size of the *F*/1.5 lens is about four times larger than the focused spot size of the *F*/3 lens. Similar to the focal length, the *F*/# is an important parameter for lens users, but it is usually not specified in lens drawings.
8. Coating type. The coating applied to the lens surface is mostly an AR coating for a certain spectral range. The coating is often specified by the maximum or average residual reflectance over a certain spectral range, e.g., an average reflectance <1% over 450–650 nm.

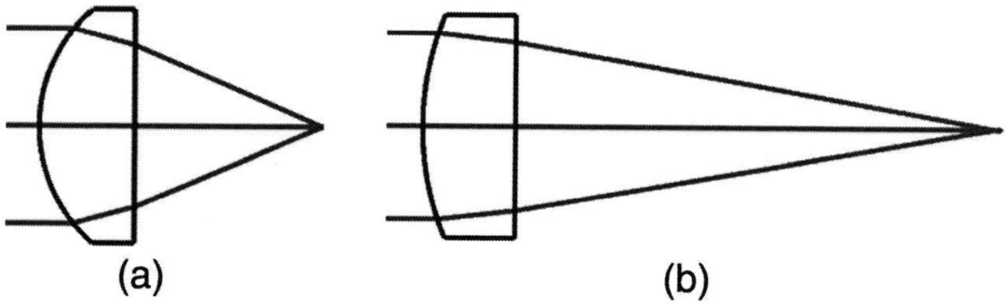

Figure 6.1 (a) *F*/1.5 lens and (b) *F*/3 lens. *F*/# is defined by the ratio of the focal length to the clear aperture of the lens. A smaller *F*/# means faster focusing, theoretically possible smaller focused spot, and more severe spherical aberration.

9. Scratch and dig. This factor defines the lens surface quality. For example, 80-50 means the maximum width of all scratches on the lens surface will be less than 80 μm and the maximum diameter of all the digs on the lens surface will be smaller than 500 μm.

   There are several grades of scratch and dig. 80-50 means low quality; these optical components are suitable for use far from image planes. Most commercial optical components have a 60-40 scratch and dig. 40-20 means moderate quality; these optical components are used in many optical devices or used to handle low-power laser beams. A small amount of laser beam energy scattered by optical surfaces can create a lot of scattering noise, and therefore laser optics requires high surface quality. Optical components with a scratch and dig of 20-10 are of high quality and can be used for high quality viewing device and to handle medium power laser beams. Optical components with a 10-5 scratch and dig are of extremely high quality and used for special research or to handle high-power laser beams.

   Scratch and dig are also used to specify the surface quality of prisms and mirrors. Letters can be used to designate specific values of scratch and dig (see Table 6.1).
10. Surface irregularity. This value is defined as the peak–valley deviation of the real lens surface to the ideal surface shape. The He-Ne laser wavelength of $\lambda = 0.633$ mm is often used as the unit. For example, $\lambda/2$ means low surface quality, $\lambda/4$ means medium surface quality, and $\lambda/8$ means high surface quality. Surface irregularity is also used to specify the surface quality of prisms and mirrors.
11. Wedge. This value is the parallelism between the two surfaces of a lens and is specified for high-precision lenses. Wedge is specified as an angle (in arc minutes) or as a sag difference (in millimeters). This topic is discussed later in Section 14.3.2.
12. For aspheric lenses, only the surface radii of curvatures are specified with tolerance; all other aspheric parameters are specified without tolerance. The tolerance of an aspheric surface is specified by the overall surface quality in terms of RMS or peak-to-valley surface fluctuations from the designed surface profile.

Table 6.1 Letter designation of scratch and dig.

Scratch Letter	Scratch Number	Dig Letter	Dig Number
A	5	A	5
B	10	B	10
C	20	C	20
D	40	D	30
E	60	E	40
F	80	F	50
G	120	G	70

6.1.2 Spherical lenses

Most lenses used have spherical surfaces that can be made by using a tool with a spherical surface to grind the lens surface. One grinding machine can often grind ten or more spherical lenses simultaneously, depending on the size of the lens and machine. Spherical lenses are relatively easy to make, inexpensive, and widely used.

The main shortcoming of spherical lenses is that rays passing through a spherical lens can have severe spherical aberration. Figure 6.2(a) reproduces the lens shown in Fig. 1.23(a). This lens is an equally convex spherical lens made by Ohara S-FPL53 glass with a 15-mm diameter, 5-mm central thickness, and 20-mm focal length. The spherical aberration is very severe.

6.1.3 Singlets, doublets, and triplets

All of the lenses discussed so far are a single lens, or singlet, which is the most common type of lens. Two properly designed singlets can be cemented together to create unique and useful properties that two separate singlets cannot possess, e.g., the ability to minimize color aberration. Such a cemented lens is called a "doublet." A doublet is usually formed by one positive singlet and one negative singlet. The inner surfaces of the two singlets have the same radii of curvature to fit each other. A thin layer of cement between the two singlets bonds them together. The two singlets are made of two different materials with different refractive indices and Abbe numbers. The two materials used must be carefully chosen to create the desired properties.

Doublets are widely used in lenses that handle light of a broad spectral band. For example, a commercial camera lens needs to have at least one doublet in order to handle the light in the visible band.

To further enhance the power of a doublet, another singlet can be joined to it to form a "triplet." A triplet usually has a positive–negative–positive or negative–positive–negative lens configuration. The materials of the three

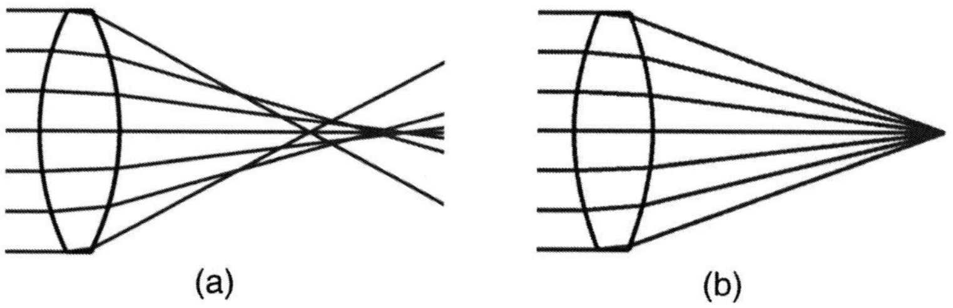

Figure 6.2 (a) An equally convex spherical lens made by Ohara S-FPL53 glass with a 5-mm central thickness, 15-mm diameter, and 20-mm focal length has severe spherical aberration. (b) An aspheric lens made by adding two equally conic terms on the two surfaces of the lens shown in (a). This aspheric lens can well focus the rays and has very small spherical aberration.

singlets must be different and properly match each other for the best results. Triplets are used less often than doublets. (Note that going beyond three joined lenses is rare because the tolerance involved can be cumulative.)

The outer surfaces of a doublet or a triplet can be either spherical or aspheric depending on the application requirements. The inner surface of a doublet or a triplet is rarely aspheric because in such a case, people must make two aspheric surfaces and cement them to make one effective aspheric surface, which is less cost effective. Figures 6.3(a) and (b) show the shapes of two typical doublets and two typical triplets.

6.1.4 Aspheric surfaces

Lenses and mirrors with an aspheric surface profile have certain unique and powerful functions, and they are widely used. Any surfaces that deviate from a spherical surface are called aspheric surfaces. A mirror or lens with one or two aspheric surfaces is an aspheric mirror or aspheric lens, respectively.

In the field of optical engineering, the profile of an aspheric surface is described by the even aspheric equation

$$z = \frac{\frac{\rho^2}{r}}{1 + \sqrt{1 - (1+k)\frac{\rho^2}{r^2}}} + a_1\rho^2 + a_2\rho^4 + a_3\rho^6 + a_4\rho^8 + a_5\rho^{10} + a_6\rho^{12} + a_7\rho^{14} + a_8\rho^{16}, \tag{6.1}$$

where z is the surface sag, ρ is the surface radial variable, r is the surface spherical radius of curvature, k is the conic parameter, and a_i ($i = 1 - 8$) is a series of aspheric parameters. The 2D illustration of Eq. (6.1) is drawn in Fig. 6.4.

Theoretically, Eq. (6.1) can contain many more ρ terms with higher powers, but ρ terms with a power up to 16 are accurate enough to describe most real aspheric surfaces. An aspheric surface can also be described by an odd aspheric equation, which can be obtained by replacing all ρ^{2i} terms by ρ^{2i+1} terms. However, the even aspheric equation [Eq. (6.1)] is more commonly used.

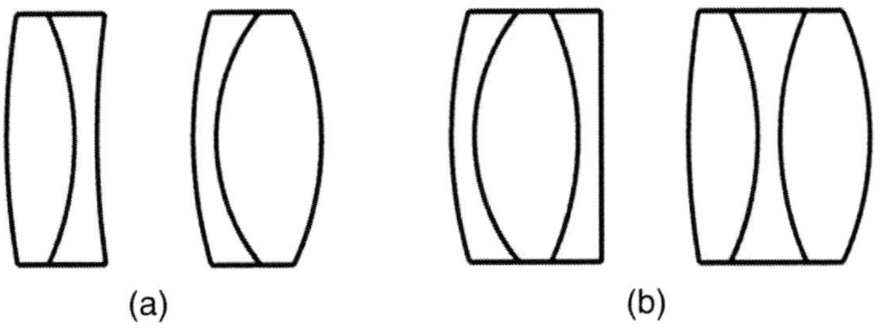

Figure 6.3 Two typical shapes of (a) doublet and (b) triplet lenses.

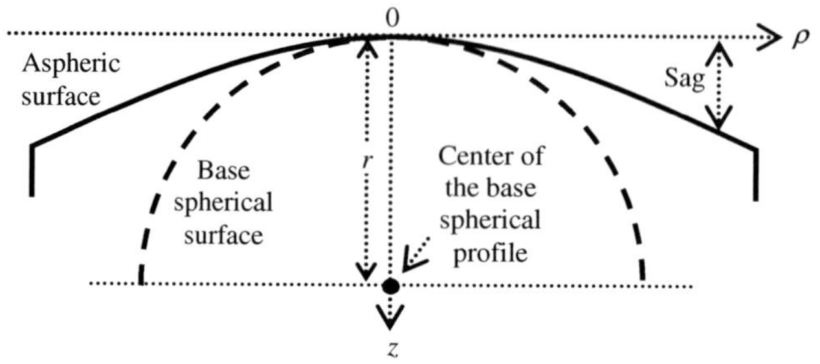

Figure 6.4 Illustration of Eq. (6.1) used to describe an aspheric surface. The solid curve is the aspheric surface, the dashed curve is the base sphere of this aspheric surface, r is the radius of the base sphere, ρ is the radial variable, and z is the lens sag. When $\rho = 0$, $z = 0$, and the local radius of curvature of the aspheric surface reduces to r.

Equation (6.1) is the general description of an aspheric surface and can be simplified to describe various simpler aspheric profiles:

1. Spherical surface. If k and all a_i are zero, Eq. (6.1) reduces to

$$z = \frac{\frac{\rho^2}{r}}{1 + \sqrt{1 - \frac{\rho^2}{r^2}}}, \tag{6.2}$$

and the aspheric surface reduces to its base spherical surface, as illustrated by the dashed curve in Fig. 6.4. Note that Eq. (6.2) can be rearranged to take the form

$$\rho^2 + (z - r)^2 = r^2, \tag{6.3}$$

which is a standard math form of a sphere with a radius r and center at $z = r$.

2. Conic surface. If $k \neq 0$ and all a_i are zero, Eq. (6.1) reduces to

$$z = \frac{\frac{\rho^2}{r}}{1 + \sqrt{1 - (1 + k)\frac{\rho^2}{r^2}}}, \tag{6.4}$$

which is a conic surface. A conic surface is a type of aspheric surface that includes three special types of aspheric surfaces:

a. If $0 > k > -1$, the surface described by Eq. (6.4) is an elliptical surface. The optical property of an elliptical surface will be discussed in Section 6.2.5.

b. If $k = -1$, Eq. (6.4) reduces to

$$z = \frac{\rho^2}{2r}, \tag{6.5}$$

which is a parabolic surface whose optical properties will be discussed in Section 6.2.6.

c. If $k < -1$, Eq. (6.4) describes a hyperbolic surface whose optical properties will be discussed in Section 6.2.7.

An aspheric lens or mirror can have various combinations of values of k and a_i. The surface can even have ripple shape profile. Since an aspheric lens has two surfaces and the lens material has dispersion, the effect of an aspheric lens surface is more complex than the effect of an aspheric mirror surface. Specific application requirements determine the best aspheric profile.

6.1.5 Aspheric lenses

An aspheric lens with proper surface shape can effectively minimize the spherical aberration. Figure 6.2(b) shows one example. This lens is made by adding two equal conics on the two surfaces of the lens shown in Fig. 6.2(a), which means that both surfaces have the same $k \neq 0$ and $a_i = 0$. The aspheric shape is weak and barely noticeable, but the spherical aberration is significantly minimized.

Large quantities of aspheric lenses can be produced by molding, although the non-recurring cost (NRC) of making molders can be high. Small-quantity aspheric lenses can be produced more cost effectively by machining a preform lens, which is a spherical lens slightly larger than the desired aspheric lens, using diamond turning or other technologies, such as CNC[1] or MRF.[2] The cost of producing one aspheric lens is about 2–4 times higher than the cost of producing a spherical lens with a similar size, shape, material, and accuracy.

Note that it is easy to create strangely shaped aspheric lenses on paper, but the fabrication of such aspheric lenses is more complicated. There are many limitations on the fabrication of glass aspheric lenses; aspheric lens vendors should be consulted during the design process (see Section 14.2.3). The fabrication of polymer aspheric lenses is relatively easier.

When a lens, such as a camera lens, must handle a large field of view, spherical aberration is likely to be the primary issue. An aspheric element can be used in such a lens to effectively reduce the aberration. If properly designed and used, the first aspheric element in a lens can perform like two to three spherical elements. The second aspheric element in a lens is usually less effective.

6.1.6 Cylindrical lenses

Cylindrical lenses have a curved surface in only one direction and can focus light in only one direction. Figure 6.5 shows the schematic of two positive cylindrical lenses and two negative cylindrical lenses.

Some special terminologies for cylindrical lenses need explanation. The term "aspheric surface" is widely used to describe a non-spherical 3D surface that is symmetrical about the optical axis. For cylindrical lenses, a new term "acylindrical" is now used to describe a non-circular 2D surface profile. "Acylindrical" is a 2D analogy of the 3D "aspheric." The term "spherical" aberration is also widely used to describe an aberration that is symmetrical about the optical axis. For cylindrical lenses, there appears still to be no 2D analogy to "spherical" aberration. So, the term "spherical" aberration is still used to describe the aberration created by cylindrical lenses, although such an aberration is not symmetric about optical axis. "Circular" aberration may be a more appropriate term.

Acylindrical lenses are relatively more difficult to fabricate and are rarely used. If acylindrical lenses are used in a design, it is important to find a lens vendor who can fabricate such lenses.

6.1.7 Gradient-index lenses

Gradient-index lenses are widely used to couple light into and out from optical fibers (and in endoscopes). As the name suggests, the refractive index of the lens material gradually varies. There are two types of index variation: radial[3] and axial.

Figure 6.6(a) shows the raytracing diagram of a radially varying index lens known as a grin lens.[4] The index profile is also illustrated. The index has its maximum value along the optical axis of the lens, decreases along the radial direction, and is symmetric about the optical axis. Since the index of the material decreases radially, the speed of the ray traveling inside such a material increases radially, which means that a divergent wavefront moving along the material will gradually become a plane wavefront and then gradually become a convergent wavefront. This type of gradient index lens is more widely used than the other type.

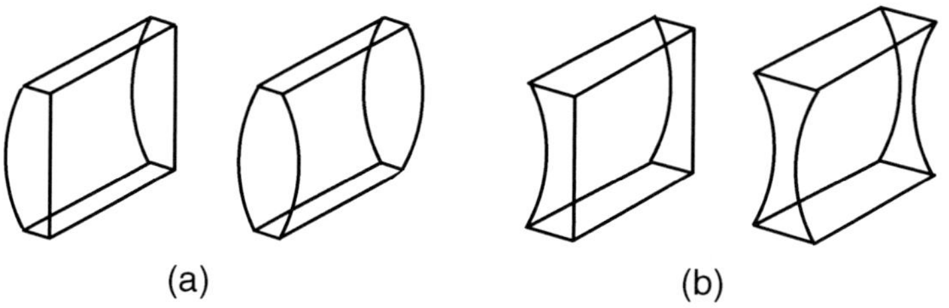

(a) (b)

Figure 6.5 (a) Planar-convex and convex-convex positive cylindrical lenses. (b) Planar-concave and concave-concave negative cylindrical lenses.

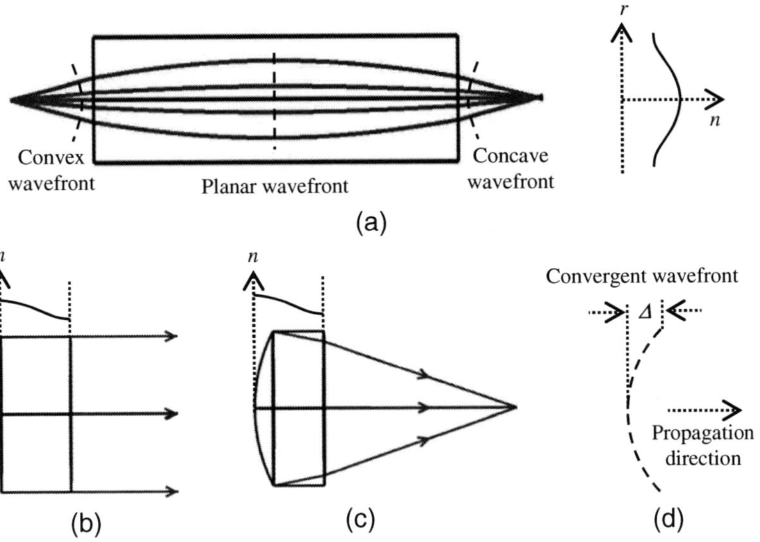

Figure 6.6 Raytracing diagrams of two types of gradient-index lenses. (a) A Go!FOTON SELFOC® grin lens that has a radially varying refractive index. The refractive index along the optical axis has the largest value, and the ray travels the slowest along the optical axis, thereby focusing the rays. (b) A LightPath Gradium lens that has an axially varying refractive index. The index decreases in the axial direction. Without a convex front surface to initiate focusing, the lens has no power. (c) A LightPath Gradium lens that has an axially varying refractive index. With a convex front surface, this lens can focus. (d) In an axially varying index lens, the edge part of a convergent wavefront propagates faster than the central part. The wave is thereby gradually focused.

Another type of gradient index lens has an axially decreasing index, as shown in Figs. 6.6(b) and (c). A planar wavefront inside such a lens will remain planar. Lenses with such an index profile need a convex front surface to generate a convergent wavefront at the start. Once a convergent wavefront is generated, the edge of the wavefront is always a delta ahead of the center part, as illustrated in Fig. 6.6(d), and it encounters an index relatively smaller than the center part of the wavefront encounters. This effect makes the edge of the wavefront always propagates faster than the center part. Thereby, the focusing power of the lens is gradually increased as the wavefront moves forward.

Compared with a fixed-index lens with the same index value and focal length, the surface curvature radius of a gradient-index lens is larger, and the spherical aberration is thus smaller, which is the main advantage of gradient-index lenses.

6.1.8 Fresnel lenses

Figure 6.7(a) shows a bulky and heavy conventional lens. A properly designed Fresnel lens (Fig. 6.7(b)) can perform tasks similar to the conventional lens but is much lighter and thinner. The Fresnel lens has a surface curvature the

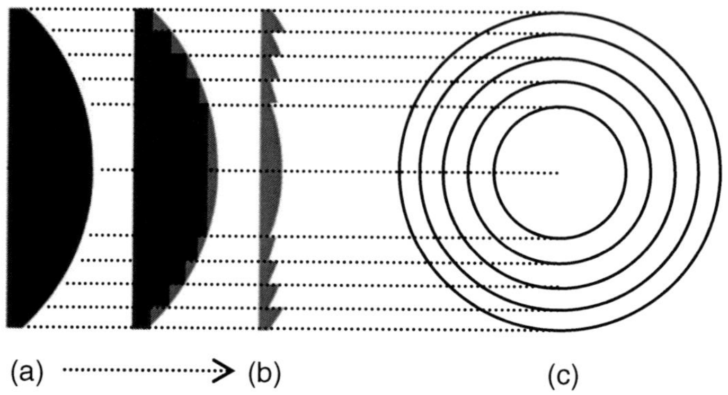

Figure 6.7 (a) A bulk and heavy conventional convex lens. (b) A Fresnel lens has the same surface curvature as the convex lens, but the curvature is broken into many sections and realigned, and thus the Fresnel lens is thin and light. (c) Top view of the Fresnel lens. Image reprinted courtesy of Thorlabs.

same as that of the conventional lens, but the curvature is broken into many sections and realigned. Thereby the Fresnel lens is thin and light. Figure 6.7 is a simplified sketch for illustration purposes; a real Fresnel lens has much more and smaller sections. Towards the edge of the lens, the surface curve is steeper, so to maintain the same section depth, the section width has to gradually decrease from the lens center to the lens edge.

Because of the complex surface structure, only optical polymers are used to mold Fresnel lenses. The lens shape can be circular or square, and the lens size can be up to hundreds of millimeters. Mass production makes the price of Fresnel lenses very low, down to a few dollars per piece. Polymer lenses are of lower quality than glass lenses, and the many edges of sections will scatter light and even cause diffraction. Therefore, Fresnel lenses are mainly used in illumination and low-resolution image applications.

6.1.9 Diffractive lenses

Diffractive lenses utilize complex microstructures in the lens to create complex interference and/or diffraction intensity patterns to perform certain tasks. Some diffractive lenses can focus or collimate a laser beam, the same as conventional lenses. Some diffractive lenses have multifocal lengths or multifocal points.

Diffractive lenses are less flexible than conventional lenses. They can only perform within the specified spatial and spectral ranges. Outside the ranges, these lenses can behave strangely. The conventional raytracing technique may not be able to correctly predict the behaviors of diffractive lenses. Caution is necessary when using diffractive lenses with conventional lenses. Figure 6.8 illustrates the potential problems. In Fig. 6.8(a), a diffractive positive lens has a focal length f_1 and can well focus collimated rays at its focal point F. A conventional negative lens with focal length f_2 is then placed a distance of f_2

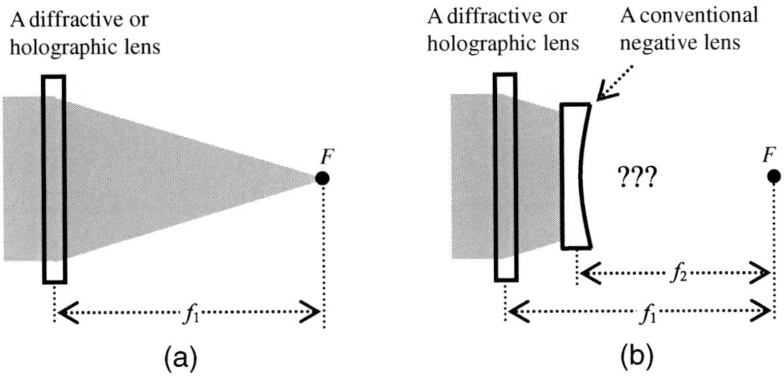

Figure 6.8 (a) A diffractive positive lens with focal length f_1 can well focus collimated rays at its focal point F. (b) A conventional negative lens with focal length f_2 is placed a distance of f_2 away from F. If the diffractive lens is a conventional positive lens, the conventional negative lens will collimate the rays.

away from F to collimate the focused rays, as shown in Fig. 6.8(b). If the positive lens is conventional, the rays will be collimated. But the positive lens here is diffractive, not conventional, so the final result is hard to predict. Because diffractive lenses utilize complex interference and/or diffractions, the conventional negative lens placed close to the diffractive lens may disturb the interference and/or diffractions. Only a real test can determine the result.

There are several types of diffractive lenses. Most diffractive lenses have their microstructures on their surfaces. Holographic lenses are a type of diffractive lens. The microstructures in a holographic lens are produced by applying a holographic image to photosensitive materials, such as dichromated gelatin, and they are inside the lens. Binary lenses are another type of diffractive lenses that have staircase type of microstructures. Diffractive lenses usually look like simple optical plates and have no convex or concave surfaces because the human eye cannot see the microstructures inside the lenses.

The design of diffractive lenses is a challenging task and requires special training and skill. Few consulting firms have the expertise to design diffractive lenses. The NRC of making diffractive lenses can be tens of thousands dollars, but once the production tool and process are set, the mass-production cost can be low, down to a few dollars per lens.

6.1.10 Lens materials

Most lenses are made of various optical glasses, particularly the inexpensive glasses, such as N-BK7 glass of Schott glass or B270 glass of Ohara Corp. All optical glasses have high transmission from about 0.4 μm to over 2 μm, covering the visible and near-IR ranges.

Various optical polymers are used to make inexpensive lenses. Most optical polymers have high transmission from about 0.4–1.0 μm, which is enough to cover the visible and the sensitive spectral range of CCD and CMOS sensors. Polymers can form strong aspheric shapes that are difficult for glasses to form. Figure 6.9 shows three examples of such strong aspheric polymer lenses. Once one sees lenses with shapes comparable to the shapes shown in Fig. 6.9, one can assume these are polymer lenses.

All of the UV and IR optical materials discussed in Sections 4.4 and 4.5 can be used to make lenses. The material types indicate the spectral range in which the lens is intended for use.

6.2 Mirrors

6.2.1 Mirror types and specifications

Mirror types are defined by three aspects: coating types, functions, and surface profiles. The other specifications concern the mirror surface quality:

1. Surface irregularity, which usually ranges from low quality ($\lambda/2$) to high quality ($>\lambda/8$), where λ is the He-Ne laser wavelength of $\lambda = 0.633$.
2. Surface scratch and dig, which usually range from low quality (80-50) to high quality (20-10), where the first number is the scratch width with a unit of microns and the second number is the dig diameter with a unit of tens of microns.

Laser mirrors often have their damage threshold specified.

6.2.2 Metallic mirrors and dielectric (dichroic) mirrors

Various types of metal coatings and dielectric coatings discussed in Sections 5.1 and 5.2 can be applied to mirror substrates to make various types of mirrors. Among these mirrors, metallic coating mirrors are of the lowest cost, have good high reflectance from UV to mid-IR range and are the most commonly used mirrors. Dielectric mirrors are relatively more expensive and mainly used for special and high performance applications, such as laser line reflection. Some dielectric mirrors are called dichroic mirrors because they have certain colors.

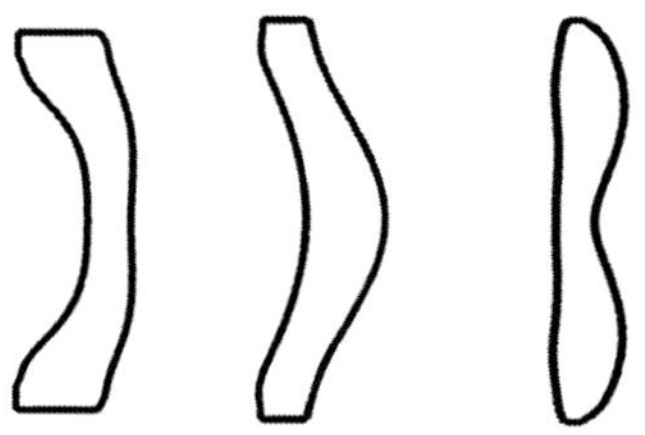

Figure 6.9 Shapes of three polymer lenses with strong aspheric surfaces. It is very difficult to make glass aspheric lenses with similarly strong aspheric surfaces.

6.2.3 Spectral mirrors

Various types spectral mirrors can be made by applying to mirror substrates various types of multilayer dielectric spectral coatings which are discussed in Sections 5.5–5.7. In terms of their function, spectral mirrors mainly include cold mirrors, hot mirrors, broadband mirrors, and laser line mirrors.

Since IR radiation is often used for heating, in this sense, a shorter wavelength is considered "colder" than a longer wavelength. Therefore, a mirror that reflects a short wavelength spectrum and lets a long wavelength spectrum pass is called a cold mirror, which is specified by its cutting wavelength. Figure 6.10(a) shows the spectra of a cold mirror with a 700-nm cutting wavelength.

Similarly, a hot mirror reflects a long wavelength spectrum and lets a short wavelength spectrum pass. A hot mirror is also specified by its cutting wavelength. Figure 6.10(b) shows the spectra of a hot mirror with a 700-nm cutting wavelength.

A broadband mirror has high reflectance inside its spectral band, and the reflectance drops fast outside the band. The spectral bandwidth can reach up to several hundred nanometers. Figure 5.4 shows the spectra of four broadband mirrors with different spectral bands in the visible and near-IR ranges, respectively. The reflectance is over 99% inside the spectral ranges.

A laser line mirror is a narrowband mirror that has high reflectance inside a spectral band of ten nanometers or so. The reflectance drops quickly outside the spectral band. The spectra of an Ar-ion laser line mirror with the intended wavelength range of 333–364 nm is plotted in Fig. 6.11. The designed angle of

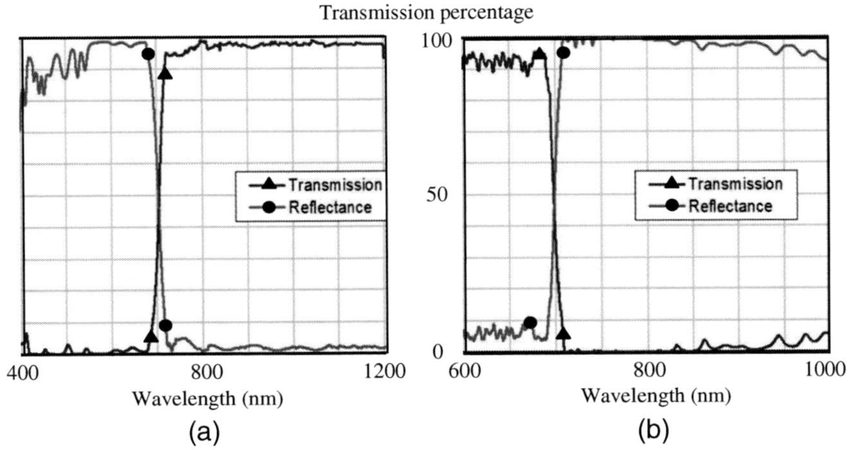

Figure 6.10 The spectrum of a (a) cold (longpass) mirror and (b) hot (shortpass) mirror with a 700-nm cutting wavelength. Image reprinted courtesy of Thorlabs.

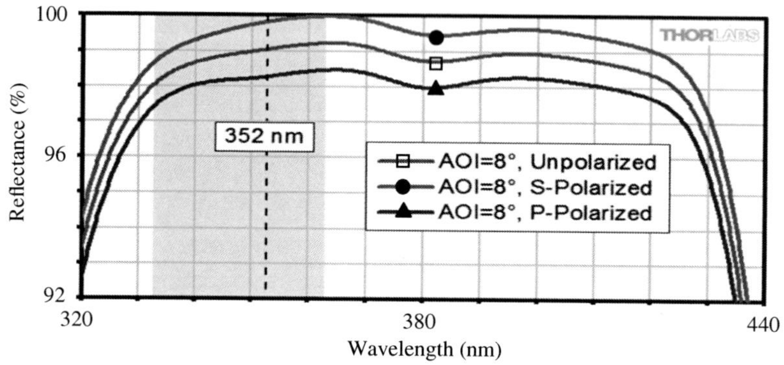

Figure 6.11 Spectra of an Ar-ion laser line mirror with the intended wavelength range of 333–364 nm. The designed angle of incidence is 8°, the reflectance for *s*- and *p*-polarizations are different, and the reflectance for the unpolarized light is the average of the reflectance for *s*- and *p*-polarizations. Image reprinted courtesy of Thorlabs.

incidence is 8°, the reflectance for *s*- and *p*-polarizations are different, and the reflectance for the unpolarized light is the average of the reflectance for *s*- and *p*-polarizations.

Two notes need to be made here:

1. All of the specific wavelengths, such as the cutting wavelength of cold or hot mirrors and the central wavelength of broadband and laser line mirrors, are specified for a certain angle of incidence (AOI). If the AOI increases, all of these specific wavelengths decrease, and vice versa.
2. The damage threshold of a laser line mirror can be important because pulsed laser beams can have high energy densities. The typical damage threshold for high-power lasers is several joules per square centimeter. The topic of laser damage threshold will be further discussed in Sections 9.1.2 and 9.2.2.

6.2.4 Spherical mirrors

Mirrors often have a surface profile that is either spherical and aspheric, which includes elliptical, parabolic, hyperbolic, etc. Flat mirrors can be considered as spherical with an infinitely large radius of surface curvature.

Flat mirrors are widely used for steering laser beams and reversing image orientation in one direction, as illustrated in Fig. 6.20(b) later in this chapter. To completely reverse an image, two flat mirrors in orthogonal arrangement are required.

Spherical mirrors have concave and convex types. Concave mirrors are often used to focus rays incident on them to form images. Figure 6.12(a) shows the Zemax-generated raytracing diagram of rays emitted by a point a distance away focused by a spherical concave mirror. However, the focus is not sharp because of the existence of spherical aberration.

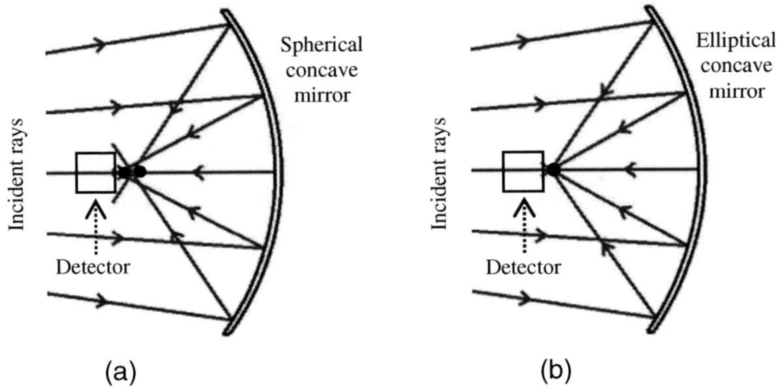

Figure 6.12 (a) A concave spherical mirror focuses incident rays with noticeable spherical aberration on a detector. The detector blocks the central portion of the incident rays. (b) An elliptical aspheric mirror focuses with minimized spherical aberration rays emitted by a point source located at another focal point of the ellipse.

6.2.5 Elliptical mirrors

Different types of aspheric mirrors have different optical properties and applications. Elliptical mirrors are one type, illustrated in Fig. 6.12(b), which is a modification of the spherical mirror shown in Fig. 6.12(a). It focuses the rays emitted from the same point source to a small spot.

Any ellipse has two focal points. Rays that are emitted from one focal point and hit the ellipse will be focused onto the other focal point of the ellipse. In this case, the elliptical mirror is only a small section of the ellipse. The point source is located at the other focal point of this elliptical mirror.

To be specific, the point source and the focused spot are 140 mm and 20 mm from the mirror vertex, respectively. The mirror has a 40-mm clear aperture, a 35-mm surface radius of curvature, and a conic of $k = -0.562$.

Elliptical cylindrical mirrors are often used in solid state lasers to focus light from a pumping light tube, which is placed along one focal line of the cylindrical mirror, onto a laser medium rod, which is placed along the other focal line of the cylindrical mirror.

6.2.6 Parabolic mirrors

The advantage of using mirrors to form images is that mirror reflection has no color dispersion. The disadvantage of using mirrors is that the detector used to receive the focused rays will block the central portion of the incident ray bundle, as illustrated in Figs. 6.12(a) and (b). To avoid the ray-blocking problem, the rays must incident on the mirror with an angle to the optical axis of the mirror, as shown in Fig. 6.13 (A). However, in such a situation, either a spherical or an elliptical mirror can only poorly focus rays.

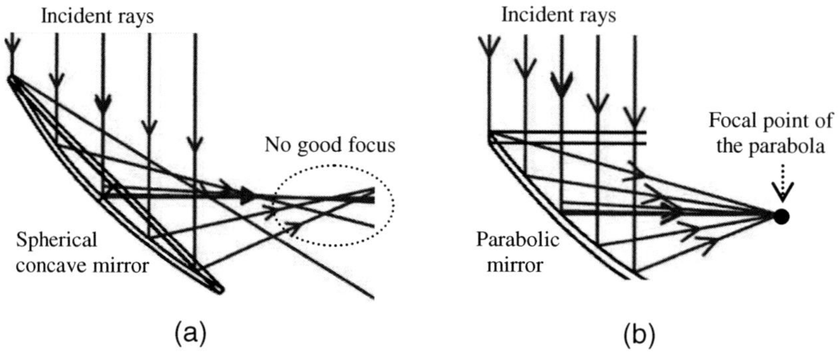

Figure 6.13 (a) A spherical concave mirror cannot well focus rays incident on it with an angle to the mirror axis. (b) A parabolic mirror can well focus rays incident on it with an angle to the mirror axis. The focused spot is at the focal point of the parabola. Shown here is a small section of a parabolic mirror.

Parabolic mirrors are another type of aspheric mirrors. One optical properties of a parabolic mirror is that it can well focus parallel rays incident on it with an angle to the mirror axis, as shown in Fig. 6.13(b). The focused spot is the focal point of the parabolic mirror. The raytracing process is reciprocal. If a point source is placed at the focal point of a parabola, the rays emitted by the point and incident on the parabolic mirror will be collimated by the mirror. If a parabolic mirror is described by Eq. (6.5), its focal length is $f = r/2$.

Parabolic mirrors are often used in car headlights, where the light bulb is placed at the focal point of the mirror. Light emitted by the bulb is collimated by the mirror and projected forward.

A real parabolic mirror is often a small section away from the center of the whole parabola.

6.2.7 Hyperbolic mirrors

Hyperbolic mirrors are another type of aspheric mirror with special applications. The standard mathematical description of a hyperbolic mirror is given by

$$\frac{z^2}{a^2} - \frac{\rho^2}{b^2} = 1, \tag{6.6}$$

where a and b are two parameters. Note that Eq. (6.6) is equivalent to Eq. (6.4) with $k < -1$ and is more convenient to illustrate the property of hyperbola.

Equation (6.6) is plotted in Fig. 6.14. Mathematically, hyperbolas appear in a pair, and their vertices are not at the origin of the coordinator, but they appear at a distance a from the origin. However, in real applications, hyperbolic mirrors are never used in pairs.

The unique optical property of a hyperbolic mirror is that it can well focus incident rays aiming at its focal point to the other focal point of the paired

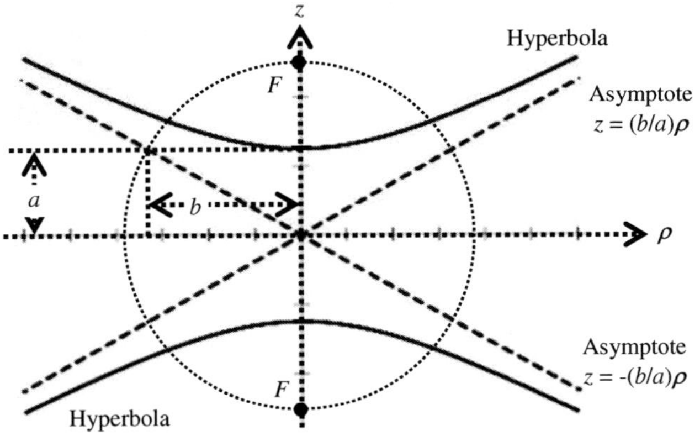

Figure 6.14 The two solid curves are a pair of hyperbolas described by Eq. (6.6). The two dashed lines are the two asymptotes of the hyperbola. The two focal points F of the two hyperbolas are marked by the solid dots. The focal length of such a hyperbola is $f = (a^2 + b^2)^{0.5}$, measured from the origin of the coordinator and not from the vertex of the hyperbola. The focal point is a distance of $(a^2 + b^2)^{0.5} - a$ from the vertex.

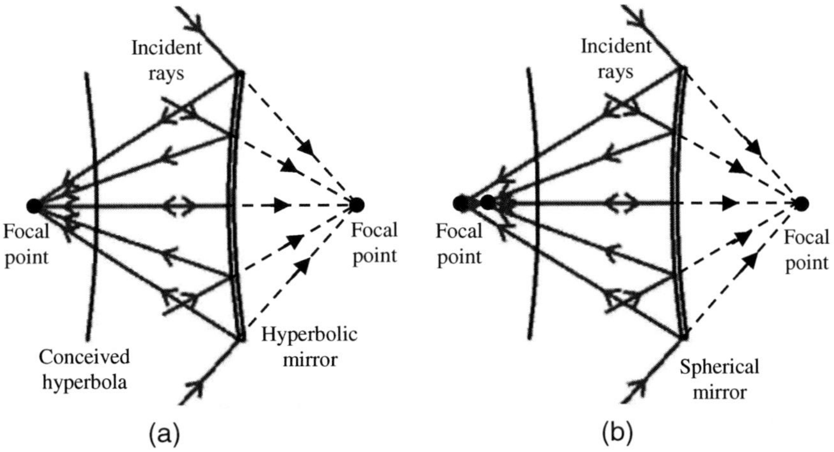

Figure 6.15 (a) A hyperbolic mirror can well focus incident rays aiming at one focal point of the hyperbola onto the focal point of the paired hyperbola. (b) A spherical mirror, with a shape and size similar to the hyperbolic mirror shown in (a), can only poorly focus the incident rays the same as the rays in (a).

hyperbola, as shown by the Zemax raytracing diagram in Fig. 6.15(a). This hyperbolic mirror has a surface radius of curvature of 180 mm, a conic parameter $k = -25$, a clear aperture of 62 mm, and a distance of 30 mm between the vertices of the pair of hyperbola.

As a comparison, the Zemax raytracing diagram of a spherical mirror with shape and size being similar to those of the hyperbola is plotted in Fig. 6.15(b). The rays are not well focused with large spherical aberration.

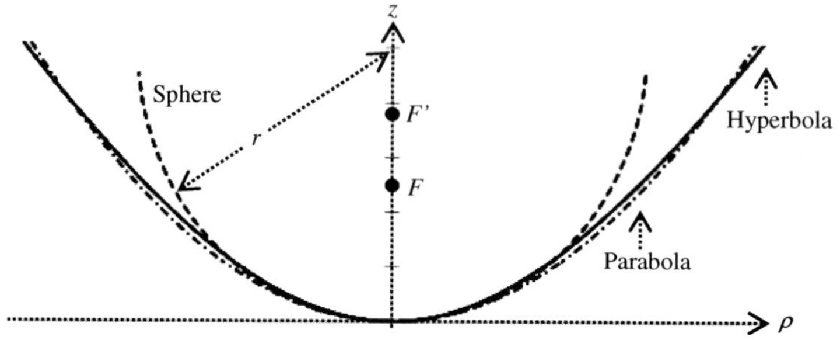

Figure 6.16 Comparison of shapes of a sphere, parabola, and hyperbola. The dashed curve is a sphere with a focal length of *r*/2 (radius of curvature is *r*). The dot-dashed curve is a parabola with a focal length of *r*/2. The solid curve is a hyperbola with a focal length of 2.38*r*. Note that the focal length of a hyperbola is measured from the focal point to the middle point of the hyperbola pair. The hyperbola is intentionally plotted here with its vertex coinciding with the vertices of the sphere and the parabola for a comparison of shapes. *F* is the focal point of the sphere as well as of the parabola. *F*′ is the focal point of the hyperbola.

Hyperbolic mirrors can be used in reflective telescopes, which will be discussed in Section 7.6.2.

Figure 6.16 plots a sphere, parabola, and hyperbola to compare their shapes. The focal point of the sphere and the parabolas are intentionally selected to coincide. The selected hyperbola has a shape similar to the shape of the parabola. Parabolas are more curved than hyperbolas.

6.3 Prisms

6.3.1 General comments

In optics, the term "prism" means a polygon, which is a 3D object with several flat surfaces. Various shapes of prisms are used to steer or shape laser beams, to reverse images, to fold the optical path, etc. The information about prism functions is sometimes included in their names. Prisms are named in several ways:

1. After its shape, such as right-angle prisms and equilateral prisms.
2. After its function, such as dispersion prisms and corner reflectors.
3. After a unique name that can be traced back years, such as Dove prisms and Pechan prisms. These prisms often have a complex shape and/or can perform complex functions that cannot be described by a couple of words.

The main specifications of prisms are

1. Surface scratch and dig, which usually range from low quality (80-50) to high quality (20-10), where the first number is the scratch width with a unit of microns, and the second number is the dig diameter with a unit of tens of microns.

2. Surface irregularity, which usually range from low quality ($\lambda/2$) to high quality ($\lambda/8$), where λ is the He-Ne laser wavelength of $\lambda = 0.633$ μm.
3. Angle tolerance and size tolerance, usually ±3 arcmin and ±0.2 mm, respectively.

There are many types of prisms, but this section introduces only some of them.

6.3.2 Dispersion prisms

The refractive index of glasses varies as the wavelength changes. This phenomenon is called dispersion. It is the root cause of color aberration and is undesired in most applications.

Glass dispersion can be utilized for good purpose, e.g., to separate different wavelengths (colors). Figure 6.17 presents an example: a single ray containing three colors RGB is incident on a right-angle prism made of Ohara S-NPH3 glass. This glass has a very small Abbe number of 17.5 and a very strong dispersion. The three wavelengths for the RGB are 0.65 μm, 0.55 μm, and 0.45 μm, respectively. The three colors are separated by the prism. The raytracing is accurately performed by Zemax, not just a sketch.

6.3.3 Right-angle prisms

Right-angle prisms can be used to reflect beams or reverse the orientation of images. One right-angle prism can reverse the image orientation in only one direction, and the ray direction is reversed as well, as illustrated in Fig. 6.18(a). To completely reverse the image orientation and keep the ray direction unchanged, a pair of right-angle prisms must be used in an orthogonal setup, as illustrated in Fig. 6.18(b). Such a setup is widely used in binoculars.

6.3.4 Corner reflectors

A corner reflector consists of three mutually perpendicular and intersecting surfaces. These surfaces are flat and highly reflective, as illustrated in Fig. 6.19(a). A corner reflector can reflect any ray incident back to its origin, as illustrated by a 2D simplification in Fig. 6.19(b). The three flat surfaces can

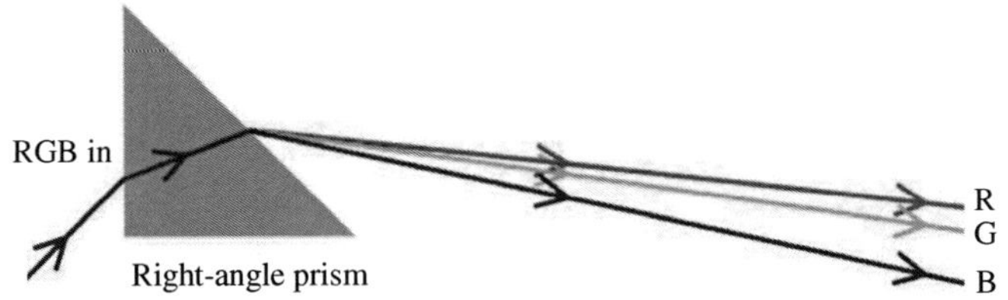

Figure 6.17 A right-angle prism disperses R(0.65 μm), G(0.55 μm), and B(0.45 μm) colors. The three colors coincide in the same ray incident on the prism and are separated when they exit the prism.

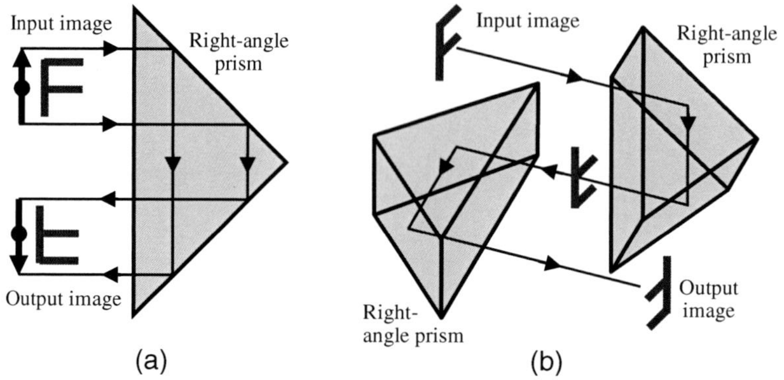

Figure 6.18 (a) One right-angle prism can reverse the image orientation in one transverse direction and the ray direction as well. (b) A pair of right-angle prisms can reverse the image orientation in both orthogonal transverse directions and maintain the ray direction.

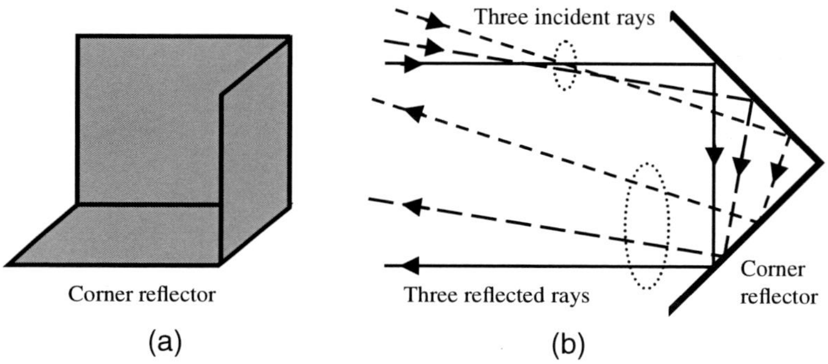

Figure 6.19 (a) A corner reflector consists of three mutually perpendicular and intersecting, flat and reflective surfaces. (b) A 2D simplification of a corner reflector. Three rays marked by the solid, dashed, and dotted lines, respectively, are incident on the reflector with different angles. The three corresponding reflected rays have the same angles as the incident rays. Therefore, each ray returns to its origin (with a small displacement).

be three mirrors or the three surfaces of a glass cube cut in half. A corner reflector is also called a corner cube or corner retroreflector.

Corner reflectors are used in surveys and ranging, where a laser beam is aimed at a corner reflector placed a distance away, and then the beam reflected by the corner reflector traces its way back before being intercepted and analyzed.

6.3.5 Penta prisms

Penta prisms are used to change the direction of rays while maintain the image orientation unchanged. Fig. 6.20 (A) shows a 2D sketch of a Penta prism to simplify the situation. The solid and dash lines denote the rays from the tip and the bottom of the arrow, respectively. Fig. 6.20 (B) shows a 2D sketch of

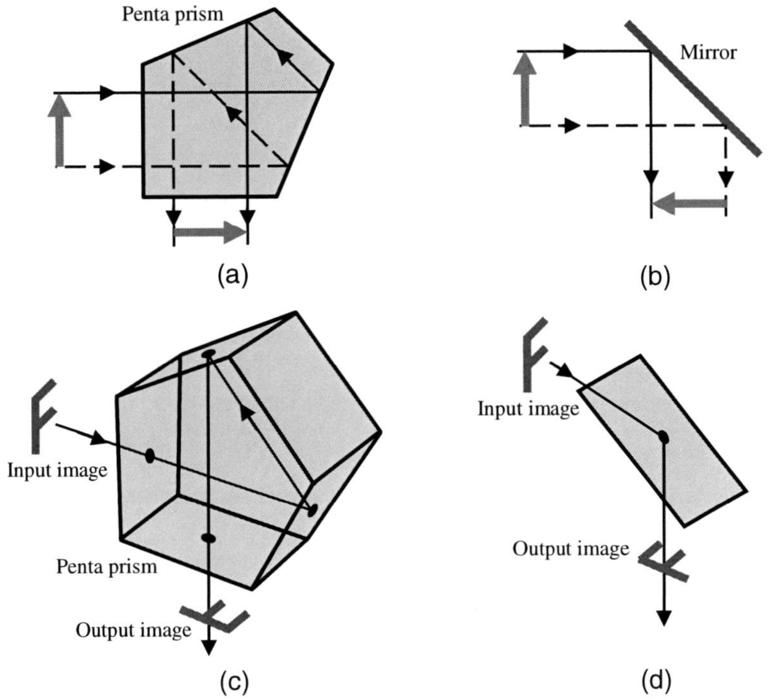

Figure 6.20 (a) A 2D sketch of a Penta prism to simplify the situation. The solid and dash lines denote the rays from the tip and the bottom of the arrow, respectively. (b) A 2D sketch of a flat mirror to compare with the Penta prism. The solid and dash lines denote the rays from the tip and the bottom of the arrow, respectively. The rays reflected by the prism and the mirror have opposite orientation. (c) A 3D sketch of a Penta prism that folds the ray while maintains the orientations of the image unchanged. (d) A 3D sketch of a flat mirror that folds the ray and reverses the orientation of the image in one direction.

a flat mirror for comparison with the Penta prism shown in (a). The solid and dash lines denote the rays from the tip and the bottom of the arrow, respectively. The rays reflected by the prism and the mirror have opposite orientation. Figure 6.20(c) depicts a 3D sketch of a Penta prism that folds the ray while maintains the orientations of the image unchanged. Figure 6.20(d) presents a 3D sketch of a flat mirror for comparison with (c). The mirror folds the ray and reverses the orientation of the image in one direction.

6.3.6 Dove prisms

Dove prisms are used to reverse the image orientation while maintaining the ray direction, as illustrated by Fig. 6.21(a). The optical axis of the prism is parallel to its base surface. The image is reversed vertically if the prism is not rotated, as illustrated by Fig. 6.21(b). The image can be rotated by rotating the prism about its optical axis. The image rotation angle is twice the prism rotation angle, as illustrated by Fig. 6.21(c).

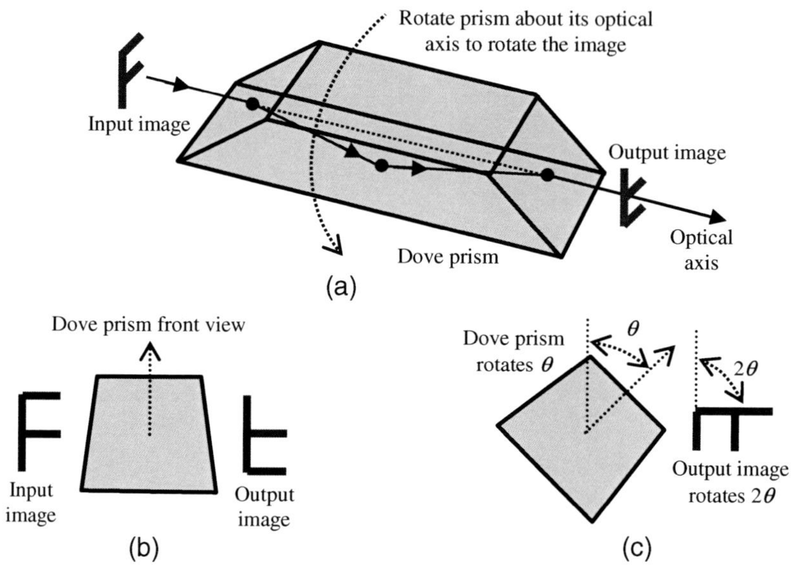

Figure 6.21 (a) A 3D sketch of a Dove prism that vertically reverses the orientation of the image exit from the prism while maintaining the same ray direction. (b) and (c) 2D sketches to further illustrate the situation. The Dove prism vertically reverses the image orientation while keeping the ray direction unchanged. The image orientation can be rotated by rotating the Dove prism about its optical axis. The image rotation angle is twice the prism rotation angle.

6.3.7 Pechan prisms

Some optical devices, such as head-mounted displays, must be small. The optical path required can be too long for the limited space. Variously shaped prisms are invented to fold multiple times the optical path. Pechan prism is one of them.

Figure 6.22 shows the Zemax-generated raytracing diagram for a Pechan prism. A Pechan prism consists of two parts. The top and bottom surfaces are high-reflection coated as marked. All of the other surfaces can be either uncoated or AR coated. The ray directions are marked by the arrows. The event sequence as the rays travel through the prism is as follows:

1. Rays transmit through the front surface.
2. Rays are reflected by total reflection at the first inter-surface.
3. Rays are reflected at the bottom surface by the high-reflection coating.
4. Rays are reflected by total reflection at the back surface.
5. Rays are reflected at the top surface by the high-reflection coating.
6. Rays are reflected by total reflection at the second inter-surface.
7. Rays transmit through the back surface.

The multiple reflections that occur inside the prism significantly shorten the physical length that the optical path takes. By carefully following the rays, the orientation of the image is shown to be reversed in one direction by the prism.

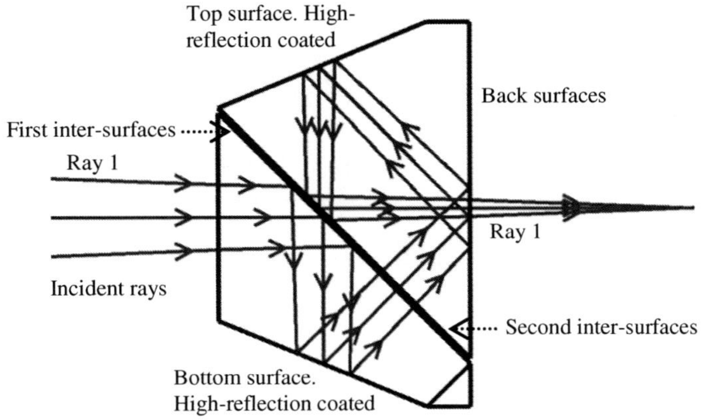

Figure 6.22 A raytracing diagram generated by Zemax for a Pechan prism. The ray directions are marked by the arrows. As the rays travel through the prisms, they are reflected five times: twice by the high-reflection coatings at the top and bottom surfaces, and three times by the total reflections that occur at the two inter-surfaces and the back surface. The multiple reflections significantly shorten the physical length that the optical path takes. The orientation of the image is reversed in one direction by the prism.

6.3.8 Anamorphic prisms

The word "anamorphic" is derived from Latin and means intentional distortion, such as unequal magnification of an image along two perpendicular axes. Anamorphic prisms are used to manipulate laser beams. Figure 6.23 sketches a pair of anamorphic prisms to expand or reduce the beam in the direction along the plane of this book page. The beam propagation direction determines whether expansion or reduction occurs. The expansion or reduction ratio, as

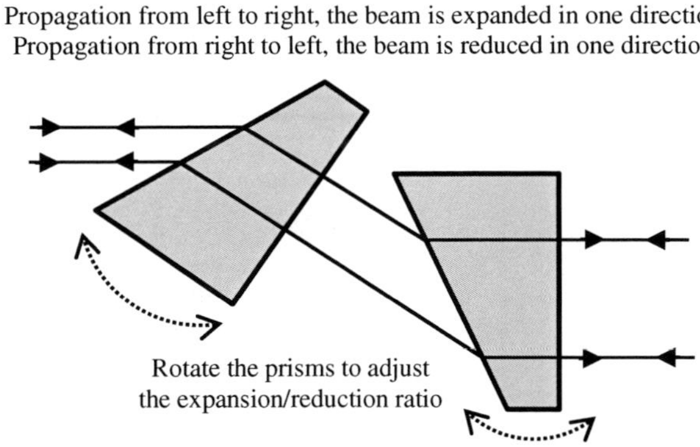

Figure 6.23 One anamorphic prism can either expand or reduce the size of a beam propagating through it in one direction perpendicular to the propagation direction, depending on the beam propagation direction. The propagation directions of the input and output beam are different. A pair of anamorphic prisms can increase the expansion/reduction ratio while keeping the propagation direction of the output beam the same as the propagation direction of the input beam. The expansion/reduction ratio can be adjusted by rotating the prisms.

well as the propagation direction of the beam after it leaves the prisms, can be adjusted by rotating the two prisms.

Figure 6.23 shows that one anamorphic prism can also expand or reduce the beam with smaller magnitude, but the often-undesired result is that the propagation direction of the input and output beam will be different.

6.4 Non-Polarizing Beamsplitters

6.4.1 General comments

There are two types of beamsplitters: non-polarizing and polarizing. A non-polarizing beamsplitter splits the intensity of a beam propagating through it by a specified ratio; the polarization state of the beam is irrelevant. Several types of non-polarizing beamsplitters are commercially available, e.g., cube, plate, pellicle, polka dot, etc. Figure 6.24 illustrates a cubic beamsplitter and a plate beamsplitter. Pellicle beamsplitters are thin films. Polka-dot beamsplitters are also plates.

Non-polarizing beamsplitters are specified by the ratio of transmission and reflection intensities, such as 10:90, 30:70, 50:50, 70:30, or 90:10, determined by the coating used. But, as will be shown in Fig. 6.25, the splitting ratio for all beamsplitters is an approximate number. This is not a quality issue but a common phenomenon because the mechanism utilized to split the beam is complex. The real transmission curves for each production run can be slightly different from the typical curves. Beamsplitter operators must include this performance variation in their tolerance budget.

Spectral filters, such as longpass filters and shortpass filters, can also be considered as a type of beamsplitter because they split the beam based on wavelength. This type of filter was previously discussed in Sections 5.7 and 6.2.3.

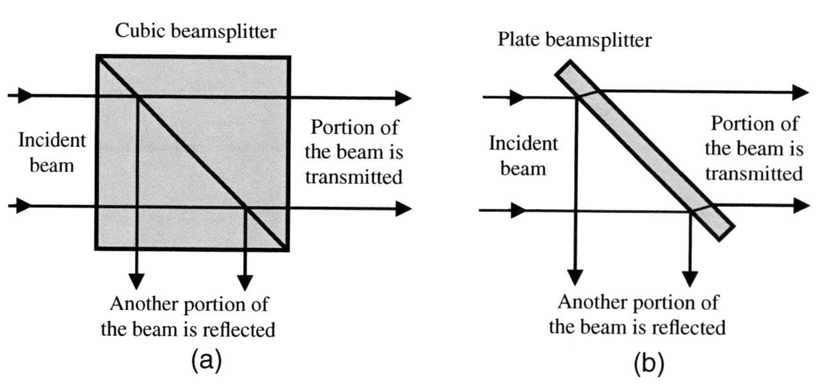

Figure 6.24 (a) A cubic beamsplitter. (b) A plate beamsplitter. Both beamsplitters can be either polarizing or non-polarizing. In the former case, the *p*-polarization (in the plane of the page) is transmitted. The *s*-polarization (perpendicular to the plane of the page) is reflected. In the case of the latter, the beam intensity is split.

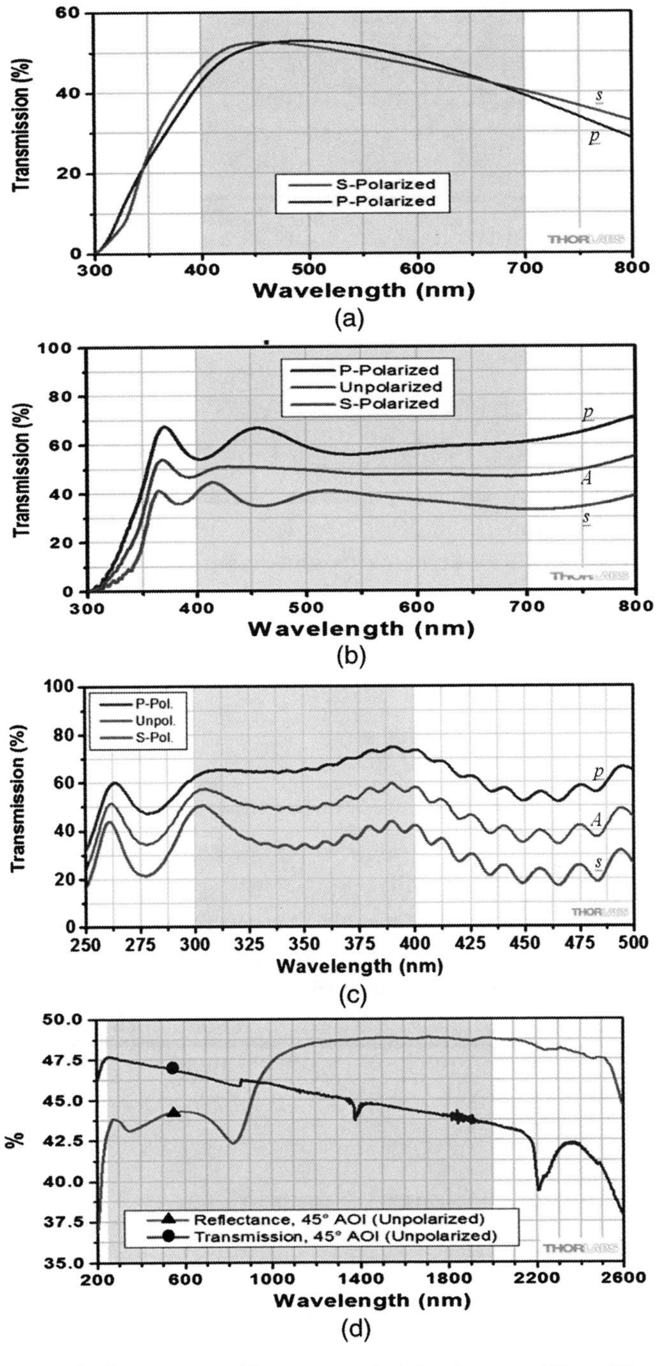

Figure 6.25 Transmission curves of four non-polarizing beamsplitters. The curves marked by *p*, *s*, and *A* are for *p*-polarization, *s*-polarization, and the average of *p*- and *s*-polarizations, respectively. Typical transmission curves between 400–700 nm for a (a) 50:50 cubic beamsplitter, (b) 50:50 plate beamsplitter, and (c) 50:50 pellicle beamsplitter. (d) Typical reflectance and transmission curves of a polka-dot beamsplitter. The splitting ratio is polarization and wavelength insensitive. Images reprinted courtesy of Thorlabs.

6.4.2 Cubic non-polarizing beamsplitters

A cubic non-polarizing beamsplitter has a special coating sandwiched in between the two right-angle prisms. The type of coating determines the splitting ratio. Figure 6.25(a) shows a type transmission curve for a 50:50 cubic beamsplitter for 400–700 nm. The splitting ratio varies apparently in this spectral range, about 12% in this case, and the transmission of *p*- and *s*-polarizations are slightly different. The advantage of cubic beamsplitters is that the two split beams have identical optical paths, which is important in applications that requires high accuracy, such as interferometers. The disadvantages of cubic beamsplitters are their large size and weight, and high cost.

6.4.3 Plate non-polarizing beamsplitters

A plate beamsplitter has a special coating on one surface of the plate. The type of coating determines the splits ratio. Figure 6.25(b) shows a typical transmission curve for a 50:50 plate beamsplitter for 400–700 nm. The transmission of *p*- and *s*-polarizations can have as much as a 22% difference, and the *s*- and *p*-polarization averaged splitting ratio marked by *A* is close to 50:50 over the entire range. The advantages of plate beamsplitters are their small size and weight, and low cost. The disadvantage of plate beamsplitter is that the two split beams have different optical paths, which will impact applications that require high accuracy, such as interferometers.

6.4.4 Pellicle non-polarizing beamsplitters

A pellicle beamsplitter is a membrane nitrocellulose material that has beamsplitting function. The membrane is only a couple of microns thick and has no coating on it. Figure 6.25(c) shows a typical transmission curve for a 50:50 pellicle beamsplitter for 400 nm to 700 nm range. The curves are similar to the curves for the plate beamsplitter, except that the curves exhibit sinusoidal oscillations as a function of wavelength. Because a pellicle beamsplitter is a thin film, interference occurs inside the film. The advantages of pellicle beamsplitters are their light weight and very low cost; however, they are fragile and easy to break.

6.4.5 Polka-dot non-polarizing beamsplitters

A polka-dot beamsplitter has a regularly repeating metal coating array on a substrate plate. The array reflects portions of the beam that is incident on the array. The uncoated part of the substrate lets the other portions of the beam pass through. The ratio of the coated area size to the uncoated area size determines the beamsplitting ratio. The advantage of polka-dot beamsplitters is that the metal coating is not sensitive to the wavelength and polarization state, as shown by the transmission and reflectance curves in Fig. 6.25(d). The disadvantage is that both the transmitted and reflected beams are transformed into a large number of subbeams, which diffract and interfere each other. Therefore, polka-dot beamsplitters can be used only in incoherent applications.

6.5 Polarizing Beamsplitters

6.5.1 General comments

A polarizing beamsplitter splits the two orthogonal polarizations of the beam propagating through the beamsplitter. If the incoming beam is randomly polarized, the intensity of the two polarizations are the same, and the splitting ratio is 50:50 or near 50:50.

Cubic and plate polarizing beamsplitters are the two types of widely used polarizing beamsplitters, as shown in Fig. 6.24. The *p*-polarization (in the plane of the page) is transmitted, and the *s*-polarization (perpendicular to the plane of the page) is reflected. Polarizing beamsplitters are specified by the extinction ratio of the transmitted beam, i.e., the ratio of the transmitted *p*-polarization intensity to the leaking-through *s*-polarization intensity in the transmitted beam. An extinction ratio of 100:1 is common, and 1000:1 or higher is frequently seen.

Both cubic and plate polarizing beamsplitters only work in a certain wavelength range, and the extinction ratio can vary inside this wavelength range. The extinction ratio of the reflected beam is usually not specified and will be lower than the extinction ratio in the transmitted beam. A polarizing beamsplitter is a type of polarizer.

6.5.2 Typical performances

Figure 6.26(a) shows typical *p*- and *s*-polarization transmission curves for a plate polarizing beamsplitter designed for 633 nm, but the *s*-polarization transmission is near zero from 615–645 nm. Figure 6.26(b) shows typical *p*- and *s*-polarization transmission curves for a cubic polarizing beamsplitter for 420–680 nm.

The linear scale of the curves in Fig. 6.26 does not provide the details of the *s*-polarization transmission; within the specified spectral range the *s*-polarization transmission is near zero and the extinction ratio is probably below 100:1. But the transmission curves in Fig. 6.26 show that within the specified spectral range, the transmission of *p*-polarization is often a few percent less than 100%, which means a few percent of the *p*-polarization are reflected. Even the *s*-polarization is 100% reflected, and the extinction ratio of the reflected beam will be much lower than 100:1. Thus the reflected beam is less "clean" than the transmitted beam. Conventionally, polarizing beamsplitters are specified by the extinction ratio only for the transmitted beam.

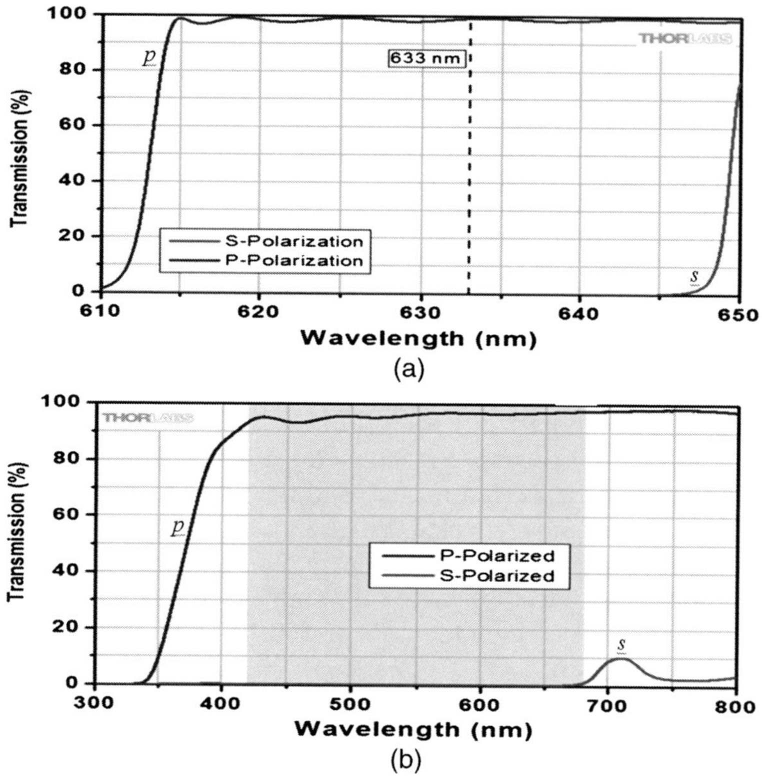

Figure 6.26 (a) Typical transmission curves of a plate beamsplitter for *p*- and *s*-polarizations for 615–645 nm. (b) Typical transmission curves of a cubic beamsplitter for *p*- and *s*-polarizations for 420–680 nm. Images reprinted courtesy of Thorlabs.

6.6 Polarizers

A polarizer is slightly different than a polarizing beamsplitter. A polarizer only provides a high extinction ratio in one beam, usually the transmitted beam, and does not affect the quality of the other beam, usually the reflected beam. A polarizing beamsplitter provides not only a high extinction ratio in the transmitted beam but also a good extinction ratio in the reflected beam. Three types of polarizers are widely used: birefringent, wire grid, and thin film.

6.6.1 Birefringent polarizers

Some crystals, such as calcite crystal, have a birefringent property[5] that can be used to make polarizers. There are two orthogonal axes in a calcite crystal. One axis is called the "ordinary axis," and the refractive index in this direction is from 1.9 at 190 nm to 1.5 at 1700 nm.[5] The other axis is called the "extraordinary axis," and the refractive index in this direction is from 1.6 at 190 nm to 1.4 at 1700 nm.

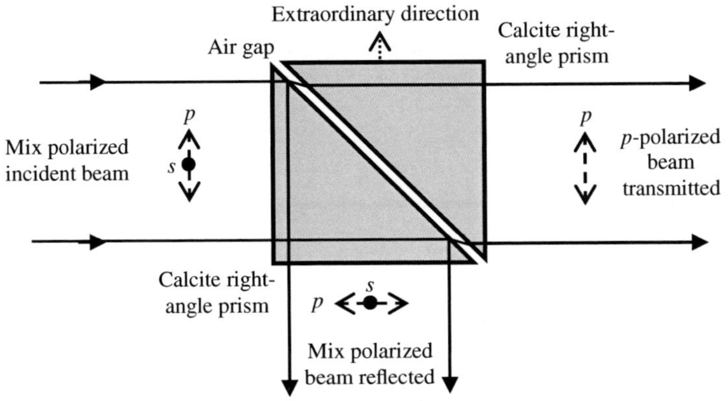

Figure 6.27 Sketch of a calcite birefringent polarizer. There are two right-angle prisms made from calcite with an air gap between them. The ordinary direction of the prisms is perpendicular to the plane of the page. In this direction, the refractive index is ~1.66. The extraordinary direction of the prisms is in the plane of the page as marked, and the refractive index is ~1.49. The *s*-polarization will be totally reflected. Most *p*-polarization will transmit through because of the index difference in two directions. Thereby, the transmitted beam is polarized.

Figure 6.27 shows a sketch of a calcite birefringent polarizer. There are two right-angle prisms made of calcite with an air gap in between them. The ordinary direction of the prisms is perpendicular to the plane of the page, and the extraordinary direction of the prisms is in the plane of the page as marked. The incident beam has mixed polarization. The *s*-polarization component is in the ordinary direction and is totally reflected at the calcite–air gap of the first prism. The *p*-polarization component is in the extraordinary direction and can transmit through because of the relatively low refractive index in this direction. Thereby, the transmitted beam has a high extinction ratio up to 100,000:1. A small portion of the *p*-polarization is still reflected by the two interfaces of calcite–air and air–calcite of the two prisms. Therefore, the reflected beam has a relatively low extinction ratio. The purpose of having the second prism there is to keep the propagation direction of the transmitted beam unchanged and to provide an optically identical path for the reflected and transmitted beams.

Figure 6.28 shows a typical transmission curve of *p*-polarization for a 10-mm-thick calcite polarizer. Birefringent polarizers have a broad working spectral range, from the visible to IR.

There are several types of calcite prism polarizers, and they all utilize the birefringent nature of calcite.

6.6.2 Wire-grid polarizers

A wire grid is another structure that has polarizing capability. A wire-grid polarizer consists of an array of parallel metal wires sandwiched between two glass plates or right-angle prisms. The polarization parallel to the wires is

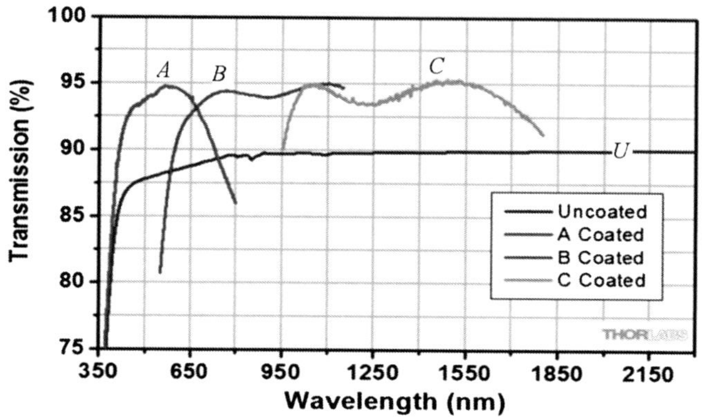

Figure 6.28 Typical transmission of *p*-polarization for a 10-mm-thick calcite polarizer. The transmission loss of an uncoated polarizer is mainly from the surface reflections. The transmission in certain spectral ranges can be raised by applying AR coatings on polarizer surfaces, as shown by the curves marked by *A*, *B*, and *C*. Image reprinted courtesy of Thorlabs.

reflected by the wires, and the other polarization can transmit through the intervals of the wires. The extinction ratio is about 1000:1 or higher. The working spectral can be extremely broad from nanometers in the UV to several microns in the mid-IR range. The limitation of wire-grid polarizers is that the wavefront of the beam transmitting through the polarizer is broken into many subwavefronts unless the applications are incoherent, otherwise the undesired diffraction and interference among the subwaves will affect the performance.

Although there are different types of wire-grid polarizers, the working principles behind them are the same.

6.6.3 Thin film polarizers

Thin film polarizers utilize complex interferences that occur in multilayer dielectric coatings to polarize. The working spectral band of thin film polarizers can cover several hundred nanometers in the visible or near-IR range. The extinction ratio can be from 1000:1 to over 10000:1. The cubic and plate polarizing beamsplitters discussed in Section 6.5 use thin films to polarize.

Some types of dichroic thin films can also polarize a beam. A dichroic film is a type of dielectric multilayer film or coating. The name "dichroic" comes from the fact that such films have certain colors; they are mostly used to reflect or transmit a certain spectral band.

6.7 Spectral Filters

6.7.1 Dielectric coating filters

Dielectric coating spectral filters are made by applying certain multilayer dielectric spectral coatings on transparent substrates and are the most commonly used spectral filters. They have different functions, such as highpass, lowpass, bandpass, band rejection, etc. Spectral coatings were previously discussed in Sections 5.7 and 6.2.3.

6.7.2 Colored-glass filters

Colored-glass filters utilize the volume absorption of various materials mixed in glasses to achieve various spectral functions. The cost of colored-glass filters is only a fraction that of dielectric coating filters and do not perform as well as dielectric coating filters. For example, colored-glass bandpass filters have much wider bands and less-clean cutting edges compared with dielectric coating filters. Figure 6.29 shows the typical spectral curves of two groups of colored-glass filters.[6]

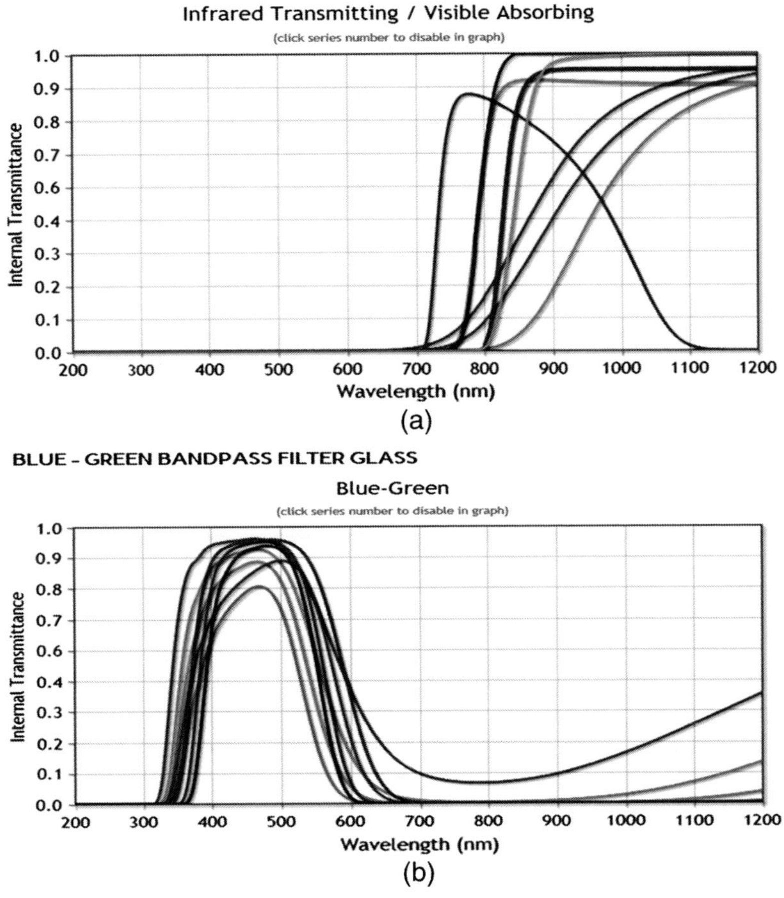

Figure 6.29 Spectral curves of two groups of colored glass filters: (a) highpass filter with a cutting edge at the near IR. (b) Blue-green bandpass filter. Images reprinted courtesy of Kopp Glass Corp.

6.8 Wave Plates

6.8.1 Working principle

Section 2.2.3 and Fig. 2.1 demonstrate that a linear polarization can be decomposed into two orthogonal polarizations. When a phase difference is created between the two decomposed polarizations, elliptical polarization appears.

A wave plate is made of a certain type of birefringent material. The polarization in the ordinary direction propagates with a speed different than the propagation speed of the polarization in the extraordinary direction. After propagating a certain distance in the material, a phase difference appears between the two polarizations in the ordinary and extraordinary directions. The plate thickness is usually so chosen that the phase difference is either a quarter-wave or a half-wave; such wave plates are called quarter-wave or half-wave plates, respectively. The phase difference created between the two polarizations can change the polarization state of light passing through the plate from linear to elliptical or vice versa, or change the polarization direction.

To make a wave plate work, the input light must be linearly polarized with the polarization direction in neither the ordinary nor the extraordinary direction. A quarter-wave plate will change a linearly polarized light to an elliptically polarized light. The elliptical ratio depends on the angle between the linear polarization direction and the ordinary (extraordinary) direction, as illustrated in Fig. 6.30. The situation shown in Fig. 6.30 is reciprocal. If an elliptically polarized light is incident on a quarter-wave plate, the exit light will be linearly polarized. For further explanation, review Section 2.2.3.

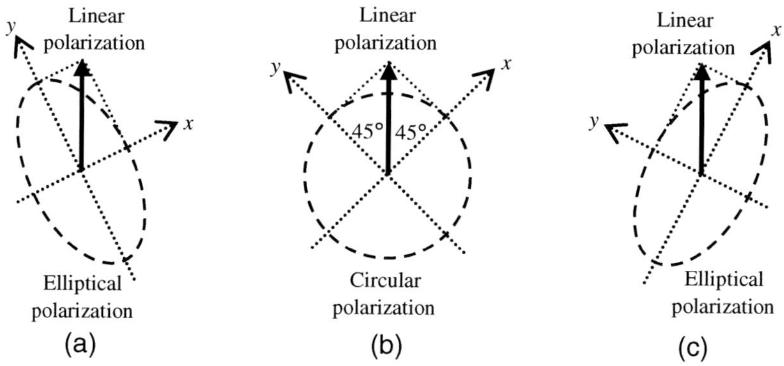

Figure 6.30 The effect of a quarter-wave plate on a linear polarizing light. *x* is the ordinary (extraordinary) direction of the wave plate, and *y* is the extraordinary (ordinary) direction of the wave plate. The solid-line arrows denote a linearly polarizing light incident on the quarter-wave plate. The dashed curves denote the polarization state of the light as it exits the wave plate. (a) and (c) The exiting light has an elliptical polarization state. (b) The exiting light has a circular polarization state. All of the situations shown in (a)-(c) are reciprocal.

Similarly, a half-wave plate will create a π/2 phase difference between the two decomposed polarizations, which rotates the linear polarization direction, as illustrated in Fig. 6.31.

Commercial wave plates come with zeroth-order and multi-order options. Mathematically, the phase difference ϕ is described by

$$\phi = \frac{2\pi L}{\lambda} \Delta n = 2\pi(a + m), \tag{6.7}$$

where L is the wave plate thickness, λ is the wavelength, Δn is the difference between indices in the ordinary and extraordinary directions, a can be either 1/4 or 1/2 for quarter-wave or half-wave, respectively, and m is an integer. $m = 0$ is zero order, and $m > 0$ is multi-order. Equation (6.7) shows that multi-order wave plates have a relative large thickness L for a given birefringent material. The phase difference ϕ is more sensitive to wavelength deviation. Therefore, multi-order wave plates only work fine for a small wavelength band. A wave plate is also called a retarder, and a half-wave plate is also called a rotator.

6.8.2 Crystal-quartz wave plates

Crystal quartz is a commonly used birefringent material to make wave plates. The refractive indices at 590 nm in the ordinary and extraordinary directions are 1.544 and 1.553, respectively. The index delta of 0.009 is smaller than the 0.173 index delta of calcite, which is not good for making polarizers but is good

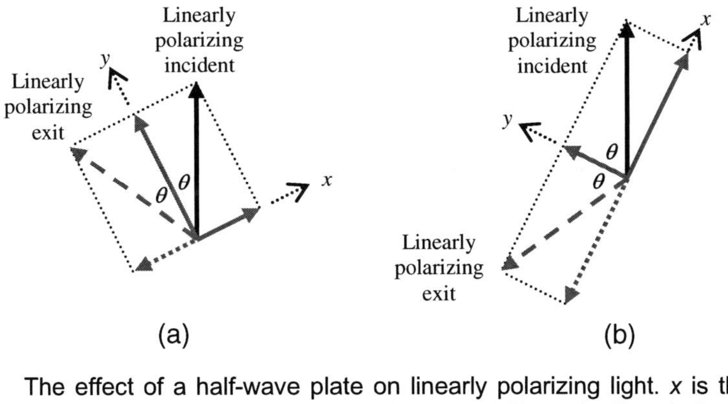

Figure 6.31 The effect of a half-wave plate on linearly polarizing light. *x* is the ordinary (extraordinary) direction of the wave plate, *y* is the extraordinary (ordinary) direction of the wave plate. The black solid-line arrows denote a linearly polarizing light that is incident on the half-wave plate with an angle θ to the *y* axis. The grey solid arrows denote the decomposed two orthogonal polarizations. The grey dotted arrows denote the polarization that has a phase of π (half-wave) change from the solid arrow. The grey dashed arrows denote the polarization of the light exiting the wave plate. The exit light is still linearly polarized, but the polarization direction is rotated by an angle of 2θ. Therefore, a half-wave plate is also called a "rotator." All of the situations shown here are reciprocal.

for making wave plates because a relatively thicker material is required to generate a given phase difference. For example, to generate a quarter-wave phase difference for a He-Ne laser of 0.633-μm wavelength, a thickness of 0.633 μm/(4 × 0.009) = 17.58 μm is required if quartz is used, whereas if calcite is used, the thickness required is 0.633 μm/(4 × 0.173) = 9.15 μm. The calcite wave plate has a thickness about half the thickness of the quartz wave plate, a thickness tolerance about twice tighter, and is more difficult to make.

6.8.3 Achromatic and tunable wave plates

The wave plates discussed previously use one type of birefringent material and functions properly only for one certain wavelength. If the wavelength changes, the phase difference will change as well, according to Eq. (6.7), the dispersion effect of the plate material will further affect the phase difference. To solve this problem people invented achromatic wave plates. An achromatic wave plate uses at least two types of birefringent materials, usually quartz and magnesium, with two properly selected thicknesses. Because these two materials have different birefringent and dispersion properties, the phase difference can be kept nearly constant over a large spectral range. Figure 6.32 shows typical phase-difference curves for an achromatic and a super-achromatic quarter-wave plate. For comparison, for a 20% wavelength increment from 1350 nm to 1650 nm, the phase difference of a simple chromatic quarter-wave plate should also have about a 20% decrement from 0.294 to 0.240, as shown by the solid line in Fig. 6.32.

Liquid-crystal variable retarders are also available. These retarders must be operated by special electronic controllers. The retardance can be tuned from zero to over one wave. The working spectral range can be in the visible or near-IR ranges.

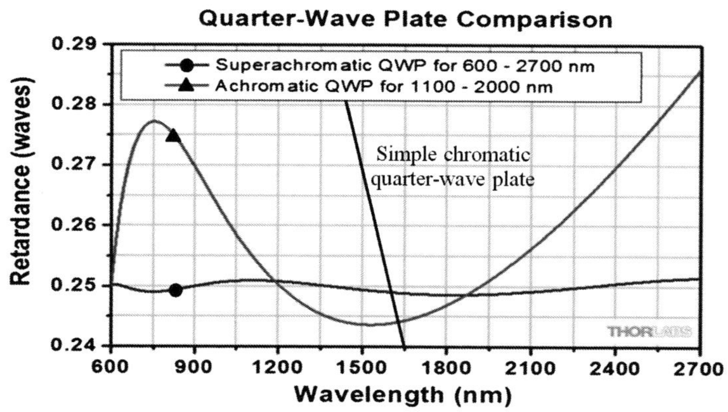

Figure 6.32 Typical phase-different curves for an achromatic and a super-achromatic quarter-wave plate. The solid line is the expected phase difference of a simple chromatic, quarter-wave plate, drawn here for comparison. Image reprinted courtesy of Thorlabs.

6.9 Attenuators

Neutral density filters are commonly used attenuators. Their name includes the word "filter," and, in some sense, they can be considered as a type of filter. They are specified by their optical density OD, defined as

$$OD = \log_{10}\left(\frac{1}{T}\right) \tag{6.8}$$

or

$$T = 10^{-OD}, \tag{6.9}$$

where T is the transmittance.

There are two types of neutral density filters: metallic and absorptive. The former are made by applying a thin layer of partially transmissive metal coating on a transparent substrate. The OD selection ranges from below 0.1 up to a single-digit number. Reflective neutral density filters have broad working spectral ranges, e.g., from the visible to the near IR, and the OD curve in the working range is flat.

Absorptive neutral density filters are made of specially selected glasses. The bulk absorption of the glass provides the attenuation. The physical thickness of the filter is also a factor determining the OD. Suitable OD values range from around 0.1 up to 1.0 or so. Because there is a limitation on feasible filter materials and thicknesses, the OD selection range is less than that for reflective neutral density filters. Absorptive neutral density filters also have broad working spectral ranges, similar to that of reflective neutral density filters, but the OD curves are less flat because the material absorption varies as the wavelength changes.

Variable-OD reflective neutral density filters can be made by applying a metallic coating with a gradually changing thickness across the substrate surface. Different portions of the filter have a different OD.

There are also fiber optic attenuators, single mode or multimode, fixed or variable. These filters are used much less frequently.

6.10 Diffusers

A diffuser is a transparent substrate plate that has a microstructure on its surface or in its volume. The microstructure can diffuse the light with certain transmission loss, usually a laser beam, incident on the diffuser. Compared with the incident light, the diffused light has much larger divergence, a more even intensity pattern, and more scattering.

Diffusers are specified by the two diffusing angles orthogonal to the light propagation direction assuming the incident light has zero divergence and incident in the normal direction of the diffuser, and by their transmissions. The intensity pattern of the diffused light is also a factor to be considered.

Four types of diffusers are commonly used:

- Opal glass diffusers have a low transmission of ~30%, a large diffusing angle of about 100° FWHM and a less-defined circular diffusing pattern. Opal diffusers usually perform well in a broad spectral range from the UV to near IR.
- Ground glass diffusers are made by sandblasting substrates with certain grit, usually from 120–1500. The transmission is from about 20% for 120 grit to around 80% for 1500 grit. The diffusing angle is from around 20° FWHM for 120 grit to around 10° for 1500 grit. Ground glass diffusers are inexpensive and have a less-defined circular diffusing pattern. Ground glass diffusers usually perform well in a broad spectral range from the UV to near IR.
- Holographic diffusers utilize complex volume micro-structures to diffuse light and are often ten times more expensive than opal glass or ground glass diffusers. Holographic diffusers have a high transmission of 80% or more, a diffusing angle from a few degrees to 20° or so, and a well-defined diffusing pattern, which can be circular, elliptical, or even linear. Holographic diffusers usually perform well with monochromatic and coherent light.
- Diffractive diffusers also utilize complex surface micro-structures to diffuse light and are as expensive as holographic diffusers. Diffractive diffusers have a high transmission of ~90%, a diffusing angle from a few degrees to over 50°, and a well-defined diffused pattern. The diffusing pattern can be circular, elliptical, square, rectangular, etc. Diffractive diffusers can perform well in a broad spectral range from the visible to near IR.

Figure 6.33(a) compares the transmission and diffusing pattern for different types of diffusers. Note that this comparison is not accurate and complete; it only serves as a general indication. Figure 6.33(b) shows the real measured diffusing intensity profile of an "engineered diffuser," e.g., a diffractive diffuser, for four wavelengths. Figure 6.33(c) shows two real images of the diffusing intensity patterns of two "engineered diffusers." One is circular and the other is square. In Figs. 6.33(b) and (c), the diffusing patterns have relatively clear edges and are well defined.

6.11 Diffraction Gratings

6.11.1 General description

Diffraction gratings are the most commonly used optical elements to disperse light. Their dispersion power is many times higher than a dispersion prism, which is discussed in Section 6.3.2.

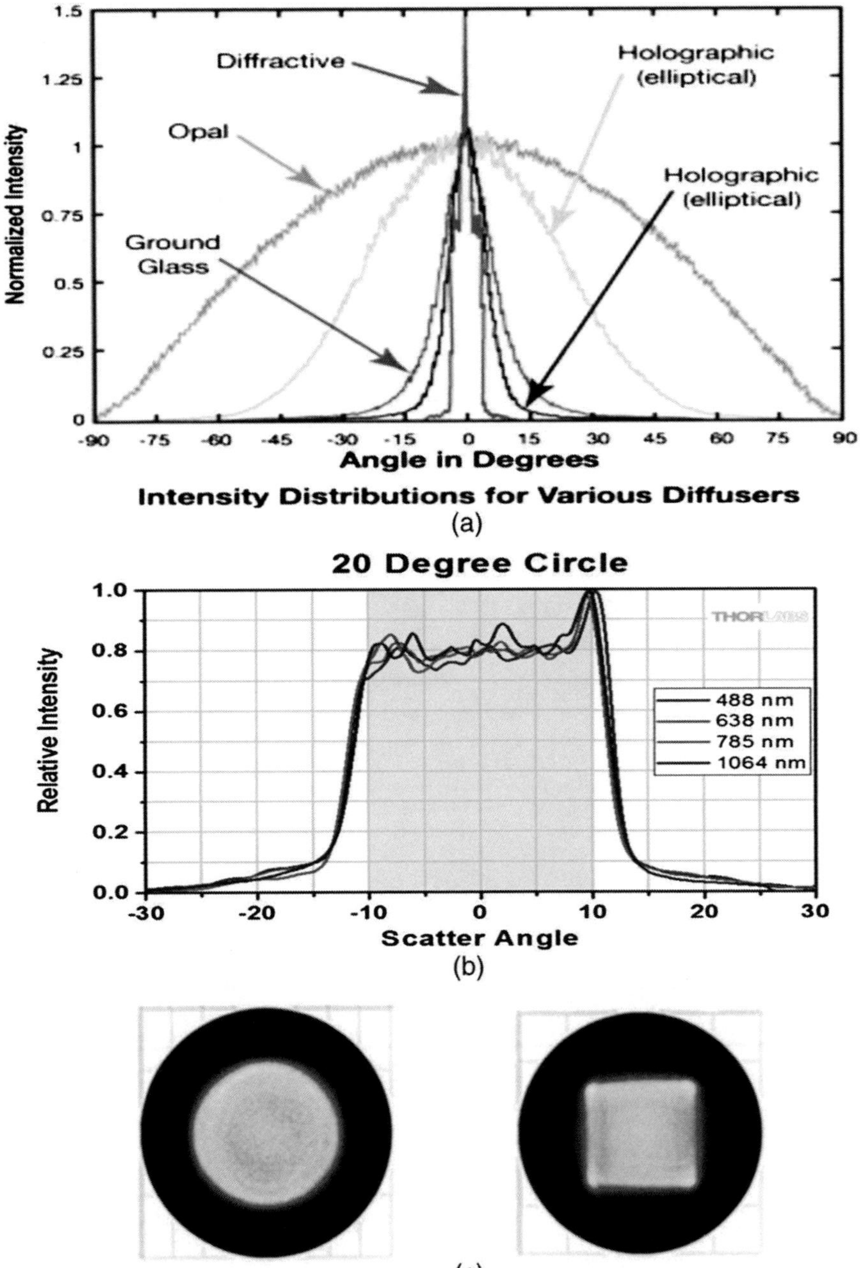

(c)

Figure 6.33 (a) Comparison of transmission and diffusing pattern for different types of diffusers. (b) Real, measured diffusing intensity profile of an “engineered diffuser” for four wavelengths. (c) Real images of circular and square diffusing intensity patterns generated by two “engineered diffusers.” Images reprinted courtesy of Thorlabs.

Several different types of diffraction gratings are commercially available. The most commonly used gratings are planar reflective gratings, which are collections of many small, identical, parallel, and slit-shaped grooves ruled on a planar high-reflective surface. Figure 6.34 shows the schematics of the

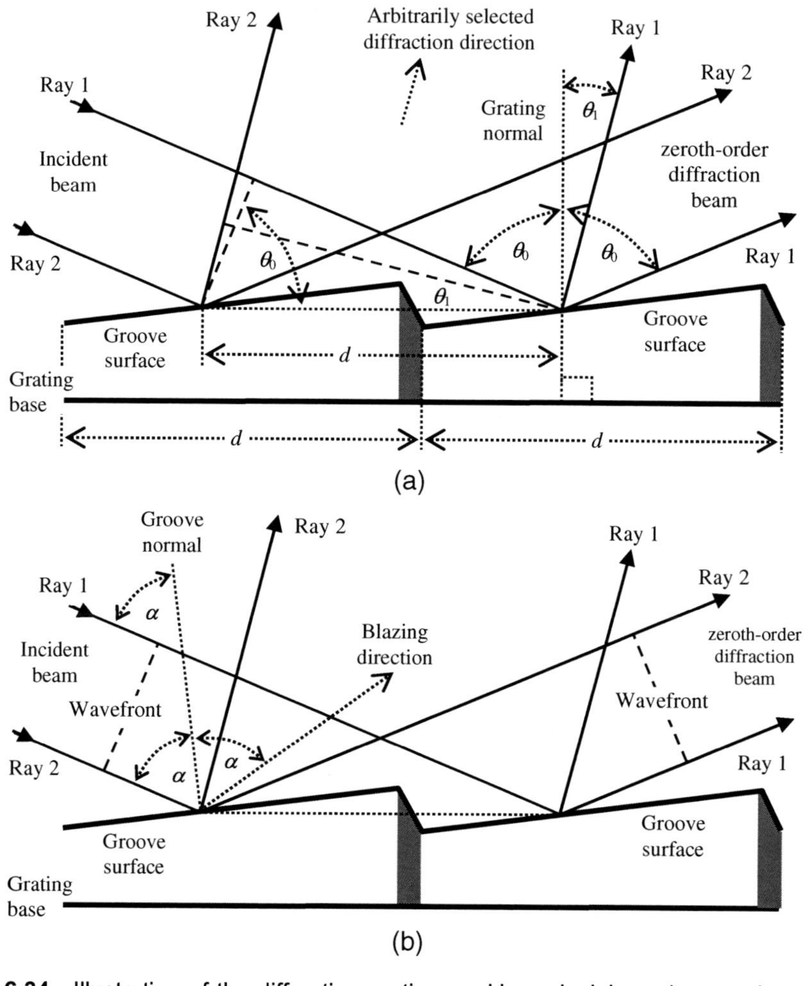

Figure 6.34 Illustration of the diffraction-grating working principle and properties. (a) The cross-section of two grooves with width d in a reflective grating. The grating normal is defined as the normal of the grating base. A beam denoted by Ray 1 and Ray 2 is incident on the two grooves at the identical locations. The grooves diffract the incident beam. The two dashed lines denote the wavefronts of the incident and diffracted beams, respectively. The optical path difference between Ray 1 and Ray 2 is a function of the diffraction angle. Different wavelengths will have a different constructive-interference direction. Thereby, the beam is dispersed. (b) The zeroth-order diffraction beam and the incident beam are symmetric about the grating normal; they meet the constructive-interference condition but not the reflection condition. The blazing direction meets the reflection condition and has the largest diffraction intensity, but only one wavelength meets the constructive-interference condition at the blazing direction.

cross-sections of two grooves, where "grating normal" is defined as the normal of the grating base, and "groove normal" is defined as the normal of the groove surfaces. The working principle of a grating is explained in the following sections using the details in Fig. 6.34.

6.11.2 Grating equation

Consider two rays, Ray 1 and Ray 2, that are incident on the identical points of the two grooves, respectively. The tiny grooves reflect and diffract the incident rays. Interference occurs among the reflected rays. In some directions, the interference is constructive, and in some other directions, the interference is destructive.

For an arbitrarily selected direction with an angle θ_1 to the grating normal, as shown in Fig. 6.34(a), the optical path difference δ between Ray 1 and Ray 2 in this direction is given by

$$\begin{aligned}\delta &= d[\sin(\theta_0) - \sin(-\theta_1)] \\ &= d[\sin(\theta_0) + \sin(\theta_1)].\end{aligned} \tag{6.10}$$

Note that the signs of θ_0 and θ_1 are defined as positive or negative at the left or right side of the grating normal, respectively. Equation (6.10) can be determined with the help of the two dashed lines shown in Fig. 6.34(a).

When δ meets the condition defined in Eq. (6.11), Ray 1 and Ray 2 will constructively interfere, where m is an integer called the "diffraction order," and λ is the wavelength. When δ meets the condition defined in Eq. (6.12), Ray 1 and Ray 2 will be destructively interfered:

$$\delta = d[\sin(\theta_0) + \sin(\theta_1)] = m\lambda, \quad \text{(constructive interference)} \tag{6.11}$$

$$\delta = d[\sin(\theta_0) + \sin(\theta_1)] = (m + 1/2)\lambda. \quad \text{(destructive interference)} \tag{6.12}$$

Equations (6.11) and (6.12) form the famous grating equation.

For a given λ, the maximum m allowed is

$$m_{\max} = \frac{2d}{\lambda}, \tag{6.13}$$

obtained by letting $\sin(\theta_0) + \sin(\theta_1) = 2$ in Eq. (6.11). Equations (6.11) and (6.12) state that for a given m and θ_0, θ_1 is a function of λ, which means that different wavelength components in an incident beam will have constructive or destructive interference in different directions after the beam is diffracted by the grating. Or in other words, the grating disperses the beam.

Any other rays in the incident beam can be grouped in pairs similar to the pair of Ray 1 and Ray 2, and analyzed in the same way. Thereby the whole incident beam and diffracted beam have the same characteristics as those of Ray 1 and Ray 2.

Note that Fig. 6.34(a) only uses two grooves to illustrate. A real grating can have several hundred up to a couple of thousands grooves in one millimeter and can generate tens of thousands of subwaves. All of these waves join the interference, and the interference fringe is very narrow.

6.11.3 Dispersion power

The angular dispersion resolution or the resolving power of a grating is defined by $d\theta_1/d\lambda$ and can be obtained by differentiating Eq. (6.11) with θ_0 being a constant. The result is

$$\frac{d\theta_1}{d\lambda} = \frac{m}{d\cos(\theta_1)}. \tag{6.14}$$

Equation (6.11) shows that when $m = 0$, $\theta_1 = -\theta_0$, that is the case for a zeroth-order diffraction beam. Equation (6.14) shows that when $m = 0$, $d\theta_1/d\lambda = 0$, which means that the zeroth-order diffraction has no dispersion effect.

Large dispersion power $d\theta_1/d\lambda$ is often desired and can be obtained by the use of a small d, a large m, and a large θ_1. The largest m value is limited by Eq. (6.13). In most real applications, $m \leq 2$. The practically largest possible θ_1 is $\leq 90°$. Commonly used diffraction gratings have a groove density from 300/mm ($d \approx 3.33$ μm) to 2400/mm ($d \approx 0.42$ μm). The corresponding dispersion resolution for $m = 1$ is $d\theta_1/d\lambda = 4 \times 10^{-4}$ rad/nm and $d\theta_1/d\lambda = 3.4 \times 10^{-3}$ rad/nm, respectively.

6.11.4 Blazing angle

Figure 6.34(a) is plotted in Fig. 6.34(b) with different notes for the convenience of illustration. Ray 1 and Ray 2 have an angle α to the groove normal. The "blazing direction" is the reflection direction of Ray 1 or Ray 2. In the blazing direction, the diffraction has the largest intensity. The wavelength that meets the constructive interference condition in the blazing direction has the largest intensity compared with any other wavelengths in any other directions. This wavelength is called the "blazing wavelength." As the wavelength shifts away from the blazing wavelength, the direction for constructive interference moves away from the blazing direction, the intensity of the diffracted beam drops.

Since the zeroth-order diffraction beam is away from the blazing direction, the zeroth-order diffraction beam is weak, although in this direction Ray 1 and Ray 2 are constructively interfered.

6.11.5 Grating efficiency

When using a grating, the goal is often to concentrate as much light energy as possible in one desired diffracted order. However, some light energy may go to other, undesired diffraction orders. Real gratings also have defects and fabrication tolerances. The groove normal direction may not be exactly the same across the groove. Light incident on the part of the grating marked by the grey areas in Fig. 6.34 is not diffracted as desired and is wasted. The coating on the groove surface may have a certain reflection loss. Therefore, there is a grating efficiency issue, defined as the percentage of incident monochromatic light that is diffracted into the desired diffraction order. Grating efficiency is very complex to calculate and should be provided by the grating vendors.

Groove density, blazing wavelength, resolving power, wavelength range, and grating efficiency are the main parameters that describe a diffraction grating. Richardson Gratings (now is a subsidiary of Newport) is the largest grating manufacturer in the world, and their *Diffraction Grating Handbook* is a great technical resource.[7]

6.11.6 Grating types

The grating illustrated in Fig. 6.34 is a reflective grating. Two types of reflective gratings are commonly used: ruled and holographic.

A ruled grating has sawtooth-shaped grooves. The production process first uses a precise mechanical engine to rule a master piece before pressing a substrate with a layer of epoxy and a reflective coating on epoxy against the master piece to generate a replica of the sawtooth grooves. The grooves shown in Fig. 6.34 are from a ruled grating.

Holographic gratings are made by first depositing a photosensitive layer on substrates and then illuminating the layer with an interference fringe pattern. The grooves generated have a sinusoidal profile. Compared with ruled gratings, holographic gratings are relatively easier to make and cost a fraction of their counterparts, but they have no clearly defined blazing angles because of the sinusoidal groove profiles.

An ion-beam etching technology was recently developed to create a sawtooth profile from the sinusoidal profile of holographic gratings. Such holographic gratings have blazing angles, cost more, and perform better than ruled gratings. However, because of the shape and size of ruled grating grooves, they tend to have more defects and errors.

There are variations of gratings, such as concave reflective gratings, which can disperse and focus simultaneously, and transmission gratings, which are the same as reflective ruled gratings except that there is no reflective coating on the epoxy.

6.12 Optical Fibers

6.12.1 General description

Optical fibers are widely used to deliver a optical signal or energy through a zigzag path, which can be over hundreds of kilometers long. Many types of optical fibers are invented for these purposes.

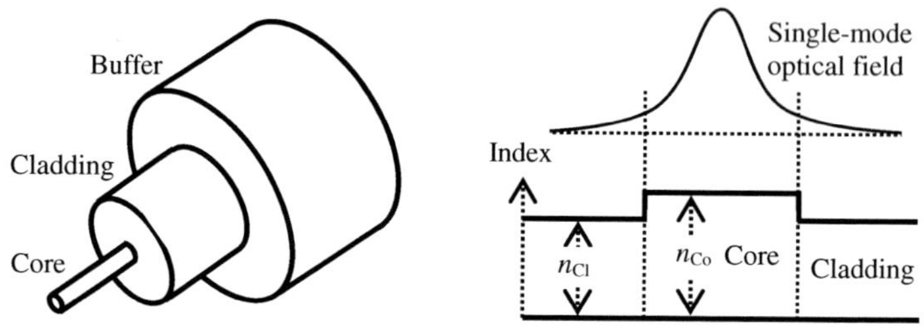

(a) (b)

Figure 6.35 Illustration of a step-index optical fiber. (a) An optical fiber, comprising a core, a cladding, and a buffer. (b) The schematic of a single-mode step-index fiber. The core has a refractive index slightly larger than the refractive index of the cladding and acts as a waveguide. Most of the optical energy is confined inside the core. A small fraction of the optical energy leaks into the cladding because of the wave nature of light.

Figure 6.35(a) shows the schematics of a widely used step-index optical fiber. The fiber consists of a core with a diameter ranging from several microns to about one hundred microns, a cladding surrounding the core with a diameter of over one hundred microns, a buffer surrounding the cladding with a diameter of a couple of hundred microns, and finally a jacket of several hundreds of microns in diameter surrounding the buffer.

The core of most optical fibers is made of doped or pure silica with a refractive index $n_{Co} \approx 1.46$. The cladding of most optical fibers is made of a material very similar to the core material with a refractive index n_{Cl}, which is about 0.01 smaller than n_{Co}, as illustrated in Fig. 6.35(b). The index difference between the core and cladding provides a waveguide mechanism and confines most of the light energy inside the core. The core and the cladding of some optical fibers are made of other types of glasses or plastics. The buffer is made of Teflon, polyurethane, or another similar material to protect the cladding and core. The jacket provides further protection.

A small fraction of the optical energy will leak into the cladding and is called cladding mode because of the wave nature of light. As a comparison, geometrical optics says 100% of the optical energy will be confined inside the core by total reflection. Figure 6.35(b) also illustrates the intensity distribution of an optical field inside a single-mode step-index optical fiber.

6.12.2 Numerical aperture

Any optical fibers have a numerical aperture or maximum acceptance angle. Figure 6.36 uses geometrical optics to illustrate the maximum acceptance angle of an optical fiber. Both the fiber core and cladding are made of silica with refractive index ~1.46 and an index delta of $n_{Co} - n_{Cl} \approx 0.01$, the reflectance at the interface of the cladding and the core $\approx 0.01^2/(1.46 + 1.46)^2 \approx 10^{-5}$. The intensity of the reflected rays is negligible.

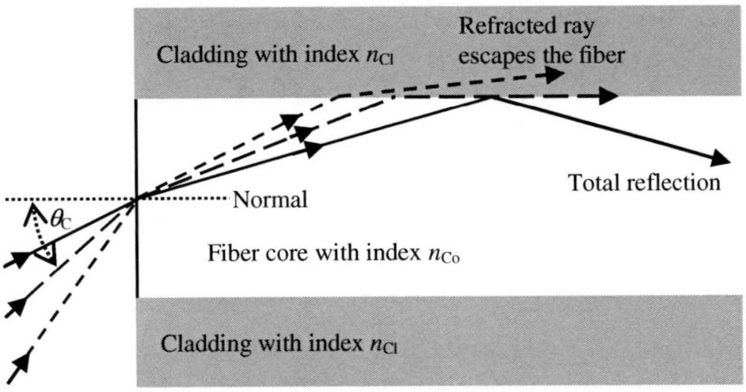

Figure 6.36 Geometrical-optics illustration of the numerical aperture or the maximum acceptance angle of an optical fiber. Three rays are drawn. For the short dashed ray, the refracted ray carries almost all of the light energy of this ray and escapes the fiber. The long dashed ray enters the fiber at the critical angle θ_C, and the refracted ray is parallel to the cladding-core interface. The solid-line ray entering the fiber with an angle $<\theta_C$ will be totally reflected and confined inside the fiber. θ_C is the maximum acceptance angle. $\sin(\theta_C)$ is the numerical aperture of this fiber.

Figure 6.36 only shows three rays as examples. The ray denoted by the short dashed line enters the fiber at an angle larger than the critical angle θ_C, is refracted by the interface of the cladding and the core, enters the cladding carrying almost all of the light energy, travels away from the fiber core, and eventually escapes the fiber.

The ray denoted by the long dashed line enters the fiber at the critical angle θ_C and is at the edge of being totally reflected by the interface of the cladding and the core.

The ray denoted by the solid line enters the fiber at an angle smaller than the critical angle θ_C and is totally reflected by the interface of the cladding and the core.

For the solid-line ray, the next internal reflection and all the follow-up internal reflections will be the same. The ray will be confined inside the fiber by total reflection as the ray travels forward along the fiber until the ray exits the fiber. All of the rays entering the fiber at an angle $<\theta_C$ will be confined inside the fiber.

Based on the geometry shown in Fig. 6.36 and using triangular geometry, θ_C is found to be

$$\theta_C = \sin^{-1}\left[\left(n_{Co}{}^2 - n_{Cl}{}^2\right)^{0.5}\right]. \tag{6.15}$$

θ_C is solely determined by n_{Co} and n_{Cl}, and has nothing to do with the fiber core diameter. θ_C is often expressed in terms of numerical aperture:

$$NA = \sin(\theta_C) = (n_{Co}{}^2 - n_{Cl}{}^2)^{0.5}. \tag{6.16}$$

For $n_{Co} - n_{Cl} \approx 0.01$, $NA \approx 0.17$.

θ_C is the maximum acceptance angle of the fiber. To couple light into an optical fiber, light must be focused onto the fiber core facet. The focused light should have a solid angle $<\theta_C$ to avoid a significant energy loss. If the fiber is severely bent, e.g., with a curvature radius of tens of millimeters, the total reflection condition for some rays will be destroyed. Those rays entering the fiber with an angle not much smaller than θ_C will escape the fiber.

6.12.3 Mode structures in a cylindrical waveguide

The optical field inside an optical fiber is traditionally called the "mode," which actually means the transverse (TE) mode, although "TE" is often omitted. For lasers, "modes" can mean either longitudinal modes or TE modes, so it is often necessary to clarify which mode is being talked about.

Optical fibers are cylindrical waveguides. Cylindrical coordinates are best for analyzing this situation. The mathematics involved for the TE mode in an optical fiber is somewhat complex. It is not necessary to dig deep into the mathematics details. Basically, it involves solving the Helmholtz equation[8]

$$\nabla^2 U(r,\phi,z) + n^2 k_0^2 U(r,\phi,z) = 0, \tag{6.17}$$

where $U(r, \phi, z)$ is the electric or magnetic fields in the fiber and can be written as $U(r, \phi, z) = u(r)\exp(-il\phi)\exp(-i\beta z)$; r is the radial variable; i is the imaginary number; l is an integer and the mode order (e.g., $l = 0$ means the basic mode); ϕ is the azimuthal angle; β is the "propagation constant;" z is the axial distance; $n = n_{\text{Co}}$ or n_{Cl} if the field inside or outside the fiber core is studied, respectively; $k_0 = 2\pi/\lambda_0$ is the wave vector; and λ_0 is the wavelength in air.

The solutions of Eq. (6.17) are[9]

$$u(r) \sim J_l(k_T r), \quad r < a \text{ (core)} \tag{6.18}$$

$$u(r) \sim K_l(\gamma r), \quad r > a \text{ (cladding)} \tag{6.19}$$

where $J_l(x)$ is the Bessel function of the first kind and order l, $K_l(x)$ is the modified Bessel function of the second kind and order l, $k_T{}^2 = n_{\text{Co}}{}^2 k_0{}^2 - \beta^2$, and $\gamma^2 = \beta^2 - n_{\text{Cl}}{}^2 k_0{}^2$.

The boundary conditions require that at $r = a$, where a is the radius of the fiber core

$$J_l(k_T a) = K_l(\gamma a), \tag{6.20}$$

$$\left.\frac{dJ_l(k_T r)}{dr}\right|_{r=a} = \left.\frac{dK_l(\gamma r)}{dr}\right|_{r=a}. \tag{6.21}$$

For a given fiber, all parameters are fixed except β, which can be manually adjusted so that the boundary condition Eqs. (6.20) and (6.21) are met. Note that the absolute value of $u(r)$ is insignificant, but the profile of the field intensity $|u(r)|^2$ is important. With the solution shown in Eqs. (6.18)–(6.21), it is not difficult to enter the optical fiber parameters into the Bessel functions and plot the mode structures. Commercial mathematics software, such as Mathcad and Matlab, can plot Bessel functions.

6.12.4 *V*-number

A parameter called the V-number is widely used to describe optical fibers.[10] Equation (6.22) defines the V-number and relates it to the wavelength λ, fiber core radius a, and fiber core and cladding refractive indices n_{Co} and n_{Cl} by

$$\begin{aligned} V &= (k_T{}^2 a^2 + \gamma^2 a^2)0.5 \\ &= \frac{2\pi}{\lambda} a\left(n_{\mathrm{Co}}{}^2 - n_{\mathrm{Cl}}{}^2\right)^{0.5} \\ &= \frac{2\pi}{\lambda} aNA. \end{aligned} \tag{6.22}$$

The V-number determines the mode number of the optical fiber.

6.12.5 Single-mode fibers

Optical fibers are usually categorized into single-mode and multimode groups. The optical field shown in Fig. 6.35(b) has only one spot and is single mode (also called basic mode). Multimode means there are more than one mode or spot in the optical field. These modes are of higher order than the basic mode.

When $V < 2.405$, the fiber can support only one transverse mode and is a single transverse mode fiber.[11] The word "transverse" is often omitted.

Most optical fiber consists of silica with an index $n_{\mathrm{Co}} \approx 1.46$ and $n_{\mathrm{Co}} - n_{\mathrm{Cl}} \approx 0.01$. Then for $\lambda \approx 0.633$ μm (He-Ne laser), Eq. (6.22) leads to $a \approx 1.42$ μm for single-mode operation. For $\lambda \approx 1.55$ μm (telecom wavelength), $a \approx 3.48$ μm for single mode. In other words, an optical fiber with a 3.48-μm core radius is single mode for a 1.55-μm wavelength but multimode for a 0.633-μm wavelength.

Single-mode optical fiber is mostly used to deliver optical signals in fiber-optic communication systems because it does not have "modal dispersion" processed by multimode fibers. Modal dispersion is discussed in Section 6.12.7. Single mode fiber is also used to clean and deliver laser diode beams. Single mode fibers have a small core of only several micron diameter. They are fragile, difficult to handle and cannot deliver a lot of energy.

6.12.6 Multimode fibers

Multimode fibers have a larger core size—up to over one hundred microns—and are used to deliver light energy or signals over a short distance.

In a multimode fiber, the optical field has multi modes or multi spots. Larger core sizes fiber supports more modes. A geometrical-optics explanation of the relation between waveguide (e.g., optical fiber) size and transverse modes is presented in Section 8.3.4 and Fig. 8.8. Figure 6.37(a) shows the mode pattern of the twelve lowest-order modes starting from the basic single mode. For step-index fiber, the mode number M is approximately related to the V-number by[12]

$$M \approx \frac{4}{\pi^2} V^2. \tag{6.23}$$

For a multimode optical fiber with a 25-μm core radius, and assuming $\lambda = 0.633$ μm and $n_{\mathrm{Co}} - n_{\mathrm{Cl}} \approx 0.01$, Eq. (6.22) leads to $V \approx 42.2$, and Eq. (6.23) leads to $M \approx 722$. All of the modes divert after they exit from a multimode fiber and overlap each other. A typical real image of the many modes from a multimode fiber is shown in Fig. 6.37(b).

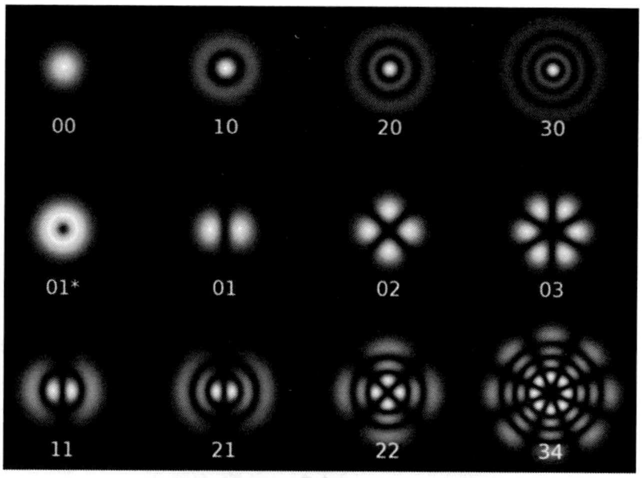

(a)

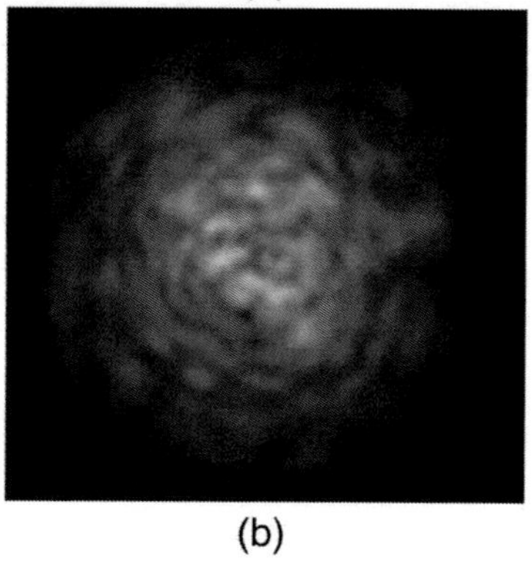

(b)

Figure 6.37 (a) Patterns of the twelve lowest-order modes of a multimode fiber.[13] The 00 mode at the upper-left corner is the basic mode or the single mode. (b) Real image of a large number of modes from a multimode fiber.

If the light from a multimode fiber is used to illuminate a small area, the light must be intentionally defocused so that the image size of every mode is significantly increased and overlap each other to form a uniform illumination pattern.

Compared with single-mode fibers, multimode fibers have a notable shortcoming: modal dispersion. Because of the modal dispersion, multimode fibers cannot be used for long-distance communication and interference-related applications.

The larger core size of multimode fibers is physically stronger and allows much more light energy to pass through. It is also much easier to couple light into multimode fibers.

6.12.7 Modal dispersion

It is awkward to use geometrical optics to explain wave optics, but it is a convention nonetheless.

For a single-mode fiber, only one mode is represented by the horizontal ray, as shown in Fig. 6.38(a). For a multimode fiber, there is more than one mode. Each mode is represented by a ray. Shown in Fig. 6.38(b) are three rays for three modes, including the basic mode. To travel the same axial distance, each ray travels a different distance. The basic mode represented by the solid line travels the shortest distance. The highest mode represented by the short dashed lines travels the longest distance. This phenomenon is called modal dispersion.

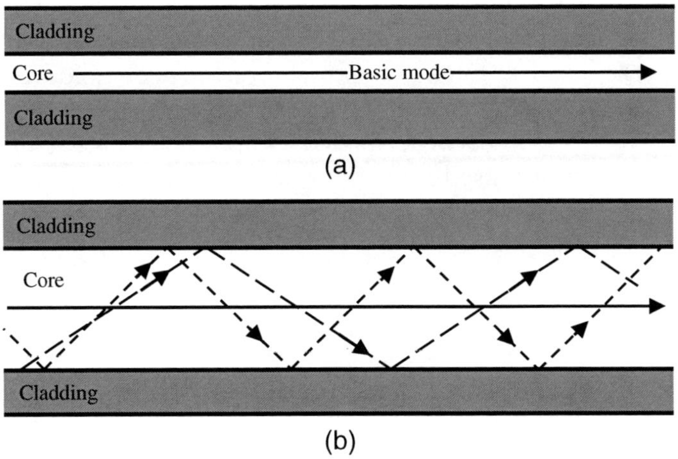

Figure 6.38 A geometrical optics explanation of mode dispersion. (a) The small core size of a single-mode optical fiber only allows the basic mode to pass through. (b) The large core size of a multimode optical fiber allows more modes to pass through. For the same axial travel distance, the higher-order mode bounces more times inside the fiber core and has a longer path. This path differences among modes is mode dispersion.

In an optical-fiber communication system, a laser beam emitted by a laser diode is modulated to carry a signal. The laser beam is coupled into an optical fiber for delivery. If the optical fiber is multimode, the laser beam inside the fiber will form many modes. Each mode carrying the same signal will arrive at the other end of the fiber at different time. In other words, modes carrying different signals and entering the fiber at different times can arrive at the other end of the fiber at the same time, thereby causing errors.

The schematic shown in Fig. 6.38(b) is not to the right proportion—it is dramatically stretched in the vertical direction for illustration purposes. The actual path length differences among modes are much smaller than those shown here. In short-distance, low-modulation-frequency communication systems, the small modal dispersion of multimode fibers may still be acceptable, and multimode fiber can still be used. But in the vast majority of optical-fiber communication systems, single-mode fiber must be used.

6.12.8 Mode field diameter

The geometrical optics model used in Fig. 6.36 to explain how some rays are confined inside an optical fiber is an approximate model. The total reflection theory, which predicts that the rays either escape the fiber or are totally confined inside the fiber, is also an approximation. Waveguide theory can accurately describe how the optical field is confined inside an optical fiber. A small portion of the confined optical field will always stay outside the core and inside the cladding, but not escape the fiber, as shown in Fig. 6.35(b). The light inside the cladding is called "cladding mode." Readers interested in waveguide theory can consult the literature.[14] However, geometrical optics can help to simplify the explanations about optical fibers and is therefore still widely used.

At the exit facet of the fiber, the light intensity pattern is the near-field intensity pattern. The $1/e^2$ intensity diameter of the near-field intensity pattern is called the "mode field diameter" (MFD) and is larger than the fiber core diameter $2a$. The relation between the MFD and $2a$ depends on the fiber type. For the most commonly used step-index single-mode fiber, the relation is[15]

$$MFD \approx 2a\left(0.634 + \frac{1.619}{V^{3/2}} + \frac{2.879}{V^6} - \frac{1.561}{V^7}\right), \tag{6.24}$$

where V is the V-number. For $V = 2.4$, $MFD = 1.08 \times 2a > 2a$.

For light exiting a single-mode fiber, the far-field light intensity pattern $I(\theta)$ as a function of the divergent angle θ is related to MFD by[16,17]

$$MFD = \frac{\lambda}{\pi}\left[\frac{2\int_0^{\pi/2} I(\theta)\sin(\theta)\cos(\theta)d\theta}{\int_0^{\pi/2} I(\theta)\sin^3(\theta)\cos(\theta)d\theta}\right]^{0.5}. \tag{6.25}$$

Since $I(\theta)$ can be measured, MFD can be back-calculated from $I(\theta)$.

Waveguide theory also states that the beam exiting a step-index single-mode fiber is approximately Gaussian and can be described by the Gaussian optics equations (3.1)–(3.3). Half of the mode field diameter $MFD/2$ can be considered as the Gaussian beam waist radius w_0, and the fiber-core exit facet can be considered as the beam waist location.

Note that for a single-mode optical fiber, the far-field divergence and NA are two different concepts and can have different values. For example, assuming $n_{Co} \approx n_{Cl} \approx 1.46$, $n_{Co} - n_{Cl} \approx 0.01$, $a = 3.5$ μm, and $\lambda = 1.55$ μm, it can be found that $NA \approx 0.17$ or $\theta_C \approx 9.8°$, $V \approx 2.4$, and $MFD = 1.08 \times 2a = 7.56$ μm. Using Eq. (3.6) with $w_0 = MFD/2$, the far-field $1/e^2$ intensity half-divergent angle of the light exiting this fiber is found to be 7.5°.

6.12.9 Attenuation

Figure 6.39 shows a typical attenuation curve for a single-mode silica optical fiber. Attenuation curves are somewhat different for multimode fibers. The attenuation in the short-wavelength side is caused mainly by scattering. The attenuation in the long-wavelength side is caused mainly by absorption. Figure 6.39 illustrates that there are three relatively-low-attenuation "windows" around 850 nm, 1310 nm, and 1550 nm, respectively. Long-distance fiber-optic telecommunication systems use the 1550-nm window for the lowest attenuation. Short-distance fiber-optic telecommunication systems also use the 850-nm and 1310-nm windows because the light sources (laser diodes) with a wavelength around 850 nm cost the lowest among the three, followed by light sources with a wavelength around 1310 nm.

6.12.10 Polarization

If the light entering an optical fiber is linearly polarized, the output light is also linearly polarized, but the polarization direction can be changed by slightly disturbing the fiber, e.g., a gentle touch on the fiber. So, in a certain sense, the

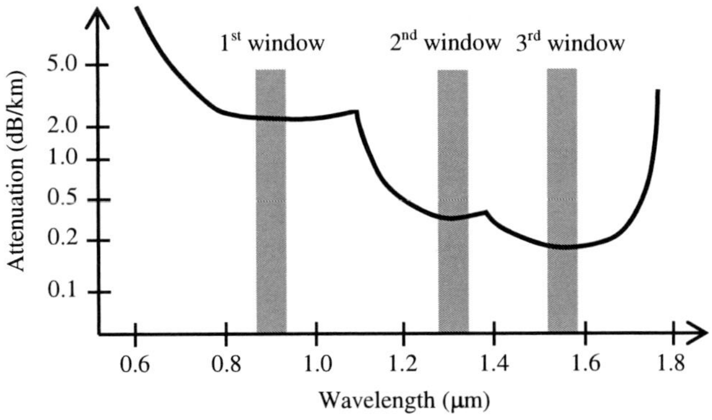

Figure 6.39 Typical attenuation curve for single-mode silica optical fiber. Three relatively-low-attenuation "windows" exist around 850 nm, 1310 nm, and 1550 nm, respectively.

light output from a fiber can be considered as having random polarization. This phenomenon is caused because the large number of random reflections that occur inside the fiber affects the polarization direction and because the bent fiber can create weak birefringence that also affects the polarization.

Specially designed and fabricated "polarization-maintaining fibers" can maintain the polarization direction as the light propagates through the fiber. When linearly polarized light is coupled into a polarization-maintaining fiber, the polarizing direction must be in the specified transversal direction of the fiber. Then the polarization direction of the light output from the fiber will remain in this specified transversal direction.

6.12.11 Fiber bundles for image delivery

Fiber bundles consist of a large number of optical fibers and are used to deliver images through a zigzag path. In applications, a lens system focuses an image onto the facet of a fiber bundle. Each fiber in the bundle acts like a pixel in a 2D sensor array. When the light exits the other facet of the fiber bundle, each fiber acts like a pixel in a small display screen. The maximum number of fibers in current fiber bundles is below one hundred thousand or so. A sensor or display array with one hundred thousand pixels is considered of very low resolution. The low resolution of an image from a fiber bundle limits its applications.

Commercially available imaging fiber bundles have a length from tens of millimeters to hundreds of millimeters, a bundle diameter of a few millimeters, and a number of fibers in the bundles from a few thousands to tens of thousands. The core diameter of the fibers is from around 10 μm to 100 μm.

6.12.12 Fiber-bundle and liquid light guides

Fiber bundles are also used to deliver optical energy in a zigzag manner. Figure 6.40(a) shows the schematics of a 19-fiber bundle with a circular cross-section for illustration purposes. The ratio of the sum of all the fiber cross-section size to the whole bundle cross-section size is the "filling factor." The typical filling factor is 70%. Light hitting the bundle in the area not occupied by the fibers is mostly scattered or absorbed and becomes heat. The filling factor decreases the transmission efficiency of the fibers.

Fiber bundles can also have linear, rectangular, or square cross-sections. The fibers in one fiber bundle can be aligned to have two different cross-section shapes at the two ends of the bundle. Figure 6.40(b) shows the schematics of a nine-fiber bundle with square and linear cross-sections at the two ends.

Most commercially available energy-delivery fiber bundles have a few to tens of fibers with a core diameter from 50–1000 μm. The typical diameter and length of fiber bundles are a few millimeters and a couple of meters, respectively.

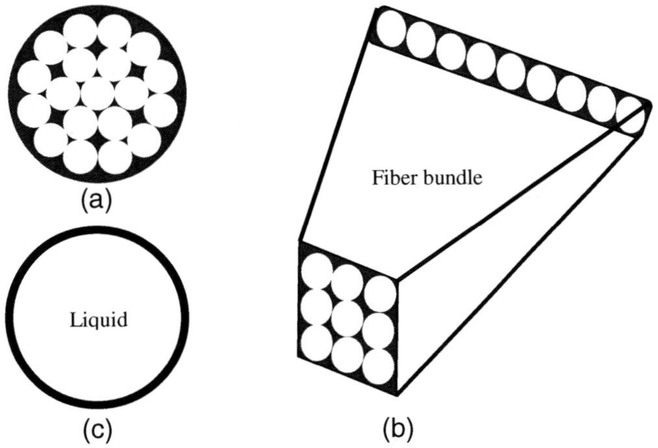

Figure 6.40 (a) Schematics of the circular cross-section of a 19-fiber bundle. The ratio of the sum of all the fiber areas to the bundle area is the "filling factor." (b) Schematics of a nine-fiber bundle with square and linear cross-sections at the two ends. (c) Schematics of the cross-section of a light guide. The filling factor is always 100%.

The limited core size and numbers of fibers in a fiber bundle, and the filling factor, limit the maximum energy that can be delivered. To deliver tens or hundreds of watts of power, liquid light guides are often a better choice. As the name suggests, a liquid light guide is a flexible tube filled with a certain type of liquid to guide light. The typical diameter and length of such guides is a few millimeters and a couple of meters, respectively. Compared with fiber bundles, liquid light guides have a much larger effective guiding area and a 100% filling factor.

The cross-section of a liquid light guide is shown in Fig. 6.40(c). The shortcoming of liquid light guide is its low transmission and limited transmission spectral range, as shown in Fig. 6.41(c).

Multimode fibers made of silica or ZrF_4 (zirconium fluoride, a type of glass) are usually used to build a fiber bundle for energy delivery because multimode fibers are bigger and easier to handle than single-mode fibers. The silica used is often doped with either a high or low hydroxyl ion (OH) for use in the 250–1200-nm or 400–2400-nm range, respectively. The attenuation curves of such fibers are shown in Fig. 6.41(a); they are different than the curves shown in Fig. 6.39. The attenuation curves of ZrF_4 fibers are shown in Fig. 6.41(b). These fibers have relatively low attenuation in the near-IR range.

6.12.13 Fiber adaptors, couplers, and lab supplies

Optical fibers are tiny, fragile, and usually placed inside a cable for protection. Fiber tips are usually mounted inside adaptors for convenience of alignment and connection. Adaptors are particularly helpful for aligning and connecting

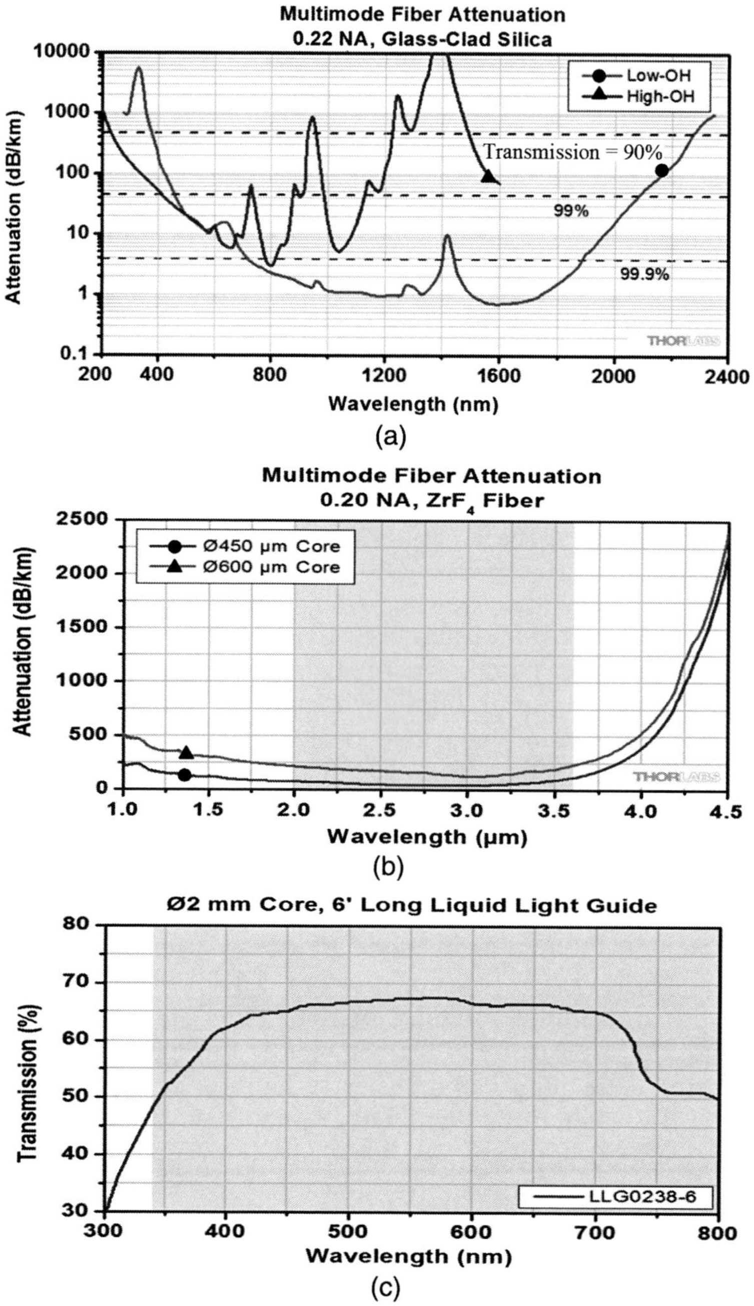

Figure 6.41 Attenuation curves of multimode energy delivery fibers made of (a) silica doped with either a high or a low hydroxyl ion (OH) for use in the 250–1200-nm or 400–2400-nm range, respectively. Three transmission levels per meter of fiber length are also marked, and (b) ZrF_4 fiber for use in the near-IR range. (c) Typical transmission of a liquid light guide. Images reprinted courtesy of Thorlabs.

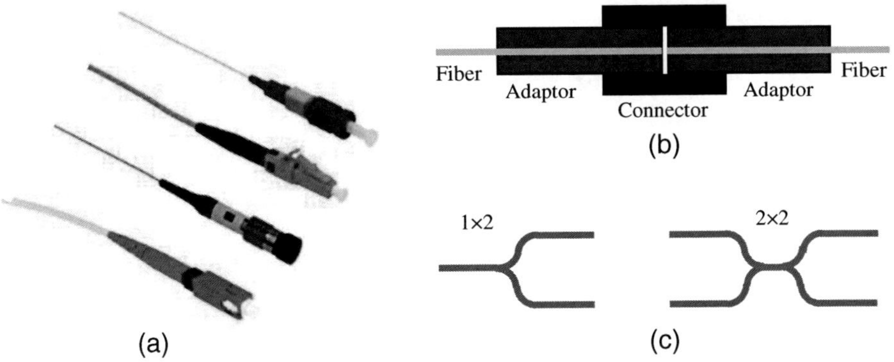

Figure 6.42 (a) Several types of fiber adaptors. (b) Schematic of aligning and connecting two fibers with two adaptors and one connector. (c) Schematic of a 1×2 and a 2×2 coupler.

single-mode fibers. There are several types of adaptor available on the market, as shown in Fig. 6.42(a). It is important to use the same type of adaptors so that different fiber cables can be easily connected. Figure 6.42(b) illustrates how two fibers are aligned and connected by two adaptors and one connector. Connectors are often part of the adaptors and come with the adaptors.

Several fiber couplers are available on the market, such 1×2 or 2×2 couplers, as illustrated in Fig. 6.42(c). These couplers are usually already mounted inside adaptors and connectors. Most fiber couplers can be used in a reciprocal way, i.e., a 1×2 coupler can be used as a 2×1, etc.

Lab tools to handle optical fibers include cleavers to cleave fiber facets, splicers to fuse two fibers together, alignment stations to precisely position fibers, and polishing and cleaning tools. Reference 18 provides information about many commercially available and commonly used optical fiber components and lab supplies.

6.13 Photodetectors

6.13.1 General comments

Strictly speaking photodetectors are not optical components, but they are frequently used in optical systems, such as cameras, to detect optical signals. This section describes the various basic properties of photodetectors for optical components.

There are many types of photodetectors, the vast majority of which are based on semiconductors. The various names of photodetectors can be confusing, such as photodiodes, photoconductive detectors, photovoltaic detectors, etc. All of them are actually certain types of photodiodes, utilizing *p-n* junctions to convert the incident photons to electrons. Even charge-coupled device (CCD) arrays can be considered a type of photodiode arrays. The differences between them indicate how they are operated. For example, photovoltaic detectors are zero biased, and solar cells are photovoltaic devices. Photoconductive detectors are often reverse biased, relatively faster, and noisy.

Photodetectors made of different semiconductor materials have different spectral response ranges. Silicon photodetectors have a response range from about 400–1100 nm with a response peak at about 800 nm. Germanium photodetectors have a response range from 800–1800 nm with a response peak at about 1500 nm. Photodetectors made of indium arsenide, indium antimonide, and mercury cadmium telluride have a spectral response range of about 1.0–3.6 μm, 1.0–5.5 μm, and 2–22 μm, respectively.

Semiconductor photodetectors in the UV, visible, and near-IR ranges are very sensitive and usually have a detection power range from about 1 nW to 50 mW for continuous wave or about 1 pJ to 1 μJ for pulsed energy. They have a short response time of ~1 ns, and their dark current is approximately doubled for every 10 °C temperature rise.

There are many types of photodetectors with many parameters and specifications. Even for the same type of photodetectors made by different companies, the characteristics can be very different. For example, CCDs usually have lower noise than CMOSs, but some scientific CMOSs have lower noise than many CCDs. Users should check the vendor's website for detail information to make sure they get what they want. References 19 and 20 provide good overviews of the characterizations of photodetectors.

6.13.2 Responsivity and quantum efficiency

A semiconductor photodetector absorbs photons incident on it and utilizes the photon energy to create free carrier pairs (electrons and holes). These free carriers form a response current in an external circuit. The responsivity R of a photodiode is given by

$$R\left(\frac{\mathrm{A}}{\mathrm{W}}\right) = \eta_D \frac{e}{h\nu}, \tag{6.26}$$

where the unit for R is the current produced by the photodiode per watt of incident photon power, η_D is the quantum efficiency, e is the electron charge, h is the Planck constant, ν is the frequency of the light, and $h\nu$ is the energy of one photon.

Quantum efficiency is one of the most important parameters about photodetectors and is proportional to the responsivity according to Eq. (6.26). In the visible range $\eta_D \geq 0.5$, $e/h\nu$ is on the order of 0.5 A/W, Eq. (6.26) leads to $R \sim 0.5$ A/W for photodetectors working in the visible range.

6.13.3 Parameters

The key parameters to describe a photodetector are

1. Spectral response range, inside which a photodetector is responsive.

2. Responsivity or quantum efficiency, usually defined in terms of how much electrical current a detector can generate for one watt of optical power incident on the detector. Responsivity is a function of wavelength; it has a peak value at a certain wavelength and falls to near zero at the two edges of the spectral range.
3. Dynamic range. A detector has a maximum optical signal level. The electrical output of the detector reaches its maximum level. Beyond this level, the detector is saturated. The electrical output of a detector is digitized by the electronics that controls the detector. The maximum electrical output is often digitized to 12–16 bits.
4. Format. Photodetectors can have only one element or many elements. The many elements in a detector can be aligned in a linear way or to form a 2D array. In a multi-element detector, an element is usually called a "pixel." Pixel size and numbers are needed to describe a multi-element detector. For example, a 640×512-pixel detector array with a 25-μm pixel size.
5. Noise. There are a few types of noise related to temperature, frequency, etc. For most applications, it is not necessary to know all of the details about the noise sources, as long as the total noise or noise equivalent power is known.
6. Speed (response time for the single-element detector or frame rate for detector array). That is about how fast a detector can respond, which is important in detecting very narrow pulses or fast varying signals, and not important for most other applications.
7. Cooling. The background noise level of photodetectors increases dramatically as the temperature of the detectors rises. Some photodetectors must be cooled for super-low noise. Some photodetectors in the IR range even require cooling for normal operation.
8. Size, weight, and price.

6.13.4 Noise equivalent power and specific detectivity

A photodetector and its operational electrical circuit have several noise sources, such as thermal noise, shot noise, Johnson noise, etc. Users of photodetector systems usually only care about the total noise in the system. A concept of noise equivalent power (NEP) is established, as illustrated in Fig. 6.43. The NEP is defined as the incident optical power level at which the magnitude of signal generated by this optical power equals the magnitude of total background noise of the detector system. In a real application, the incident optical power should be larger than the NEP to avoid being submerged in the noise and to be detectable.

Since the responsivity of a photodetector is a function of wavelength, for example, the curve shown in Fig. 6.44, the NEP is also a function of wavelength. The value of the NEP at a certain wavelength λ can be calculated by

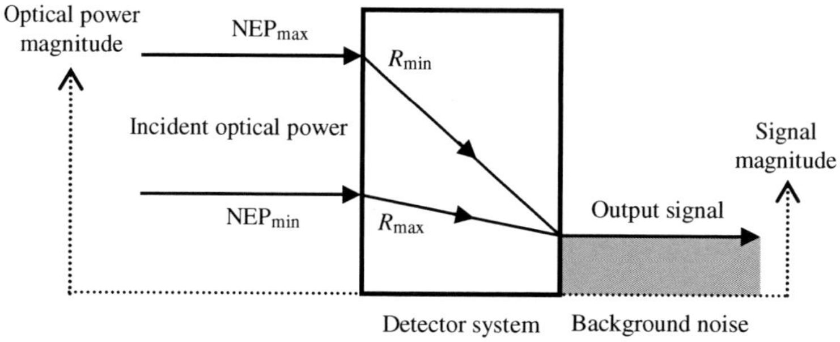

Figure 6.43 Noise equivalent power (NEP) is defined as the incident optical power level at which the magnitude of the signal generated by this optical power equals the magnitude of total background noise of the detector system. λ_{max} and λ_{min} are the wavelengths at which the detector system has the maximum and minimum responsivities, respectively.

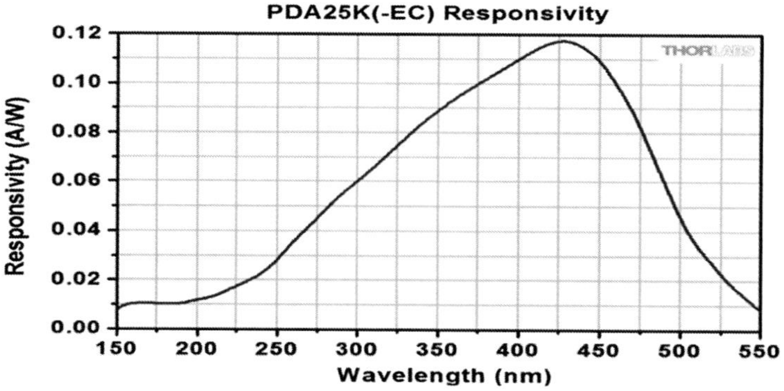

Figure 6.44 A typical responsivity curve of a gallium phosphate detector. Image reprinted courtesy of Thorlabs.

$$NEP(\lambda) = NEP_{\min}\frac{R(\lambda_{\max})}{R(\lambda)}, \tag{6.27}$$

where λ_{max} is the wavelength at which the responsivity of the detector has maximum value, and NEP_{min} is the minimum NEP that occurred at λ_{max}.

The NEP value is approximately proportional to the bandwidth. Therefore, it is necessary to normalize the NEP to 1 HZ to avoid confusion. Traditionally, the NEP is normalized to 1 $Hz^{0.5}$. Then the unit of the NEP is $W/Hz^{0.5}$. The minimum detectable optical power at wavelength λ is $P_{min} = NEP(\lambda)BW^{0.5}$, where BW is the bandwidth involved in the application.

For the same type of photodetectors, the sensing area size affects the minimum detectable optical power. For this reason, another parameter called specific detectivity D^* is defined as

$$D^* = \frac{A^{0.5}}{NEP}, \tag{6.28}$$

where A is the detector area size, and the unit of D^* is $cm \times Hz^{0.5}/W$.

References 21 and 22 provide more detailed descriptions about NEP and D^*.

6.13.5 UV detectors

Photodetectors can be categorized into materials used or responsivity, etc. To the users of photodetectors, the responsivity appears to be more important than the type of materials.

Few detectors are available in the UV range. UV detectors are mainly single-element detectors. Array detectors are rare because of the technical difficulties, high cost, and low need of UV imaging. The two commonly used UV detectors are gallium phosphate (GaP) and silicon (Si) detectors, as listed in Table 6.2. A typical responsivity curve of a GaP detector is shown in Fig. 6.44.

6.13.6 Visible and near-IR detectors

CCD and CMOS arrays, either linear or 2D, are widely used semiconductor sensors for visible imaging. A CCD consists of an array of photodiode, each of which has a capacitor built into it. Each photodiode accepts the optical power incident on it, converts the optical power to electrical charge and save the charge in its capacitor. The control circuit moves the charge from one capacitor to the next towards the outlet at one corner of the array and takes the charges from there in a sequential way before processing them.

A CMOS consists of an array of photodiodes. Each photodiode has a few transistors built into it. Each photodiode accepts the optical power incident on it and converts the optical power to an electrical signal; the transistors amplify the single; and the signals are output individually.

Traditionally, CCDs have higher quantum efficiency, lower noise, higher signal fidelity, consume more electrical power, and cost more compared with CMOSs. However, the differences between these two types of sensors are continuously shrinking.

As noted in Section 6.13.1, the responsive curves for the same type of detector, such as CCDs, made by different companies can have significant differences. The same phenomenon applies to the responsivity curves of CMOSs as well. Generally speaking, the responsivity curve of CMOSs is similar to the responsivity curve of CCDs. Figure 6.45 shows typical responsivity/quantum efficiency curves for a Si detector, a CCD sensor, and a CMOS sensor. Table 6.3 further describes these three commonly used detectors/sensors in the visible to near-IR range.

Table 6.2 Commonly used UV detectors.

Type	Spectral Range	Note
GaP	150–550 nm	Single element. Up to several mm^2 sensing area.
Si	200–1100 nm	Single element. Up to several mm^2 sensing area.

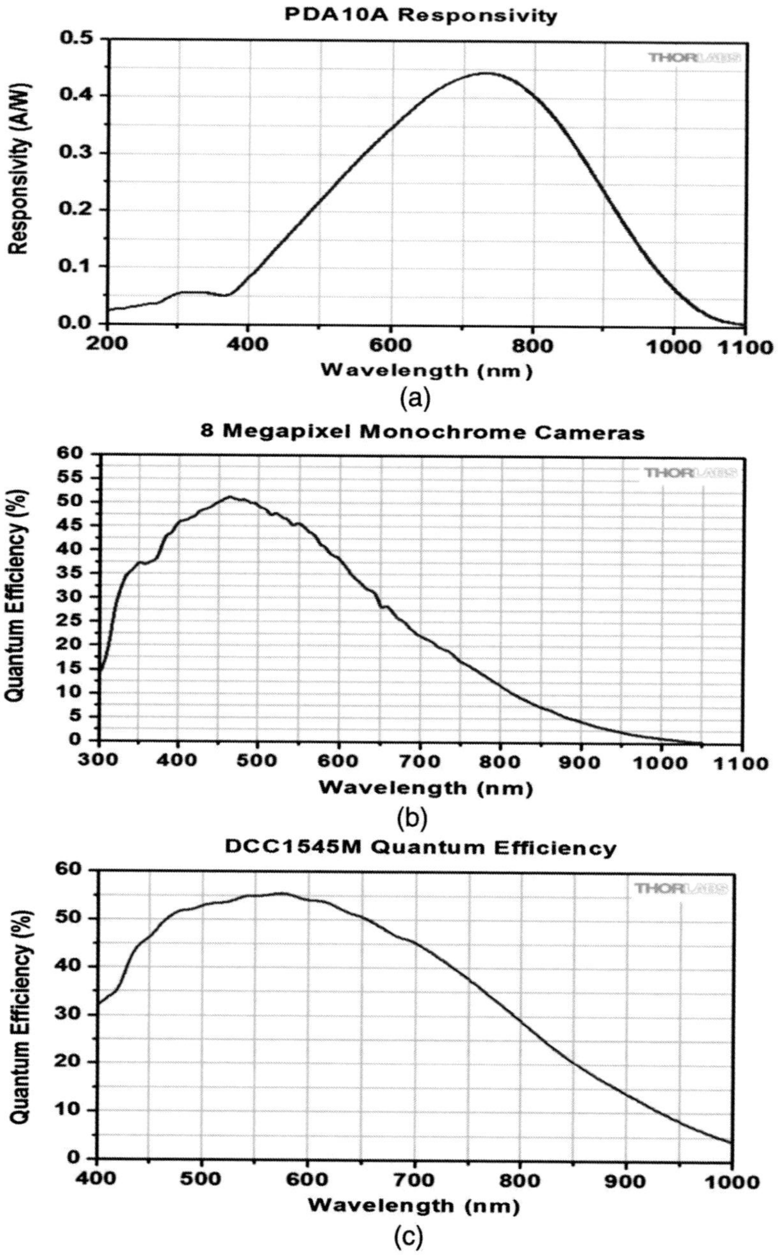

Figure 6.45 A typical responsivity/quantum efficiency curve for a (a) Si detector, (b) CCD sensor, and (c) CMOS sensor. Images reprinted courtesy of Thorlabs.

Some UV CCDs and CMOSs have a responsivity extended to ~200 nm, and some IR-enhanced CCDs and CMOSs push the spectral limit to 1200 nm or so.

Table 6.3 Commonly used visible detectors/sensors.

Type	Spectral Range	Note
Si detector	300–1000 nm	Single element. Up to a 100-mm^2 sensing area
CCD	300–1000 nm	2D array with up to several megapixels and down to pixels
CMOS	300–1000 nm	several μm^2 in size. Widely used in cameras.

6.13.7 RGB sensors

Red, green, and blue (RGB) sensors are one special type of sensor used in the visible range for sensing color images. Every modern camera, including cellphone cameras, has a RGB sensor in it. In a RGB sensor, a special optical component is placed in front of a colorblind sensors, such as a CCD array. The optical component separates the RGB components in the incident light so that the RGB can be recognized and sensed separately. Figures 6.46 and 6.48 show the schematics of two such optical components.

Shown in Fig. 6.46 is a Bayer filter array that is placed on top of a sensor array. The filter array elements have the same size as the sensor pixel size, and these two elements are aligned. Some filter elements only pass through R, some only pass through G, and some only pass through B. Then the sensor pixels behind the filter elements can only receive R, G, or B, respectively. Thereby, the black/white sensor becomes a RGB sensor.

The advantage of the Bayer filter approach is its negligible additional size to the base black/white sensor. Many existing imaging lenses originally designed for black/white imaging can be used directly for RGB imaging with a Bayer filter RGB sensor. The disadvantages are its reduced spatial resolution

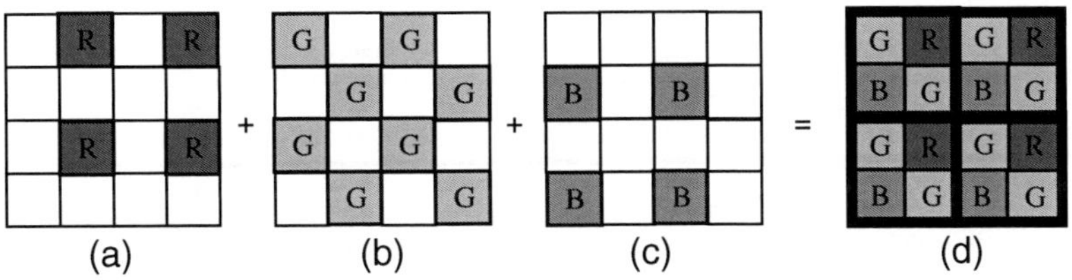

Figure 6.46 A Bayer filter array consists of three types of micro-filter arrays that only pass RGB colors, respectively. (a) A 4 × 4 array with four red filters. The other part of the array is transparent. (b) A 4 × 4 array with eight green filters. The other part of the array is transparent. (c) A 4 × 4 array with four blue filters. The other part of the array is transparent. (d) The combination of the three arrays shown in (a)–(c) form a Bayer filter array with four identical cells marked by the thick black line frames. Each cell contains one R filter, two G filters, and one B filter to imitate the human-eye spectral response, which is most sensitive to G. The micro-filter size is the same as the detector pixel size. By placing a Bayer filter on top of a detector and aligning every micro-filter to a pixel, a colorblind detector is converted to a RGB color detector. Each RGB detector unit consists of four pixels. The spatial resolution of the RGB detector is one-fourth the spatial resolution of the colorblind detector.

(because four pixels only form one RGB sensing element) and the possible high cost if small-quantity sensors are developed. A typical relative sensitivity curve of a RGB detector is shown in Fig. 6.47.

Figure 6.48 portrays another approach to turn three black/white sensors into one RGB sensor. Three dichroic prisms separate the RGB components in the incident beam and guide the RGB components to the three sensors, respectively. The advantage of this prismatic approach is its low development cost and high spatial resolution. The disadvantage is its large size, which increases the back working distance of imaging lenses and complicates the imaging lens design. Many exiting black/white imaging lenses may not work with this RGB sensor.

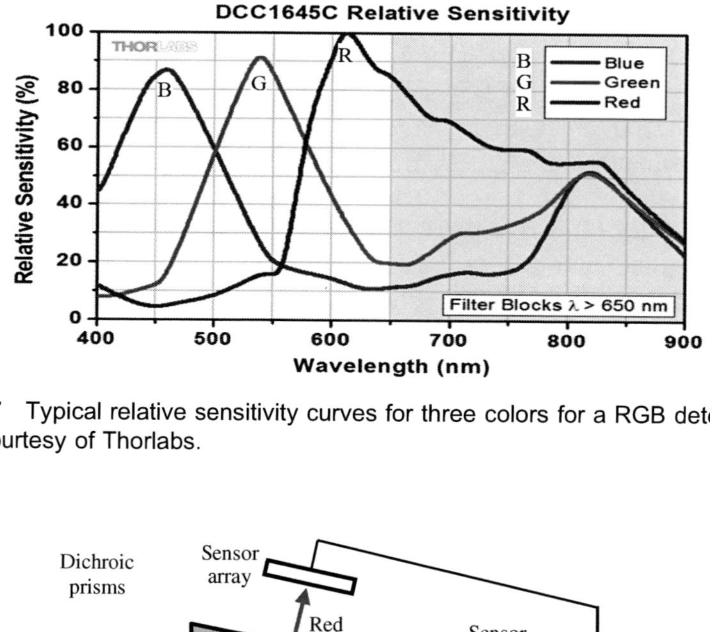

Figure 6.47 Typical relative sensitivity curves for three colors for a RGB detector. Image reprinted courtesy of Thorlabs.

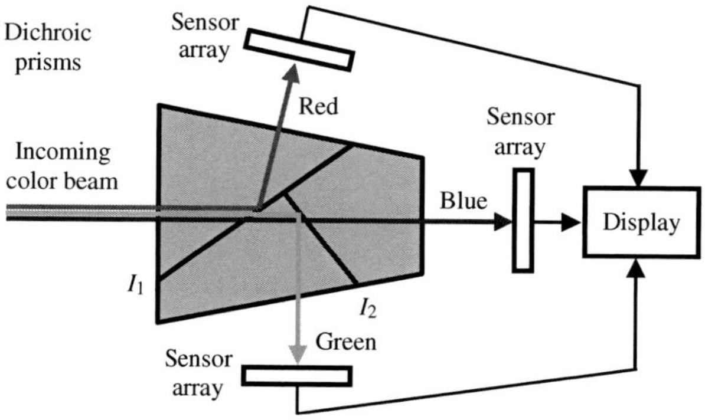

Figure 6.48 Three dichroic prisms split the incoming color beam into three RGB beams and guide the beams to three sensor arrays. A lowpass filter is coated at interface I_1 to let the green and blue light pass through and reflect the red light. An even-lower-pass filter is coated at interface I_2 to let the blue light pass through and reflect the green light. Thus the three colorblind sensors become one RGB sensor.

6.13.8 IR detectors

IR detectors without cooling to suppress noise often do not perform up to their potential. IR sensor arrays are usually available with pixel numbers much less than the pixel numbers available in visible sensor arrays, e.g., 640 × 512 pixels, and have a pixel size larger than the pixel size of visible sensor arrays, e.g., 15 μm. IR detectors with a sensitive wavelength over 10 microns are also called thermal detectors.

Table 6.4 lists six commonly used IR detectors. Note that some of them are photovoltaic detectors (the responsivities have a unit of V/W), some are photoconductive (the responsivities have a unit of A/W), and the three detectors covering longer wavelengths must be cooled to –30 °C to operate.

Figure 6.49 shows the typical responsivities of these six IR detectors.

6.13.9 Photomultiplier tubes

Photomultiplier tubes (PMTs) are vacuum tubes that have a built-in amplification mechanism. This mechanism can dramatically amplify the number of electrons generated by the optical signal (photons) received by the tube. Thereby, PMTs have a very high sensitivity and can detect as low as a single photon in the spectral range from the UV to the near IR. PMTs are single-element detectors mainly used in scientific research. Reference 23 provides a detailed review. Figure 6.50 shows the typical radiant sensitivities of five PMTs sold by Thorlabs.

6.14 Light Sources

Light sources are not conventional optical components, but they are frequently used to illuminate objects for imaging or sensing by optical systems, such as for spectral imaging, or used as wavelength standards. Three types of light sources are discussed in this section.

6.14.1 Illumination light sources

Four illumination light sources are introduced in this section. Among these light sources, deuterium, xenon, and mercury vapor (Hg) sources utilize the

Table 6.4 Commonly used IR detectors

Type	Spectral Range	Note
Ge	0.8–1.8 μm	Single element or array
InGaAs	0.85–1.70 μm	Single element or array
PbS	1.0–2.9 μm	Single element or array
PbSe	1.5–4.8 μm	Single element or array
InAsSb	1.0–5.8 μm	Single element or array
HgCdTe (MCT) photovoltaic	2.5–10.6 μm	Single element or array

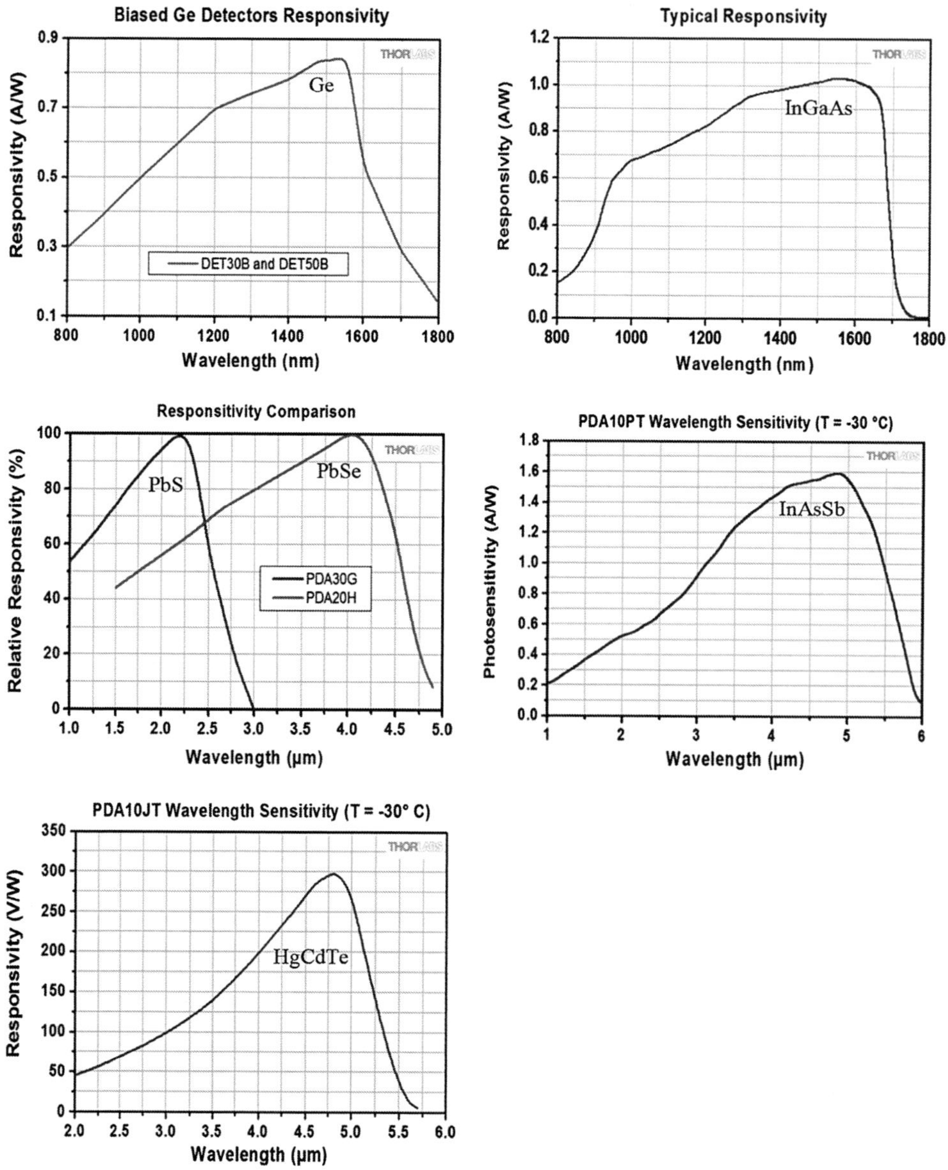

Figure 6.49 Typical responsivity curves for six IR detectors listed in Table 6.4. Images reprinted courtesy Thorlabs.

electrical discharge of an arc in a certain type of gas to light. Xenon lamps can be operated in either flash or continuous mode.

These lamps are very hot—up to six thousand degrees when being operated. The emission spectra are mixtures of blackbody spectrum and some molecular spectrum lines. Halogen lamps are incandescent lamps that are very

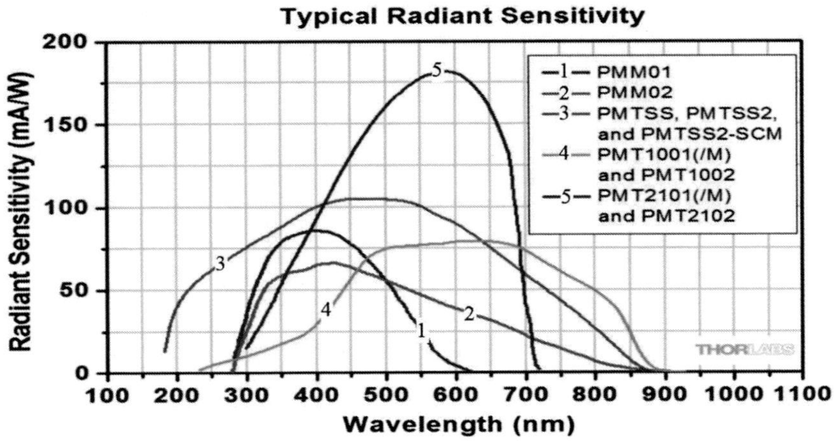

Figure 6.50 Typical radiant sensitivities of five types of PMTs. Image reprinted courtesy Thorlabs.

hot—up to five thousand degrees when being operated—and emit light like a pure blackbody.

Generally speaking, discharge lamps have higher efficiency, longer lifetime, and more complex structures compared with incandescent lamps. Discharge and incandescent light sources have been used for more than one hundred years.

There are many variations of these light sources: power ranging from a few watts to over ten thousand watts, bulb size and shape, arc size and shape, etc. More than one gas can be mixed in one lamp to emit light with a mixed spectrum. Several other gases are used to make lamps, too, such as argon, neon, krypton, sodium, metal halides, etc.

The spectra of xenon, deuterium, and mercury lamps are plotted in Fig. 6.51. The spectrum of halogen lamps is the blackbody spectrum, depending on the temperature of the specific lamp. The temperature range is usually from 3000–5000 K. The blackbody spectra for eleven temperatures that were plotted in Figs. 1.41 and 1.42 are not plotted again here.

Table 6.5 lists the five types of lamps according to their spectra. References 24–26 provide more information about light-source characteristics.

6.14.2 Spectral lamps

When atoms are excited, they emit light of certain wavelengths or spectral lines. Atoms of different elements emit different spectral lines. These lines have narrow linewidths, very stable wavelengths, and are ideal for spectral calibration or optical testing. For these applications, the output power is a secondary concern. In fact, most spectral lamps are small and output below one watt of optical power.

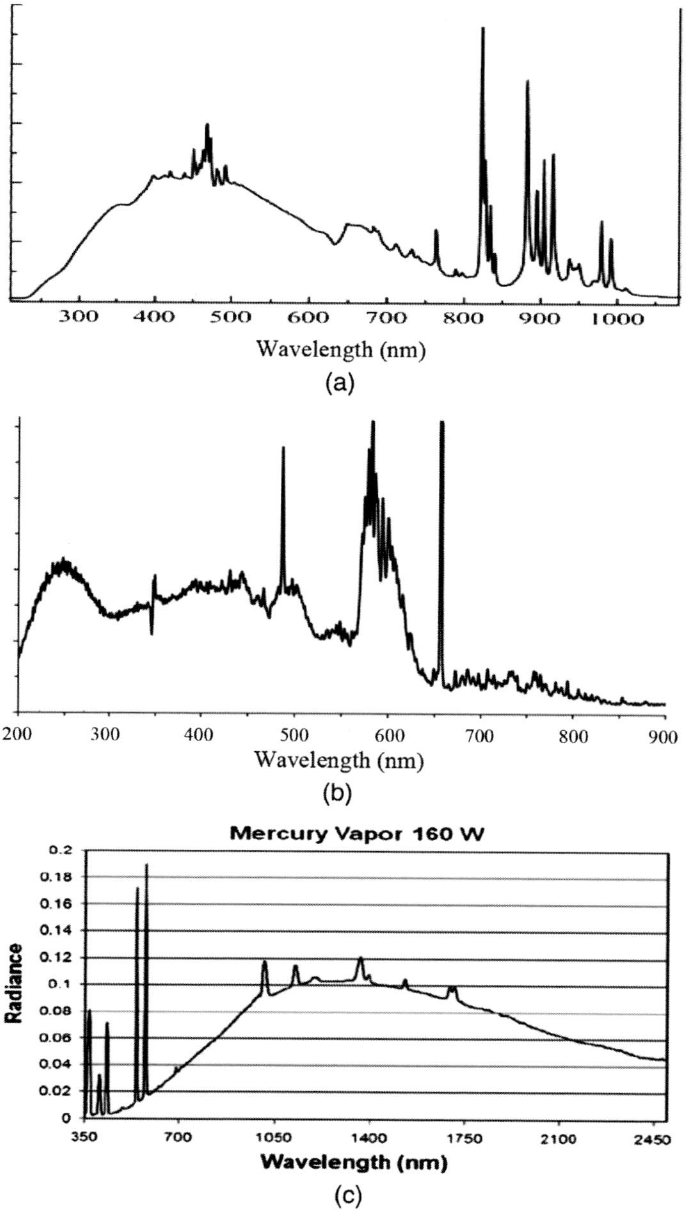

Figure 6.51 The spectrum of three lamps. (a) The spectrum of xenon lamps is a mixture of blackbody spectrum and some spectral lines. (b) The spectrum of deuterium lamps is a mixture of blackbody spectrum and some spectral lines. The light is emitted in one direction. (c) The spectrum of mercury lamps is a mixture of blackbody spectrum and some spectral lines.

A bandpass filter is often used with a spectral lamp to block the undesired lines. The commonly used spectral lamp elements and their spectral lines are listed in Table 6.6. Reference 27 provides a more complete list.

Table 6.5 Spectra of some commonly used light sources.

Spectra	Light Source
0.2–0.4 μm (UV)	Deuterium, xenon arc, Hg arc
0.4–0.7 μm (visible)	Halogen, LED, Hg arc, xenon arc
0.7–1.0 μm (SWIR)	Halogen, LED, xenon arc
1.0–3.0 μm (NIR)	Halogen, LED, xenon arc
>3.0 μm (MIR)	Blackbody

Table 6.6 Commonly used spectral lines.

Element (Symbol)	Main Spectral Lines (nm)
Cadmium (Cd)	326, 468, 480, 509, 644
Helium (He)	588, 668, 707
Mercury (Hg)	313, 334, 365, 404, 407, 435, 436, 546, 577, 579
Sodium (Na)	589, 590
Neon (Ne)	638, 640, 651, 660, 668
Zinc (Zn)	308, 328, 330, 335, 468, 472, 481, 636

6.14.3 Light-emitting laser diodes

LEDs are semiconductor devices that generate light using the same mechanism as laser diodes or a reverse mechanism of photodiodes. The differences between LEDs and laser diodes are that LEDs do not have a resonant cavity, and the light emitted by LEDs is not coherent, have larger linewidths, and divergent angles.

Compared with traditional light sources, LEDs have several important advantages. They

- Are as small as a penny, lightweight, and robust.
- Do not release much heat when being operated.
- Have a very high electrical-to-optical-power conversion efficiency.
- Have many wavelength selections, from UV (245 nm) to IR (4600 nm).
- Have a long lifetime.
- Can be battery operated and switched quickly.
- Emit light that is directional and can be focused to spots much smaller than the spots produced by traditional lamps. The can also achieve high power densities, up to hundreds of watts per square centimeter.

Because of these advantages, LED applications are rapidly increasing and replacing many traditional light sources.

RGB and white LEDs are available. A RGB LED consists of three LEDs with three colors, and a white LED has a phosphor coating on a blue LED. The coating absorbs blue light and emits yellow or green or even red light. The mixture of these color lights appears to be white to human eye, but the spectrum of this white light is not continuous but only contains several discrete bands.

Similar to other semiconductor devices, a high temperature will first reduce the efficiency and then the lifetime of LEDs. Some type of thermal management is necessary, at least to contact a LED to a heat sinker. The power ranges of LEDs are from a few mW to hundreds mW for a single emitter and hundreds watts for an emitter array.

LEDs are sensitive to electric discharge and must be handled by grounded tools or hands. LED wavelengths have a thermal shift of ~0.2 nm/°C and should not be used as wavelength standards.

Many commercial LED packages come with driven electronics, including an protection from electric discharges. Some LED packages come with a light-collecting lens that can concentrate the light.

References

1. *CNC Optical Grinding Machines*, www.optipro.com/cnc-grinding.html
2. *MRC Technologies*, mrctechnologies.webs.com
3. *Gradient-Index Optics*, en.wikipedia.org/wiki/Gradient-index_optics
4. *GoFoton*, welcome.gofoton.com/passive-optics/selfoc-lenses
5. Wikipedia, "*Calcite*," wikipedia.org/wiki/Calcite
6. *Colored Filter Glass*, www.koppglass.com/filter-catalog/filter-index.php
7. *Diffraction Grating Handbook*, www.gratinglab.com/Information/Handbook/Handbook.aspx
8. B. E. A. Saleh and M. C. Teich, *Fundamentals of Photonics*, 2nd ed., John Wiley & Sons, Inc., Hoboken, NJ, p. 332, Eq. (9.2-1) (2007).
9. *ibid*, p. 333, Eq. (9.2-6).
10. *ibid*, p. 334, Eq. (9.2-10).
11. *ibid*, p. 337, Eq. (9.2-15).
12. *ibid*, p. 338, Eq. (9.2-18).
13. By Dr Bob at English Wikipedia, CC BY-SA 3.0, en.wikipedia.org/wiki/Transverse_mode#/media/File:Laguerre-gaussian.png
14. B. E. A. Saleh and M. C. Teich, *Fundamentals of Photonics*, 2nd Ed., John Wiley & Sons, Inc., Hoboken, New Jersey, p. 333, Fig. 9.2-2 (2007).
15. C. D. Hussey and F. Martinez, "Approximate analytical forms for the propagation characteristics of single-mode optical fibres," *Electron. Lett.* **21**, 1103 (1985).
16. K. Petermann, "Constraints for fundamental mode spot size for broadband dispersion-compensated single-mode fibers," *Electron. Lett.* **19**, 712 (1983).
17. C. Pask, "Physical interpretation of Petermann's strange spot size for single- mode fibres," *Electron. Lett.* **20**, 144 (1984).
18. *Fiber Components*, www.thorlabs.com/navigation.cfm?guide_id=29
19. *Photodiode Characteristics and Applications*, www.osioptoelectronics.com/application-notes/an-photodiode-parameters-characteristics.pdf

20. W.-C. Wang, *Optical Detectors*, depts.washington.edu/mictech/optics/me557/detector.pdf.
21. V. Mackowiak, J. Peupelmann, W. Ma, and A. Gorges, *NEP – Noise Equivalent Power* www.thorlabs.com/images/TabImages/Noise_Equivalent_Power_White_Paper.pdf
22. P. C. Datskos and N. V. Lavrik, *Detectors-Figures of Merit*, 123.physics.ucdavis.edu/week_6_files/detectors_figures_of_merit.pdf
23. *Photomultiplier Tubes, Basics and Applications*, 3rd Ed, Hamamatsu Corp, www.hamamatsu.com/resources/pdf/etd/PMT_handbook_v3aE.pdf
24. T. Iguchi, A. Patel, and M. Lares, *A Guide to Selecting Lamps*, www.photonics.com/Article.aspx?AID=44487
25. C. D. Elvidge, D. T. Tuttle, and K. E. Baugh, "Spectral Identification of Lighting Type and Character," *Sensors* **10**, 3961–3988 (2010) [doi: 10.3390/s100403961].
26. *Light Source* www.hamamatsu.com/resources/pdf/etd/LIGHT_SOURCE_TLSZ0001E.pdf
27. *Oriel Pencil Style Calibration Lamps*, assets.newport.com/pdfs/e5395.pdf

Chapter 7
Commonly Used Imaging Lenses

7.1 Specifications

Imaging lenses are the most widely used lenses. An imaging lens is described by a series of specifications. It is necessary to understand the meanings of these specifications in order to understand and evaluate imaging lenses.

7.1.1 Focal length

An imaging lens can consist of many elements, but it has only one focal length, which is the collective effects of all of the elements. A focal length can be either positive or negative.

An imaging lens with a positive focal length can focus rays passing through the lens. The thin lens equation [Eq. (1.15), reproduced here]

$$\frac{1}{o} + \frac{1}{i} = \frac{1}{f} \tag{7.1}$$

states that when the object point is at infinity $o \to \infty$, the image distance i equals the focal length f, $i = f$; then the rays from the object point are focused at the focal point, and the focal length of the lens is the axial distance between the principal plane of the lens and the focal point. The principal plane can be located by forward extending the incoming rays and back extending the focused rays; the cross-point of these two extended rays is the principal plane location. The situation is illustrated in Fig. 7.1(a). If the object is not at infinity $i \neq f$ according to Eq. (7.1), the rays will not be focused at the focal point of the lens, as illustrated in Fig. 7.1(b).

An imaging lens with a negative focal length does not focus the rays passing through the lens, but rather it diverts the rays. When the object point is at infinity $o \to \infty$, the image distance i equals the focal length f, $i = f < 0$. The focal point of the lens is the cross-point of the back extension of the diverted rays and the optical axis, as illustrated in Fig. 7.2(a). The focal length of the

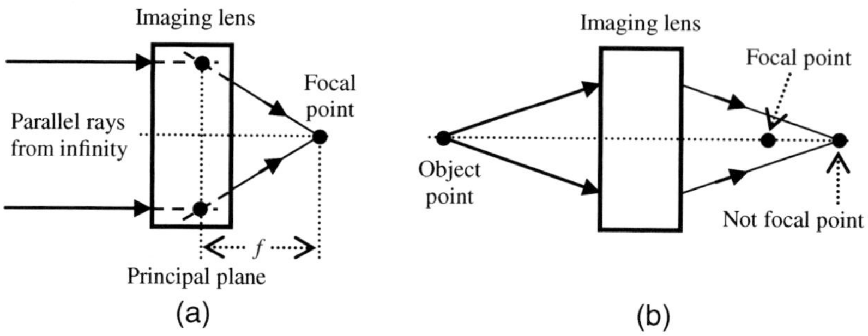

Figure 7.1 (a) An imaging lens with a positive focal length focuses parallel rays from infinity to the focal point of the lens. (b) If the incident rays are not parallel, the rays will not be focused at the focal point of the lens. Such a lens can form real images on a sensor ray.

lens is the axial distance between the principal plane of the lens and the focal point. The principal plane can be located by forward extending the incoming rays and back extending the focused rays; the cross-point of the two extended rays is the principal plane location, as illustrated in Fig. 7.2(a). If the object is not at infinity $i \neq f$ according to Eq. (7.1), the cross-point of the back extension of the diverted rays and the optical axis is not the focal point of the lens, as illustrated in Fig. 7.2(b).

Sections 1.5.2 and 1.5.3 explain the concept of focal length in more detail.

7.1.2 Field of view, image size, and working distance

The double Gauss lens plotted in Fig. 1.10 is re-plotted in Fig. 7.3(a) for illustration. It illustrates the field of view in terms of the field angle θ or in terms of the object size O, the working distance W, the object distance o, the image size I, and the lens optical length L.

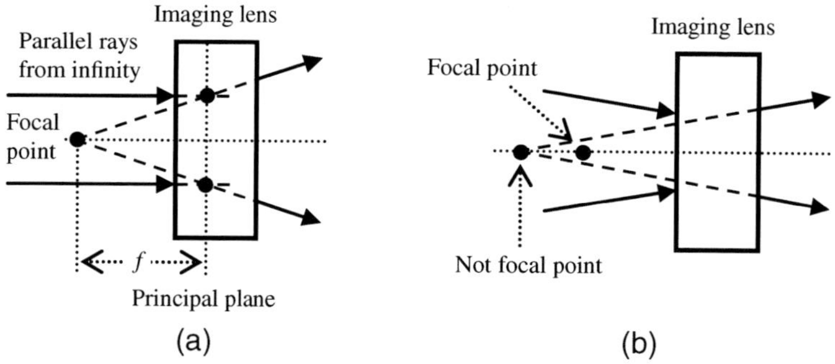

Figure 7.2 (a) An imaging lens with a negative focal length diverts parallel rays from infinity. The cross-point of the back extension of the diverted rays and the optical axis is at the focal point of the lens. (b) If the incident rays are not parallel, the cross-point of the back extension of the diverted rays and the optical axis is not at the focal point of the lens. Such a lens can form virtual images for the human eye to view.

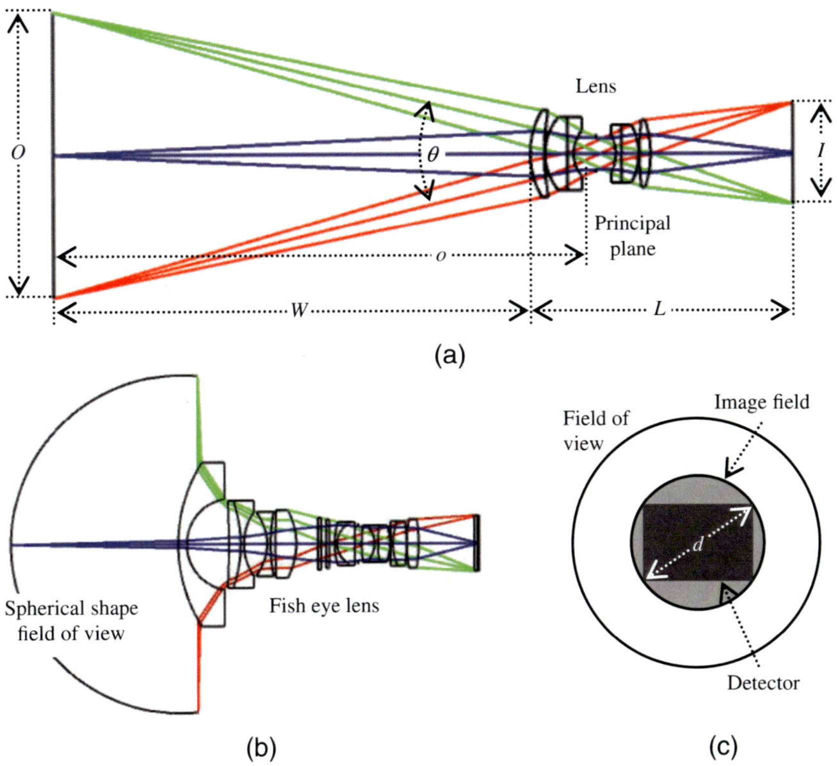

Figure 7.3 (a) The field of view in terms of the field angle θ is defined as the angle between the two edge chief rays. The field of view in terms of the object size *O* is the largest object size that the lens can see. The working distance *W* is the distance from the object to the front lens vertex. The object distance *o* is from the object to the principal plane of the lens. *I* is the image size, and *L* is the optical length of the lens. (b) Zemax raytracing diagram of a fisheye lens with a 180° field of view and a spherical object surface. (c) Illustration of the shapes of the field of view, image field, and detector.

The field of view is about how large the lens can see and is related to the lens focal length and sensor size. When the working distance is not much longer than the focal length, the field of view can be defined by either the field angle or the object size. When the working distance is much longer than the focal length, the field of view is usually defined by the field angle because when viewing a distant object, such as the moon or a mountain, object size cannot be easily determined. The complexity of an image lens increases significantly as the field angle increases. The relation is nonlinear and cannot be simply defined.

As shown in Fig. 7.3(a), the working distance *W* is the axial distance from the object plane to the vertex of the front lens. The object size is

$$O \approx 2W \tan\left(\frac{\theta}{2}\right). \tag{7.2}$$

For an object point at the edge of the object plane, the distance from this specific point to the lens is $W/\cos(\theta/2)$.

An image lens needs to form a sharp image for all points on the object plane with distances from W to $W/\cos(\theta/2)$. As the value of θ increases, both values of $W/\cos(\theta/2)$ and O significantly increase. When $\theta = \pi$, the value of $W/\cos(\theta/2) \to \infty$ and $O \to \infty$, which is impossible to handle. In such a case, the object plane must be made spherical to avoid an infinitely large object size, as shown in Fig. 7.3(b). Such a lens is called a "fisheye lens."

Image size I is related to the field of view angle θ and the lens focal length f by

$$I = \frac{2f\tan(\theta/2)}{1 - 100\Delta}, \tag{7.3}$$

where Δ is the percentage distortion of the image. For $\Delta = 0$, Eq. (7.3) reduces to the standard tangent relation.

Most lenses are symmetric about their optical axis. The object field they can see and the image field they produce have circular shapes. The image size I usually equals the diagonal size of the rectangular detector used to see the image, as illustrated in Fig. 7.3(c).

When working distance $W \gg L$, $W \approx o$, where o is the object distance measured from the object to the principal plane of the lens. For most cameras, their working distance can be from 1 m or so to infinity.

7.1.3 Image quality

The quality of images produced by a lens mainly includes two aspects: resolution and distortion. Generally speaking, the image resolution of a lens should match the pixel resolution of the detector used to see the image. The distortion of an image produced for the human eye to see is usually about 2–5%. Images with a distortion >5% will look uncomfortable. A distortion <2% will not be easily noticeable to human eye. For serious inspection, the image distortion should be below 1%. (See Section 1.13 for more details.)

7.1.4 Spectral range

The spectral range of lenses is determined by both the transmission of the glass used to make the lens and the spectral range of the detector used to see the image. Most optical glasses have high transmission from below 400 nm to over 2 μm. Most optical polymers have high transmission from 400 nm to about 1 μm. As a comparison, most widely used silicon-based detectors have decent responsivity from about 400 nm to over 1 μm. Therefore, in most cases, the spectral range of detectors is the limitation.

Most lenses are AR coated. The spectral range of the coating used can also be an issue. Outside the spectral range, the reflectivity of the coating will dramatically increase.

7.1.5 Transmission and illumination uniformity

Lens surface reflections are the main cause of low transmission. The rough estimation is 5% reflection loss per uncoated lens surface. Surface reflectance increases a little bit as the incident angle increases, as shown in Fig. 2.3.

Unless being otherwise specified, AR coatings on lenses are optimized for normal incidence. For a larger incident angle, the residual reflectivity of AR coatings increases. Furthermore, many lenses are designed to intentionally vignette (clip) some rays at image corners because these rays cannot be well focused and will ruin the image quality. The vignetting will reduce the brightness at the image corners.

Given these three factors, it is common for many imaging lenses, such as camera lenses, that the images they produce have dimmer corners. The image or illumination uniformity often must be specified. Image corners with illumination power density half of that at image center may still be acceptable for commercial use because the human-eye responsivity to brightness is not linear. Half illumination power looks like more than half to human eyes.

7.1.6 Magnification

For a lens, such as an objective, when its object and image sizes are clearly known (as in Fig. 7.3(a)), the magnification is also known as I/O. For camera lenses, the object size is approximately proportional to the object distance, and the object distance can change by thousands of folds, so it does not make much sense to define a magnification by I/O.

Binoculars and telescopes are used to look at distant objects. The rays incident on the devices are almost parallel. The rays output from the devices are not focused onto a detector but are almost parallel for human eyes to directly view. In such a case, an angular magnification can be defined as the ratio between the incident ray angle and output ray angle.

7.1.7 Depth of field

Equation (7.1) shows that once the object distance o is changed, the image distance i is also changed, and a detector that was placed at the correct position to take the focused image is no long at the correct position. The image is more or less blurred.

The concept "depth of field" (DOF) is defined as a range of object distance that can lead to images with "acceptable focus," a subjective and ambiguous term. DOF can be numerically defined at image space as the image distance range around the best focusing image distance. At the two edges of the range, the focused spot size is $2^{0.5}$ times larger than the best-focused spot size, as illustrated in Fig. 7.4. But this definition can be misleading, e.g., when the best-focused spot size a is much smaller than the detector pixel size. An increment of $2^{0.5}$ times the spot size will not result in

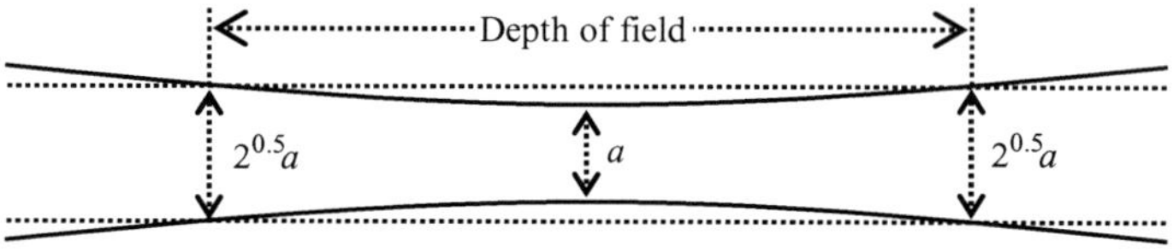

(a)

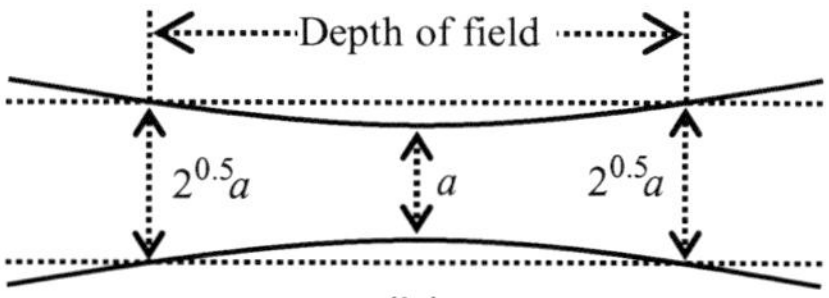

(b)

Figure 7.4 Illustration of depth of field, where *a* is the best-focus spot size. The DOF shown in (a) is longer than the DOF shown in (b), although both have the same focused spot size *a* because the focusing in (a) is slower (a larger *F/#*).

any noticeable change in image sharpness. If the lens's best focused spot size matches the pixel size of the detector used, the DOF in image space is useful.

The DOF can also be defined in object space. However, the DOF in object space for a $2^{0.5}$-times increment of the image spot size is not a constant but approximately proportional to the object distance o. Therefore, to define the DOF in object space, the object distance o must first be specified.

7.1.8 Focus adjustment

For a given lens, such as a camera lens, the image distance i is fixed. Equation (7.1) shows that to change object distance o, while maintaining focus, the focal length f must be changed or, in other words, the focusing of the lens must be adjusted.

After differentiating Eq. (7.1) with i as a constant, the result is

$$\Delta f = \frac{\Delta o}{o^2} f^2, \tag{7.4}$$

which states that as o increases, Δf decreases significantly.

Here are two numerical examples. Assuming $f = 30$ mm, $o = 2$ m, $\Delta o = 1$ m, and $\Delta o/o = 0.5$, Eq. (7.4) results in $\Delta f = 0.225$ mm, i.e., a 0.225 mm/30 mm = 0.75% change of the focal length. As a comparison, assuming $f = 30$ mm, $o = 1000$ m, $\Delta o = 500$ m and $\Delta o/o = 0.5$, Eq. (7.4) results in $\Delta f = 0.00045$ mm, i.e., a 0.00045 mm/30 mm = 0.0015% change of focal length, which is negligible. These two examples show that adjusting the focus at a close object distance is much more difficult to achieve than adjusting it at a long object distance. Roughly speaking, there is no need to adjust the focus if the object distance is beyond several hundred times the focal length.

Focus adjustments from a near distance of about tens of focal lengths to infinity can usually be achieved by moving some elements in the lens, usually the front element(s). Once an element or some elements in a lens are moved, the principal plane position of the lens can also be moved, as does the imagine distance *i*. In reality, the adjustment of most lenses is achieved by changing both the focal length and the image distance.

7.1.9 Zooming

Equation (7.3) shows that for a given image (detector) size *I*, a reduction of the focal length *f* increases the object field angle θ, and vice versa. Thus, zooming can be achieved by adjusting the focal length *f*. Equation (7.1) demonstrates that for a fixed object distance *o*, a change in *f* must be accompanied by a change in the image distance *i*. When zooming, two groups of optical elements usually must be moved, and the principal plane position of the lens is also moved so that *i* is changed but the detector position is not.

Zoom range is defined as the ratio of the focal lengths at the two edges of the zoom range. For example, at one edge of the zoom range $f = a$, and at another edge of the zoom range $f = 3a$. Then the zoom range is 3 times, or 3X. Zooming is a relative concept that has nothing to do with absolute magnification.

Zooming and focus adjustment are two different issues that are independent of each other. For the latter, the object distance is changed. The focal length and image distance are changed by only a fraction. While zooming means that the object distance is fixed, the focal length and image distance have significant changes (in particular, the focal length can change multiple times). Focus adjustments keep an image sharp for different object distances. Zooming changes the image size at a given object distance. Commercial camera lenses usually have these two functions built in.

7.1.10 *F*-number and power-collection capability of lenses

Every imaging lens has an *F*-number determined by the ratio of f/D_{En}, where *f* is the focal length, and D_{En} is the entrance pupil diameter, as illustrated in Fig. 1.10(d). Equation (2.18) shows that the Airy (diffraction-limited) spot size is proportional to the *F*-number. Thus a lens with a smaller *F*-number has the potential (only the potential) to produce higher-resolution images. Whether such potential is realized is another issue. A smaller *F*-number also leads to larger spherical aberration and other aberrations. If these aberrations are not well minimized, a smaller *F*-number can result in lower image resolution.

Lenses that can collect more optical power from the object are suited to produce bright images. For a given focal length, a smaller *F*-number means a larger entrance pupil diameter and more incoming optical power being collected, as shown in Fig. 1.10(d). There is a common misconception that the size of the front element in a lens determines the optical power collected; however, as can be seen from Fig. 7.3(a), for any field angle only a small portion of the front element collects optical power.

7.2 Magnifiers

Magnifiers are the most commonly used lens to optically magnify objects for clear viewing. Any positive lenses can be used as magnifiers. Figure 7.5 illustrate the working principle of a magnifier using Zemax-generated raytracing diagrams. The magnifier is a planar-convex lens made of N-BK7 glass with a 50-mm diameter, 20-mm central thickness, 40-mm back surface radius of curvature, and 77-mm focal length.

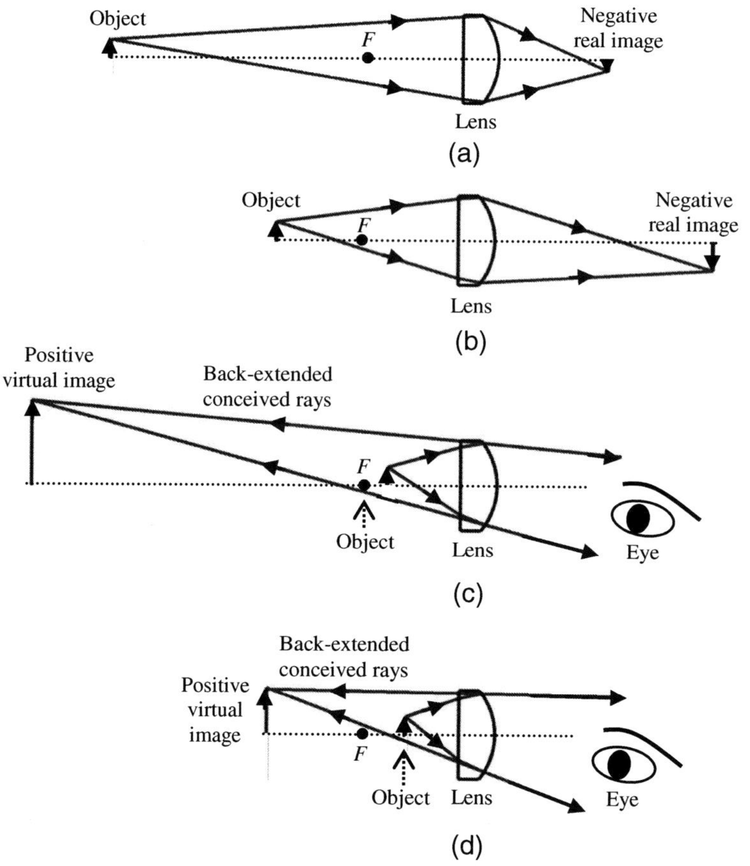

Figure 7.5 Principle of a magnifier. The left-side focal point *F* of the lens is marked by the solid dot. (a) When the object is placed far outside the focal length, the lens will form a negative and real image. Human eyes cannot directly view such an image. (b) When the object is moved closer to the lens but still outside the focal length, the size of the negative and real image increases. (c) When the object is placed slightly inside the focal length, the rays passing through the lens are not focused to form any real image. However, the back extension of the rays will form a positive and virtual image. Human eyes can directly view such an image. (d) If the object is placed even closer to the lens, the positive and virtual image size decreases.

If the object to be viewed is placed outside the focal length, the magnifier will form a negative real image, as shown in Figs. 7.5(a) and (b). Such an image cannot be directly viewed by the human eye but can be accepted by a 2D sensor array and displayed on a screen for viewing.

When the object is placed inside the focal length, a positive virtual image can be formed by the magnifier for direct viewing, as shown in Figs. 7.5(c) and (d). The magnifications can be adjusted by changing the distance between the object and the magnifier. Most magnifiers consist of a single element, and they have severe spherical and color aberrations and low image quality.

7.3 Eyepieces

7.3.1 General description

Eyepieces are high-quality magnifiers that consist of several elements and have the color and spherical aberrations minimized. Eyepieces should be used in the designed conditions to generate high-quality virtual and positive images for direct viewing, similar to the situations shown in Figs. 7.5(c) or (d).

Figure 7.6 shows the Zemax-generated raytracing diagram of a six-element eyepiece. The rays passing through the eyepiece are not focused, but the back extensions of the rays are focused to form a virtual positive image that is much larger than the object being viewed.

This eyepiece has 25-mm focal length, 3-mm working distance, 15-mm eye relief distance, 6-mm exit pupil size, 20-mm maximum field of view, 200-mm maximum image size, and 250-mm image distance. The designed magnification is 200 mm/20 mm = 10X. The real magnification of the eyepiece can be adjusted by changing the object distance.

This eyepiece can produce an image with $CTF > 0.3$ at 7 line pair/mm on the image for most part of the field. Such an image resolution matches the human-eye resolution limit when watching an image at a distance of 250 mm, as will be explained in Section 7.13.4.

7.3.2 Specifications

An eyepiece is specified by several parameters:

1. Focal length, which is an indication of the power of the lens. A shorter focal length means a more powerful lens. This eyepiece has a focal length of 25 mm. Note that the real magnification of the eyepiece depends not only on the focal length but also the object (working) distance.
2. Working distance, which is defined as the distance between the object being viewed and the frontmost plane of the eyepiece, as explained in Fig. 7.6(b). This eyepiece is optimized for a working distance of 3 mm. If an object being viewed is placed at a distance not 3 mm away from

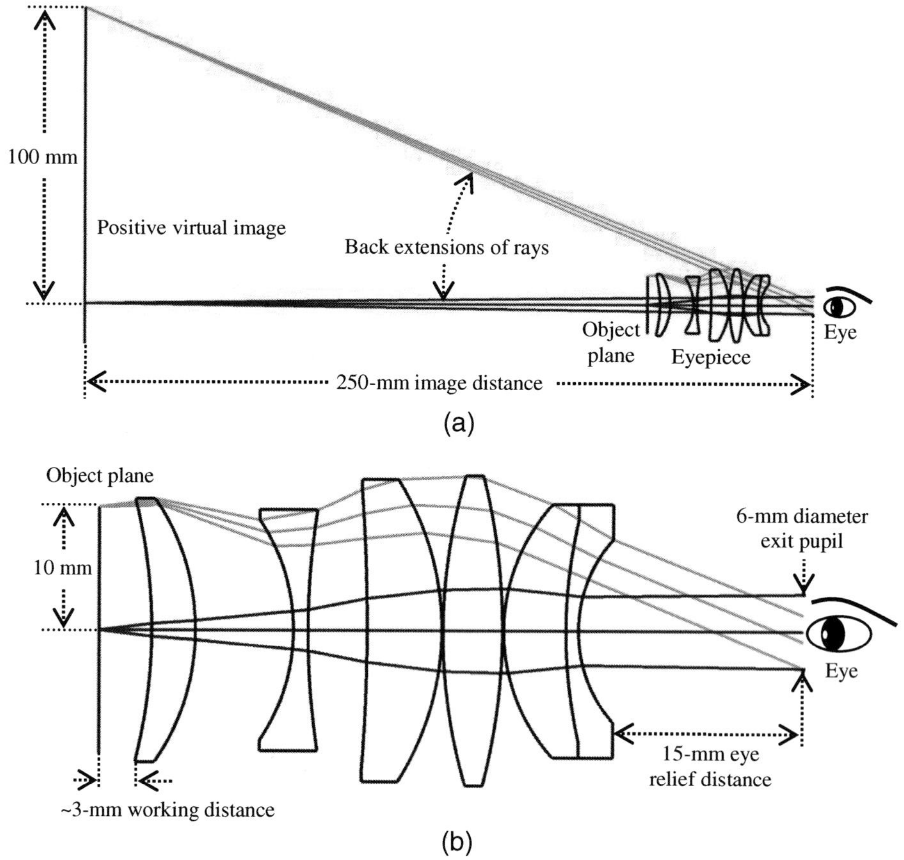

Figure 7.6 (a) Raytracing diagram of a six-element eyepiece. (b) A detailed view of the raytracing diagram. When the user's eye is placed at the exit pupil, only the eyeball must move to view the whole image.

the eyepiece, the object can still be viewed, but the image quality will be lower than the best possible, and the magnification is also different than the specified magnification. A large working distance is desired for easy use of the eyepiece. From a design perspective, large working distances are difficult to achieve, so a balance must be struck.

3. Magnification, which is defined as the ratio of the image size to the object size at the designed working distance. This eyepiece has a magnification of 10X magnification. For a given focal length, the magnification of an eyepiece can be altered by changing the working distance. Similar to working distances, a large magnification is difficult to achieve.
4. Eye relief distance, which is defined as the distance between the rearmost physical plane of the eyepiece and the user's eye, as shown in Fig. 7.6(b). A large eye relief distance is desired for easy and

comfortable use of the eyepiece. Large eye relief distances are difficult to achieve. The eyepiece shown in Fig. 7.6 has an eye relief distance of 15 mm.

5. Exit pupil size, which is the size of the exit ray bundle as shown in Fig. 7.6(b). The exit pupil should not be smaller than the human-eye pupil diameter, which is about 2–8 mm, to easily align the viewer's eye with the exit ray bundle. A large exit pupil is difficult to achieve. This eyepiece has an exit pupil size of 6 mm.
6. Field of view, which is defined at the designed working distance. A change to the working distance will change the field of view. Any object larger than the field of view will be clipped. A large field of view is difficult to achieve. This eyepiece has a field of view of 20 mm.
7. Image distance, which is usually chosen to be about 250 mm, which is the most comfortable viewing distance for most human eyes. The image distance can be altered by changing the working (object) distance as described by Eq. (7.1).

7.4 Objectives

7.4.1 General description

Objectives are high-quality, multi-element positive lenses that have the color and spherical aberrations minimized. Objectives are preferred to be used in the designed conditions to generate high-quality real images for further processing, similar to the situations shown in Figs. 7.5(a) and (b). Figure 7.7 shows the Zemax-generated raytracing diagram of an eight-element objective. The rays passing through the objective are focused. The real image is negative, has a size much larger than the object size, and can be adjusted by changing the distance between the object and the objective.

7.4.2 Specifications

An objective is specified by several parameters:

1. Focal length, which is an indication of the power of the lens. A shorter focal length means a more powerful lens. However, the real magnification of the lens also depends on the working distance. This objective has a focal length of 4.6 mm.
2. Working distance. Similar to eyepieces, the working distance is defined as the distance between the object being viewed and the most front plane of the objective. The magnification of objectives can be adjusted by changing the working distance. This objective is optimized for a working distance of 1 mm, as shown in Figs. 7.7(b) and (c), such a short working distance is normal for objectives. If the object being viewed is placed at a distance a little off 1 mm from the objective, a real image will still be generated but with a quality lower than the best

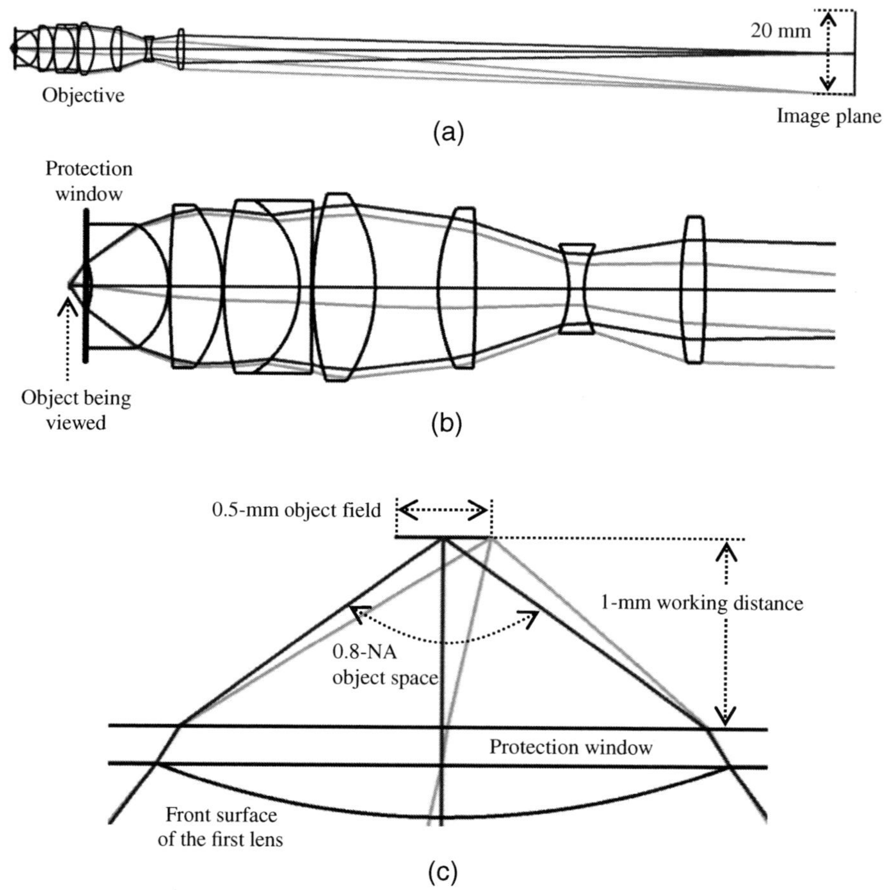

Figure 7.7 (a) Zemax-generated raytracing diagram for an eight-element objective. (b) The detailed view of the objective. (c) A more-detailed view of the most front part of the objective which is rotated clockwise by 90°. This objective has a focal length of 4.6 mm, a working distance of 1 mm, a maximum object field of view of 0.5 mm, a corresponding image size of 20 mm, an objective NA of 0.8, and a magnification of 20 mm/0.5 mm = 40X.

possible, and the magnification is also different than the designed version. A large working distance is desired for easy use of the objective, but it is difficult to achieve from a design perspective, so a balance must be found.

Although the working distance is only 1 mm smaller than the focal length, the objective can still focus rays similar to that shown in Fig. 7.5(b) because the left-side principal point is inside the objective, the distance between the object and that point of the objective is larger than the focal length.

3. Magnification, which is defined as the ratio of the image size to the object size at the designed working distance. Changing the working

distance can change the magnification. A large magnification is difficult to achieve. This objective has a magnification of 20 mm/ 0.5 mm = 40X.
4. Field of view, which is defined at the designed working distance of 1 mm. Any object larger than the field of view will be clipped. A large field of view is difficult to achieve. This eyepiece has a maximum field of view of 0.5 mm. Such a small field of view is normal for objectives.
5. Objective numerical aperture, as illustrated in Fig. 7.7(c). A large numerical aperture is ideal for collecting more light power from the object; however, a large numerical aperture is difficult to achieve. This objective has a numerical aperture of 0.8, which is common for objectives.

7.5 Microscopes

A microscope consists of an objective and an eyepiece. The objective generates a magnified negative real image for the eyepiece to view, as illustrated in Fig. 7.8. The objective and eyepiece described in two previous sections can be used to form a microscope. The magnification of the microscope as a whole is the production of the magnifications of the objective and the eyepiece.

As shown in Fig. 7.8, for the objective, if the object distance a is changed, the image distance b changes, as does the magnification of the objective. A change of b means that the object distance of the eyepiece is changed, and the magnification and the image distance d of the eyepiece are changed too.

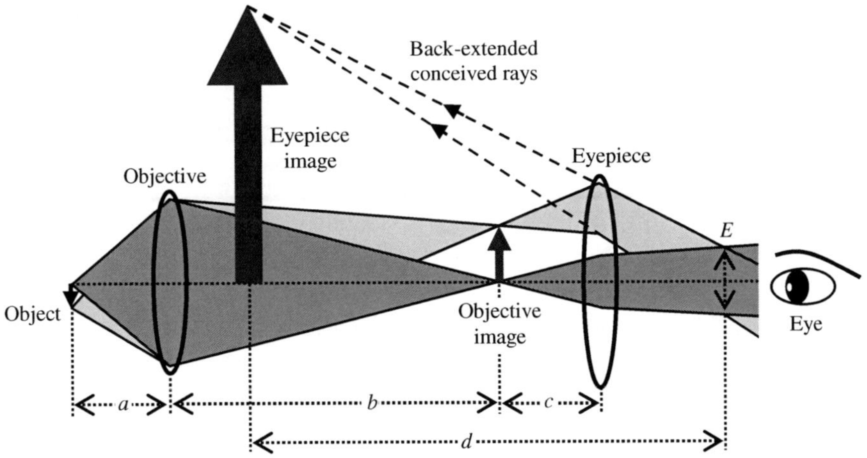

Figure 7.8 Schematic of a microscope. Objective working distance *a* must be smaller than the focal length of the objective. The objective forms a negative real image usually about $b = 160$ mm from the objective for the eyepiece to view. The eyepiece working distance *c* must be smaller than the focal length of the eyepiece. The eyepiece forms a positive virtual image usually about $d = 250$ mm from the exit pupil *E*. The user's eye is placed at the exit pupil to view the image.

If the eyepiece is moved, its object distance c changes, so the magnification and the image distance d of the eyepiece are also changed.

The image of the microscope has an orientation opposite to the orientation of the object, as illustrated in Fig. 7.8. This is not a severe issue in microscope applications and not worth the cost and effort to correct. Different objectives and eyepieces can often be paired together to change the magnification and field of view of a microscope.

Microscope specifications consist of the specifications of the objective and the eyepiece. In many applications, large magnification is highly desired. That leads to a very small working distance, very small eye relief distance, very small exit pupil size, and uncomfortable use.

7.6 Telescopes

Similar to microscopes, a telescope consists of an objective and an eyepiece. There are mainly two types of telescope: refractive (i.e., use refractive lenses) and reflective (i.e., use reflective mirrors).

7.6.1 Refractive telescopes

Refractive telescopes can be constructed using lenses. There are two basic types of refractive telescope: Galilean and Keplerian, the latter of which is an improvement of the former.

Figure 7.9(a) illustrates a Galilean telescope, which consists of a positive objective with focal length f_O and a negative eyepiece with focal length f_E. The focal point of the eyepiece is placed slightly inside the focal length of the object so that the rays output from the eyepiece are slightly divergent, and the back extensions of these rays form a positive virtual image, as shown in Fig. 7.9(a).

In Fig. 7.9(a), θ_{Ac} is the "actual angle" that is imposed by the object being viewed to the telescope optical axis and is usually only a few degrees, and θ_{Ap} is the apparent angle that is imposed by the virtual image formed by the eyepiece to the telescope optical axis. The magnification of the telescope is defined as θ_{Ap}/θ_{Ac}. Telescopes are used to view distant objects of unknown size, so it is not realistic to define the magnification by the ratio of the object size to the image size, as a microscope objective does.

The incident rays from these distant objects are virtually parallel. The image distance of the objective of a telescope can be considered as f_O and does not have any meaningful change when viewing different objects since the objects being viewed are far away. The object distance of the eyepiece is slightly smaller than f_E and can be treated as equal to f_E. With the help of some geometrical optics, it can be shown that $\theta_{Ap}/\theta_{Ac} \approx f_O/f_E$. Note that f_O/f_E is a constant, whereas θ_{Ap}/θ_{Ac} can be slightly adjusted by moving the eyepiece.

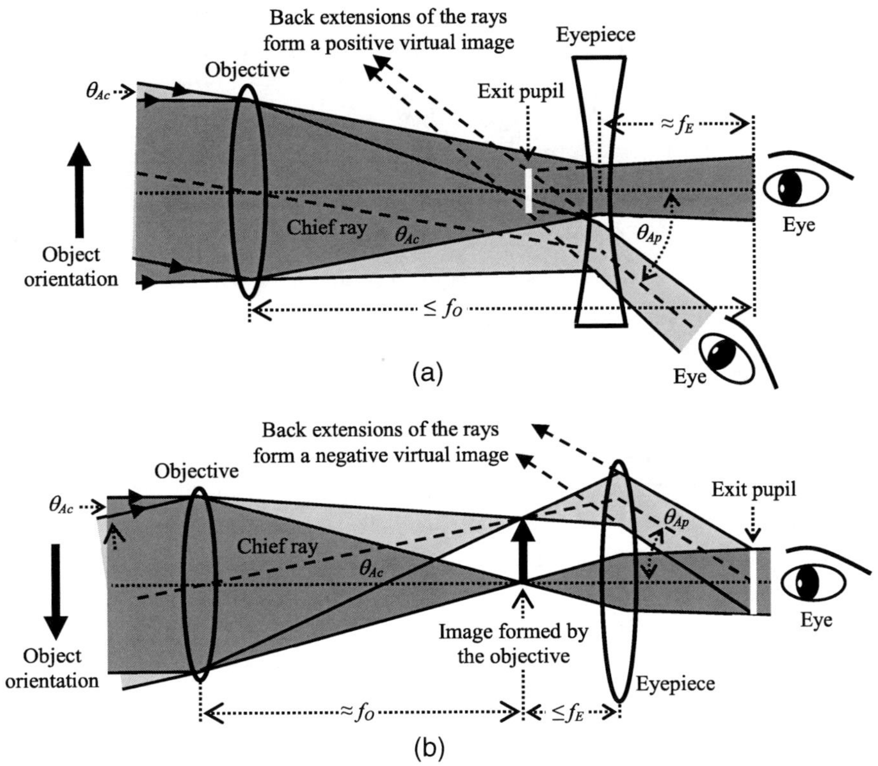

Figure 7.9 (a) Schematic of a Galilean telescope. (b) Schematic of a Keplerian telescope. In both (a) and (b), the eyepiece views the image formed by the objective. θ_{Ac} and θ_{Ap} are the actual and apparent angles, respectively, and f_O and f_E are the focal length of the objective and eyepiece, respectively. The magnification of the telescope is defined by $\theta_{Ap}/\theta_{Ac} \approx f_O/f_E$. The θ_{Ac} shown here are exaggerated for illustration purposes.

The exit pupil of a Galilean telescope is somewhere inside the telescope, as marked by the white vertical line in Fig. 7.9(a). Viewer must move their head to see the images of different field angles, as illustrated by the "eye" in Fig. 7.9(a). This is the main shortcoming of Galilean telescopes and the main reason why they are no longer used.

Figure 7.9(b) shows the schematic of a Keplerian telescope that consists of a positive objective and a positive eyepiece with focal lengths f_O and f_E, respectively. The advantage of Keplerian telescopes is that the exit pupil is outside the telescope. Viewers only need to move their eyeball to comfortably view images of different field angles. The disadvantage of Keplerian telescopes is that the image formed is negative, which does not matter much if the telescope is used for astronomy. Keplerian telescopes are still in use. The magnification of a Keplerian telescope is also $\theta_{Ap}/\theta_{Ac} \approx f_O/f_E$.

A large magnification over 50 is highly desired for telescopes and can be achieved by using a very long f_O, often over 1 m, and a very short f_E, often only a few millimeters. Because of the long focal length, telescope objectives are often much larger but much simpler than microscope objectives. Keplerian telescope eyepieces are similar or identical to microscope eyepieces.

The real objective and eyepiece of both Galilean and Keplerian telescopes can have more than one element to minimize various aberrations.

7.6.2 Reflective telescopes

Reflective telescopes can be constructed using mirrors. There are several similar versions of reflective telescopes. The Zemax raytracing diagram of a basic structure of a Cassegrain telescope is plotted in Fig. 7.10. There are two mirrors. A primary concave mirror focuses incoming rays from the object being viewed onto a secondary convex mirror. The secondary mirror further focuses the rays through a hole at the center of the primary mirror onto a sensor to form an image. Cassegrain telescopes do not create an image for direct viewing by eyes. The orientation of the image is still opposite the orientation of the object, as marked in Fig. 7.10.

Similar to refractive telescopes, the magnification of a Cassegrain telescope is given by $\theta_{Ap}/\theta_{Ac} \approx f_P/f_S$, where θ_{Ap} and θ_{Ac} are the apparent and actual angles, respectively, as marked in Fig. 7.10, and f_P and f_S are the focal lengths of the primary and secondary lenses, respectively.

Since spherical mirrors have spherical aberration and no color aberrations, at least one of the two mirrors must have aspheric surfaces to minimize the spherical aberration. For example, the primary mirror can have a parabolic surface and the secondary mirror can have a hyperbolic surface.

Compared with a refractive telescope, the advantage of reflective telescopes is that the mirrors have no color dispersion. The disadvantage of reflective telescopes is that the secondary mirror blocks a portion of the incoming light energy.

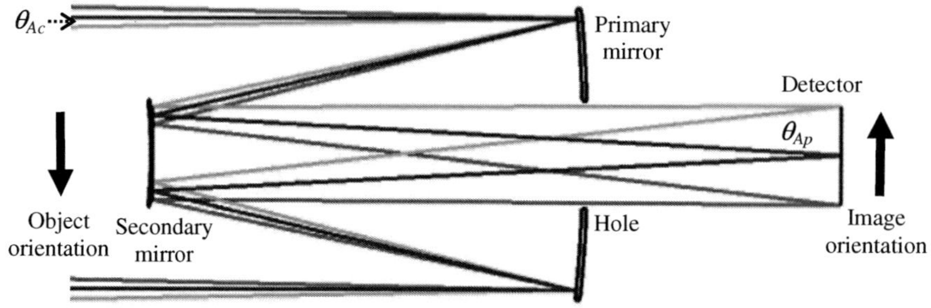

Figure 7.10 Raytracing diagram of a Cassegrain telescope. The concave primary mirror focuses rays from a distant object onto a convex secondary mirror. The secondary mirror further focuses the rays through a hole at the center of the primary mirror onto a detector.

7.6.3 Catadioptric telescopes

The term "catadioptric" means that the refractive and reflective elements are combined. Since some telescopes use a catadioptric approach, this topic is included in the telescope section.

Figure 7.11 shows the Zemax raytracing diagram of a catadioptric telescope that consists of a spherical primary mirror and a glass spherical doublet. The back surface of the doublet is high-reflection coated to act as the secondary mirror. Rays reflected by the primary mirror pass through the doublet, are reflected by the back surface of the doublet, pass through the doublet again, and finally reach the detector. The doublet helps to reduce the spherical aberration, and thus the primary does not have to be aspheric and the total cost of the telescope is reduced.

7.7 Binoculars

Binoculars consist of an objective and an eyepiece, and are used to view objects far away, similar to telescopes. The main difference between binoculars and telescopes is that the image generated by the eyepiece of binoculars must have the same orientation as that of the objects because binoculars are used to view objects in daily life, so a reversed image would be annoying. Binoculars also have a magnification of only several times, much smaller than that of telescopes.

Figure 7.12 shows a Zemax-generated raytracing diagram for a binocular. Note that the diagram still shows an image orientation opposite the object orientation. In real binoculars, a pair of Porro (right-angle) prisms are used to reverse the image orientations in both the vertical and horizontal directions, as previously illustrated in Fig. 6.18. When used in binoculars, these right-angle prisms are called Porro prisms. When designing binoculars, the Porro prisms pair is modeled by two glass blocks that have no image-reversing functions,

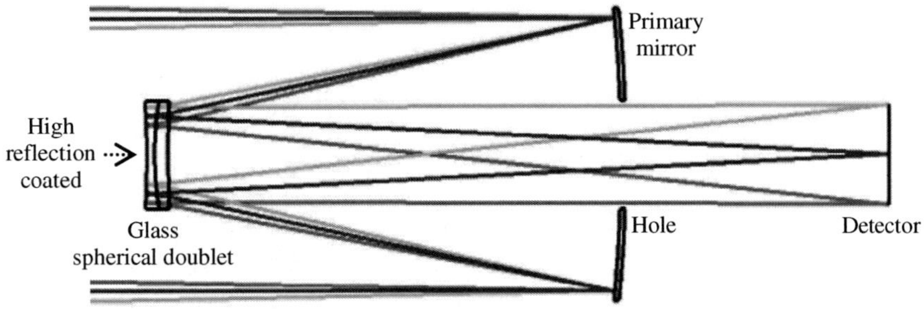

Figure 7.11 Raytracing diagram of a Catadioptric telescope that consists of a spherical primary and a glass spherical doublet whose back surface is high reflection coated to act as the secondary mirror.

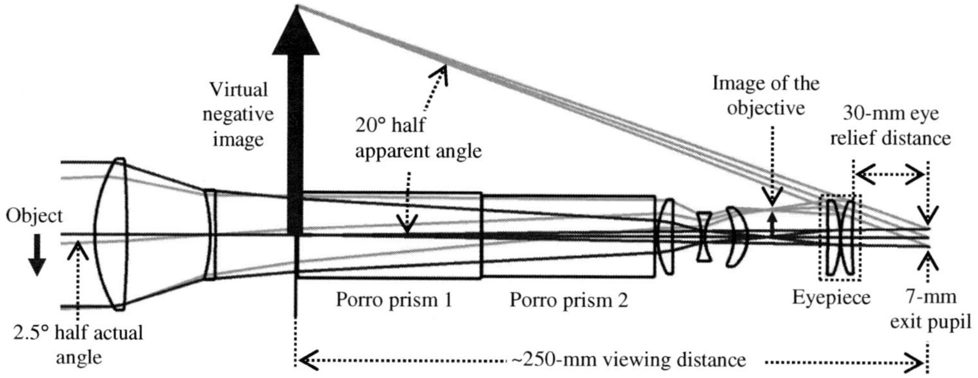

Figure 7.12 Zemax-generated raytracing diagram of a binocular that has a 2.5° half actual angle, 20° half apparent angle, a magnification of 20/2.5 = 8X, a 30-mm eye relief distance, a 7-mm exit pupil size, and a 250-mm image distance. The two Porro prisms used to reverse the image orientations are modeled by two glass blocks.

like those shown in Fig. 7.12. The Porro prisms are very thick glass blocks that have strong dispersion effects and should be included in the optical design process for accurate results.

The last two elements inside the dotted rectangle is the eyepiece; all other elements belong to the objective. Just like telescopes, the magnification of a binocular is the ratio of the apparent angle to the actual angle and is approximately equal to the ratio of the objective focal length to the eyepiece focal length. The magnification can be changed by adjusting the spacing between the objective and the eyepiece.

This binocular has $CTF > 0.3$ at 8 line pair/mm for images 250 mm from the viewer's eye. Such image quality matches the human-eye resolution limit, as will be explained in Section 7.13.4.

7.8 Gun Scopes

Like binoculars, gun scopes are used to view distant objects, and the image generated must have the same orientation as the object. However, the Porro prism pair used in binoculars folds the optical path twice and makes binoculars short and fat, and the optical axes of the objective and the eyepiece are not along the same line. These characteristics of binoculars are not ideal for gun scopes. A gun scope should be a thin and long tube that can be comfortably mounted on a rifle. The objective and eyepieces should be along the same ling. Therefore, a gun scope uses a relay element group to reverse the image rather than a Porro prism pair. This approach complicates the scope's optical structure.

Figure 7.13 shows a Zemax-generated raytracing diagram for a gun scope. The relay element group is inside the dotted rectangle. The orientations of the

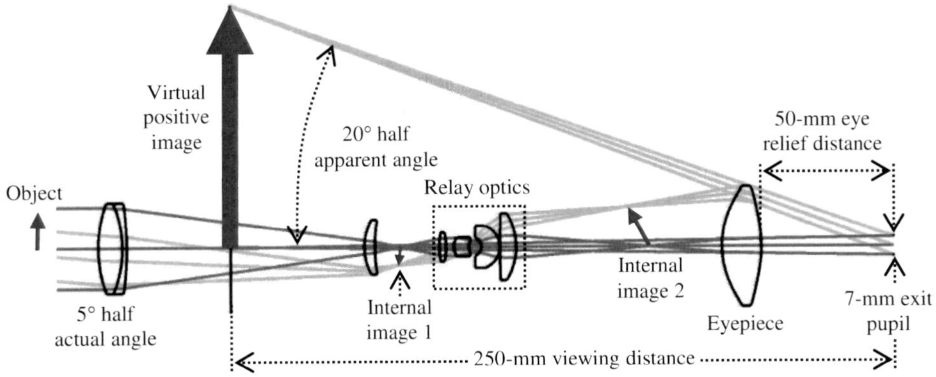

Figure 7.13 Zemax-generated raytracing diagram of a gun scope that has a 5° half actual angle, a 20° half apparent angle, a magnification of 20/5 = 4X, a 7-mm exit pupil size, a 50-mm eye relief distance, and a 250-mm image distance.

two internal images before and after the relay lens group have opposite directions that confirm the reversal of the image. The two internal images are tilted because of the field curvatures, as shown in Fig. 7.13. This is not a problem so long as the final image generated by the eyepiece has high quality. The elements in front of and behind the relay element group are the objective and the eyepiece, respectively.

The long eye relief distance of 50 mm reduces the impact of recoil while shooting. As can be seen in Fig. 7.13, the size of the eyepiece is proportional to the eye relief distance. Larger eyepieces tend to have larger spherical aberration and are more difficult to fabricate. The eyepiece used in this gun scope has two aspheric surfaces.

This gun scope has $CTF > 0.3$ at 8 line pair/mm for images 250 mm from the viewer's eye. Such image quality matches the human-eye resolution limit, as will be explained in Section 7.13.4.

7.9 Camera Lenses

Camera lenses are the most widely used lenses. Every cell phone has a miniature camera lens in it. Most of the commonly used lens specifications described in Section 7.1 apply to camera lenses. This section plots raytracing diagrams for four camera lenses and describes the basic features of these camera lenses.

7.9.1 *F*/2.8, 12-mm-focal-length lens

Figure 7.14 shows a Zemax-generated raytracing diagram for a *F*/2.8, 12-mm-focal-length camera lens. This lens uses one aspheric element (the front one), three spherical singlets, two spherical doublets, and one spherical triplet.

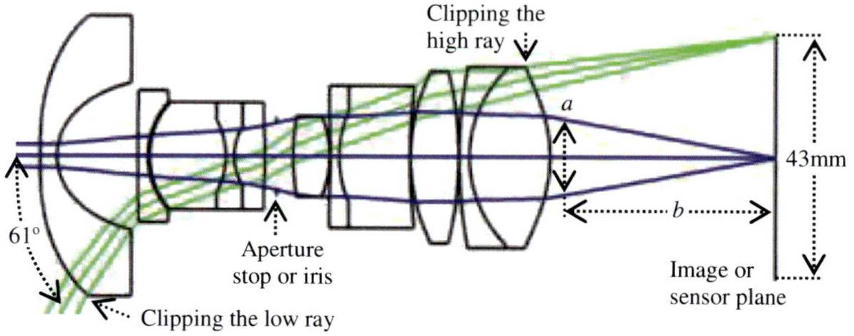

Figure 7.14 A Zemax-generated raytracing diagram of a camera lens that has a 12-mm focal length, $F/2.8 = b/a$, ±61° field of view, and 43-mm full image size.

All of these elements are made of several types of optical glasses. The lens has a very large field of view of ±61°. An aspheric element is very effective at handling such a large field of view. The edge portions of the full-field ray bundle are clipped by the first and last element, respectively, because these rays have large aberrations and will ruin the full-field image quality. This is a common practice in camera lens design. An adjustable iris is mounted at the aperture stop position of the lens to control the amount of the light reaching the sensor or adjusting the *F/#*. This camera lens produces images with $CTF \geq 0.3$ at 50 line pair/mm, and an image distortion ≤5%.

7.9.2 Cellphone camera lens

Figure 7.15 shows a Zemax-generated raytracing diagram for a cellphone lens. All five elements used are aspheric and made of polymers. Four of the five elements have strong aspheric profiles that are not suitable for glasses. The front surface aperture of the first element serves as a fixed aperture stop of the lens. The imaging brightness is automatically adjusted by the electronics that controls the sensor.

This cellphone lens has a ±32.5° field of view, 4.3-mm focal length, 5.7-mm image size, *F*/2.4, images with $CTF \geq 0.3$ at 250 line pair/mm, and an image distortion ≤1%.

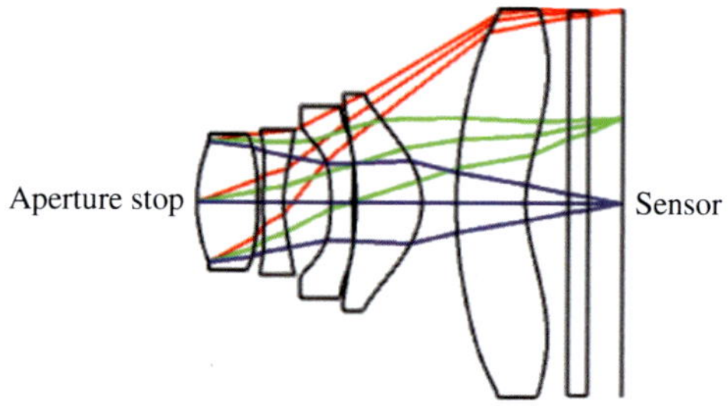

Figure 7.15 A Zemax-generated raytracing diagram of a cellphone camera lens.

7.9.3 Focusable camera lens

Figure 7.16 shows a Zemax-generated raytracing diagram for a focusable lens. Most focusable lenses change the focus by moving the front element (group). To focus at a closer distance, the element (group) must be moved forward. When focused at 0.5 m, the focal length of the lens is 44.8 mm; when focused at 10 m, the focal length is 59.1 mm. The *F*/4 lens produces images with $CTF > 0.3$ at 150 line pair/mm and an image distortion <3%.

7.9.4 6X zoom camera lens

Figure 7.17 shows a Zemax-generated raytracing diagram for a 6X zoom lens. To change the zoom, two element groups are usually moved simultaneously with different distances. This zoom lens has the two groups moving in the opposite directions, whereas some other designs have the two groups moving in the same direction. This lens produces images with $CTF > 0.3$ at 150 line pair/mm and an image distortion <3%.

7.10 Projection Lenses

A projection lens is mostly a camera lens used in a reverse way: it focuses rays from a small and bright object (or a display device or computer screen) onto a large display a few meters away, as illustrated in Fig. 7.18.

However, there are differences between a camera lens and a projection lens. In a camera lens, the light energy from a large object is concentrated onto a much smaller and very sensitive sensor. The brightness of the image is rarely a concern. In a projection lens, the light energy from a small object is projected onto a much larger screen and is significantly diluted. The brightness of the

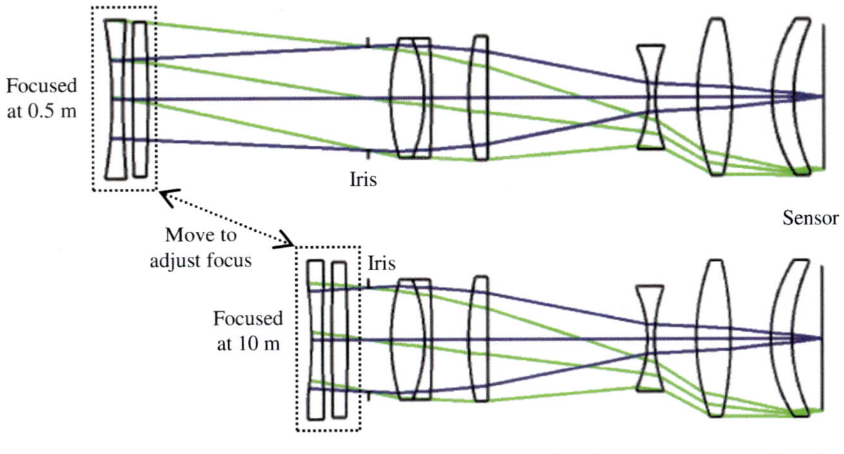

Figure 7.16 A Zemax-generated raytracing diagram of a focusable lens. The focusing adjustment is achieved by moving the front element group, marked by the dotted-line frame. This lens consists of one glass spherical doublet and six glass spherical singlets.

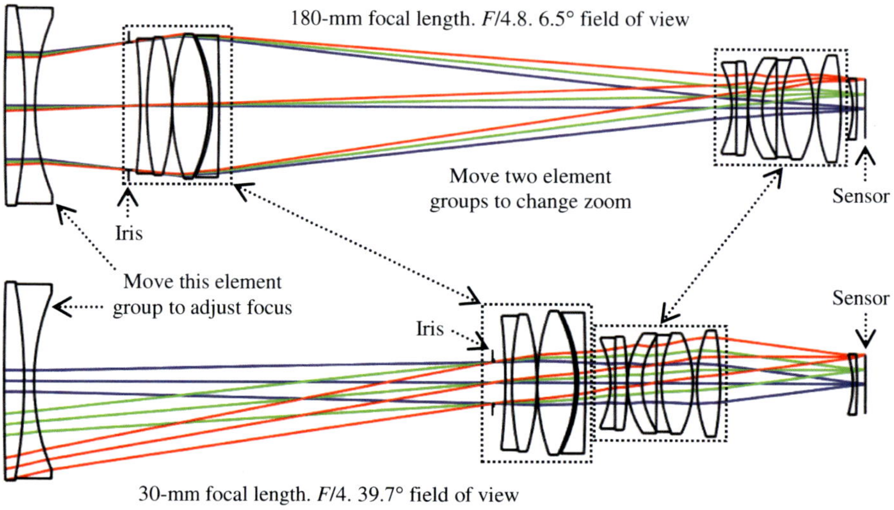

Figure 7.17 A Zemax-generated raytracing diagram of a 30 mm/180 mm focal-length 6X zoom lens. This lens consists of four glass spherical doublets and five glass spherical singlets. The changing of zoom is achieved by simultaneously moving two groups of elements. The iris is moved together with the left-side moving group. The iris size is adjustable to maintain the *F*-number for different focal lengths. The front element group can be moved independently of the two zoom-element groups to adjust the focus.

image is a major concern. The light-collecting power and throughput of a projection lens must be good.

In most projection lenses, the object is often illuminated by a powerful illumination source that produces a lot of uneven heat. The projection lens must be thermal stabilized well for a large temperature range and large temperature gradient across the lens. In addition, additional illumination optics may be needed to generate even illumination on the object.

To maximize the light-collecting power, the project lens should have a "telecentric" feature, i.e., the chief rays of all fields from the object and entering the lens are in the normal direction of the object, as illustrated in Fig. 7.18, because most objects have their largest radiance in the normal direction.

7.11 Inspection Lenses

There are many types of inspection lenses, depending on the applications. Generally speaking, inspection lenses are of very high quality. They look at an object at a close distance and produce images with high resolution and low distortion. Figure 7.19 shows a Zemax-generated raytracing diagram of a 5X inspection lens. This lens has a near-diffraction-limited (nearly perfect image resolution) $CTF > 0.3$ at 60 line pair/mm, allowed by a large *F*-number of *F*/18 and a small image distortion of 0.2%.

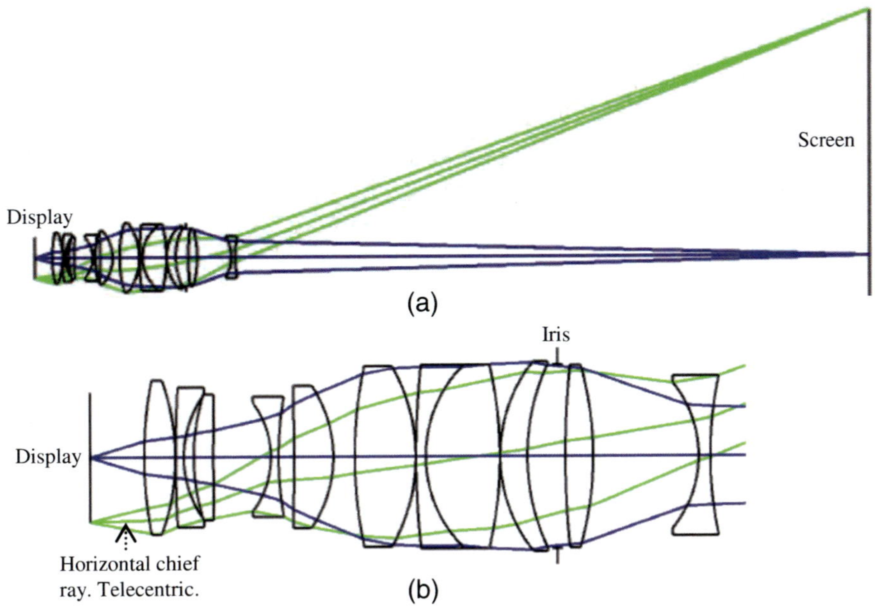

Figure 7.18 (a) Zemax-generated raytracing diagram of a projector lens that forms the image of a display on a large screen. (b) The detailed view of the lens layout. The lens has a relatively small *F*-number of *F*/1.6 to collect more radiance from the display. The light-collecting cone is about in the normal direction of the display since the maximum radiance of most displays is in the normal direction. An iris is placed at the location of the aperture stop to adjust the *F*/# of the lens.

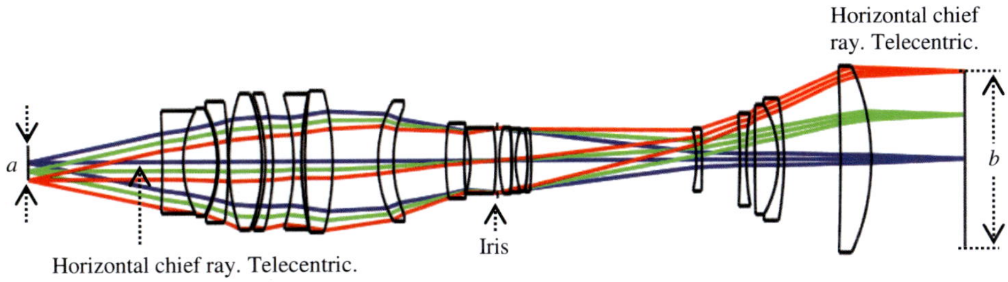

Figure 7.19 Zemax-generated raytracing diagram of a $b/a = 5$X inspection lens. This lens is telecentric in both object and image spaces.

Because the image size b is five times larger than the object size a, a spot on the object plane will have a five-times-larger image spot on the image plane. Based on the concept of etendue, the focusing cone angle at the image space should be one-fifth of the light-collecting cone angle at the object space. Such a cone angle difference can be visually seen in Fig. 7.19. This inspection lens is telecentric at both object and image spaces and can be used in a reverse way with a magnification of 0.2X. In such a case, the image resolution will be $CTF > 0.3$ at 300 line pair/mm.

7.12 Endoscopes

Endoscopes are imaging lenses that have a long tube shape and can go through a small opening and see the object behind the opening. Endoscopes are used in industrial and medical applications. There are two types of endoscopes: rigid and flexible.

7.12.1 Rigid endoscope

A rigid endoscope is a long metal tube several millimeters in diameter and a few hundred millimeters long with a lens system mounted inside the tube. The layout of an endoscope lens system is plotted in Fig. 7.20, which consists of an objective, four relay lenses, and an eyepiece. The relay lenses have very small diameters and are very thick to fit inside a thin tube and increase the image delivery distance, respectively. The large thickness is also beneficial to limit the lens tilt when mounting the lenses inside the metal tube. The endoscope length can be adjusted by adding or removing relay lenses.

An objective feeds the light from the object into the endoscope. An eyepiece picks up the output light from the endoscope and forms a virtual

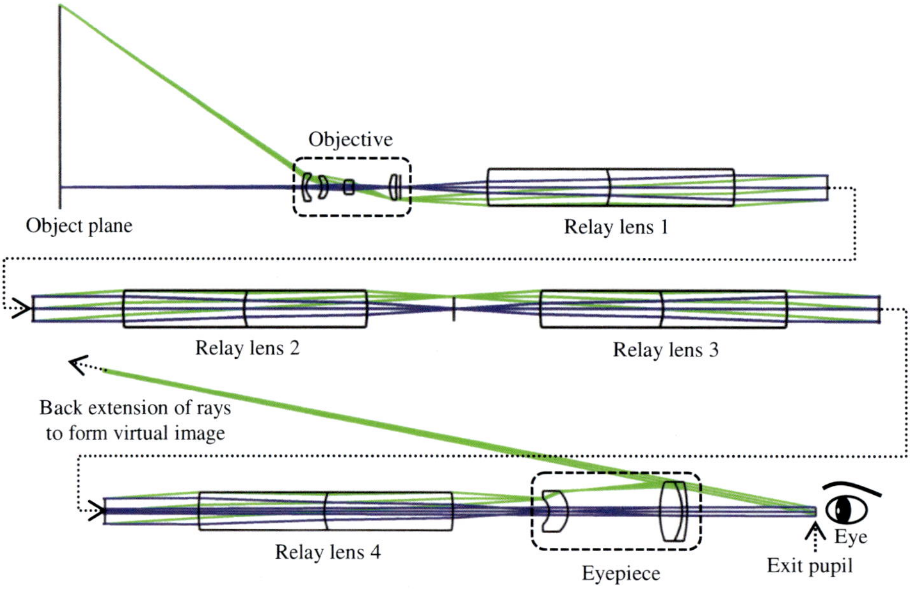

Figure 7.20 Zemax-generated raytracing diagram of a rigid endoscope. Because the diagram is too long to fit this book page, it is broken into three sections. An objective focuses the light from the object being viewed onto a plane, and a relay lens group picks up the light. Several identical relay-lens groups transmit the image through the long tube. An eyepiece picks the output light from the last relay lens group and generates a magnified positive virtual image for the user to view. Both the objective and the relay lens must have a small diameter of a few millimeters for easy insertion into a small opening. The number of relay lenses used is determined by the relay distance. The eyepiece diameter can be a little larger.

image for the user to view. The orientation of the image can be made the same as the object orientation by carefully designing the relay lens groups and selecting the right number of relay lenses. The eyepiece can be replaced by a focusing lens to focus the light from the endoscope onto a 2D sensor for display on a computer screen.

The difficult part of designing an endoscope is that all of the lens elements used must be very small, i.e., only a few millimeters in diameter. The small size of the relay lenses and the multiple relays inside the metal tube reduce the image quality. Generally speaking, the endoscope image resolution is below the human-eye or sensor-pixel resolution limit.

7.12.2 Flexible endoscope

The schematic of a flexible endoscope is shown in Fig. 7.21. An objective focuses the light from the object being viewed onto a bundle of optical fibers. The fiber bundle consists of up to tens of thousands fibers. At the input end, one fiber acts like a sensor pixel. At the output end, one fiber acts like a display pixel.

Thereby, the fiber bundle delivers an image. An eyepiece magnifies the image delivered by the fiber bundle for the user to view. The eyepiece can also be replaced by a focusing lens if a 2D sensor is used to accept the image instead of a human eye.

The advantage of a flexible endoscope is that the flexible fiber bundle can pass through a zigzag path. The disadvantage is the low image resolution limited by the fiber number. Even up to tens of thousands, the fiber numbers is still not large enough to deliver a high resolution image. As a comparison, most 2D sensor arrays and 2D displayers nowadays have megapixels.

7.13 Human Eyes

7.13.1 Photopic curve

With a lighting condition of >3 luminance, human eyes have a spectral response as shown in Fig. 7.22. The spectrum covers a range from violet (~400 nm) to red (~700 nm) with the peak at green (~555 nm). This spectral range is the visible range, and such a curve is called the photopic curve.[1]

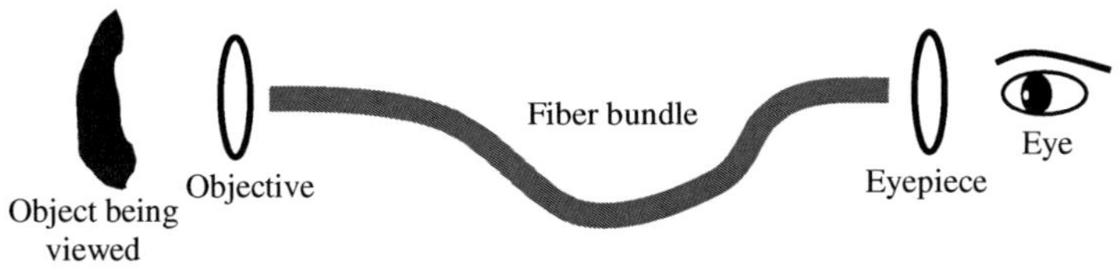

Figure 7.21 Schematics of a flexible endoscope that consists of an objective, an imaging fiber bundle, and an eyepiece.

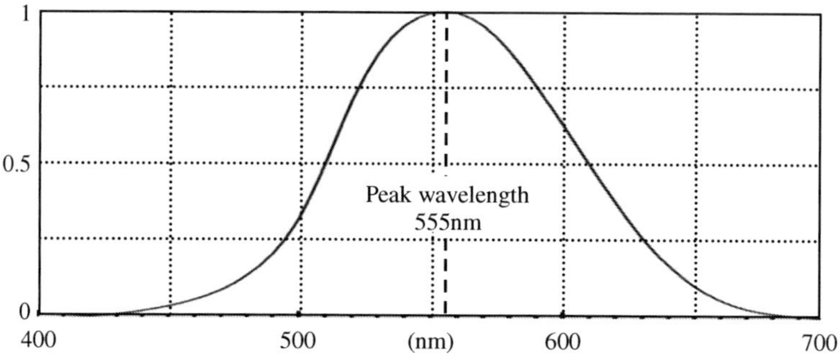

Figure 7.22 Photopic curve: human-eye spectral response.

With a dim light condition of <0.03 luminance, the human-eye spectral response is called the scotopic curve.[2] It has a shape and width similar to those of the photopic curve, only the peak is shifted from ~555 nm to ~505 nm.

7.13.2 Focal length and pupil size

Human eyes are the absolutely most widely used optical device and are focus-adjustable positive lenses with a basic focal length of ~17 mm. Some literature cite 22 mm as the focal length of human eyes because they are filled with water with a refractive index of 1.33, and a 17-mm focal length in air is 1.33×17 mm $\approx$ 22 mm in water,[3] as illustrated in Fig. 7.23. The comfortable viewing distance of human eyes is much larger than its focal length, usually considered to be >250 mm.

A human eye can only work as a positive lens to focus divergent rays entering the eye onto the retina to form image, but it cannot divert convergent rays entering the eye to focus onto the retina. Any lenses designed to generate images for human eyes to see must output divergent ray bundle. Conversely,

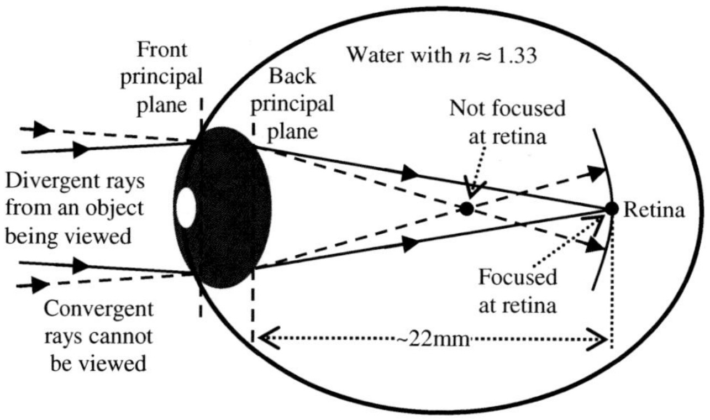

Figure 7.23 A human eye is a focus-adjustable positive lens with an in-water focal length of ~22 mm. A human eye can focus a divergent ray bundle (denoted by the solid lines) entering the eye onto the retina to form an image, but it cannot divert a convergent ray bundle (denoted by dashed lines) entering the eye to focus onto the retina to form an image.

any lenses designed to generate images for a sensor must output convergent rays that are focused onto the sensor.

Human eye pupil sizes vary from ~2 mm at bright illumination to ~8 mm in dark,[4] which is similar to a camera lens. When a camera is used to take pictures under bright illumination, its iris (aperture stop) size must be reduced, and vice versa. Any lenses that generate images for a human eye to view should have an exit pupil size larger than 8 mm for easy viewing. Aligning a human eye to an exit pupil smaller than 2 mm is not comfortable.

7.13.3 Visual field

Human vision can cover a total of about 200° of horizontal field by rotating the eyeballs, i.e., about 100° outward and 60° inward (towards nose), as illustrated in Fig. 7.24(a). Vertically, human eyes can cover a total of ~135° visual field by rotating the eyeballs, which is ~60° upward and ~75° downward.[5] The rectangular ratio of the human visual field is 200:135 ≈ 1.5:1. Since it is relatively easier for human beings to turn their head horizontally than vertically, most TV or computer display screens nowadays have a rectangular ratio of 16:9 ≈ 1.8:1 > 1.5:1.

7.13.4 Visual acuity

The nominal visual acuity of human eyes is about 1 arcmin at the forward direction and falls rapidly as the visual direction moves away from the forward direction,[6] as shown in Fig. 7.25 with a reciprocal unit of resolution. Visual acuity falls as the contrast of the object being viewed falls. Reference 7 provides an excellent illustration. Visual acuity also falls as the illumination level falls, and vice versa.

When designing a lens for a human eye to view, the most effective image resolution is the resolution that matches the human-eye visual acuity. Any finer image generated by the lens cannot be appreciated by human eyes and is wasted. For example, the binocular shown in Fig. 7.12 generates a virtual image at a distance of 250 mm. 1-arcmin eye acuity means 2 arcmin per line pair angular resolution, which leads to $(2/60) \times (\pi/180) \times 250\ \text{mm} = 0.146$ mm/line pair linear resolution at an image plane 250 mm away, or expressed in a reciprocal way, 1/0.146 mm ≈ 7 line pair/mm.

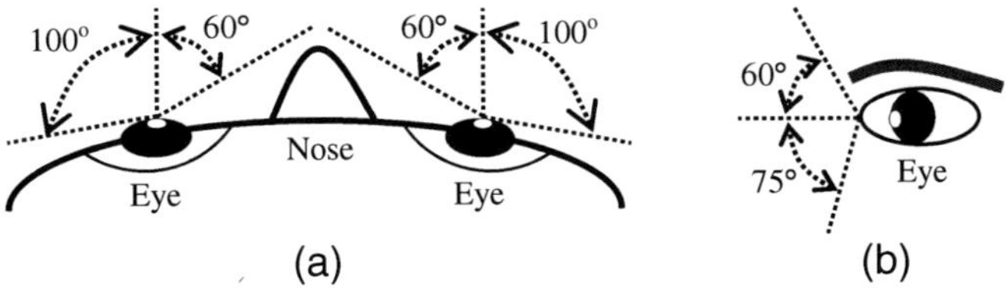

Figure 7.24 Illustration of human eye visual fields: (a) ~200° horizontal total and (b) ~135° vertical total.

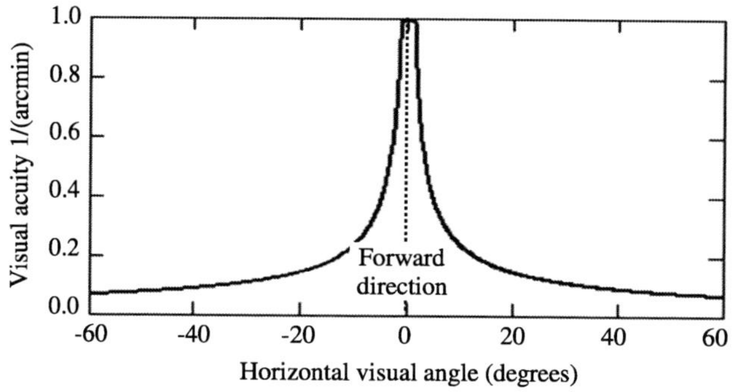

Figure 7.25 Average human-eye acuity expressed in a reciprocal way.

References

1. Wikipedia, "Photopic vision," wikipedia.org/wiki/Photopic_vision
2. J. Schwiegerling, "Photopic $V(\lambda)$ and Scotopic $V'(\lambda)$ Response," *Field Guide to Visual and Ophthalmic Optics*, SPIE Press, Bellingham, WA (2004) [doi: 10.1117/3.592975.p12].
3. D. A. Atchison and G. Smith, "The Equivalent Power and Focal Length," *Optics of the Human Eye*, Butterworth-Heinemann, Oxford, UK, pp. 7–8 (2000).
4. D. A. Atchison and G. Smith, "Pupil Size," *Optics of the Human Eye*, Butterworth-Heinemann, Oxford, UK, p. 22 (2000).
5. R. H. Spector, "Visual Fields," Chapter 16 in *Clinical Methods: The History, Physical, and Laboratory Examinations*, 3rd ed., Butterworth, Boston (1990).
6. *Acuity*, michaeldmann.net/mann7.html
7. *3 Illumination*, Fig. 18, webvision.med.utah.edu/book/part-viii-gabac-receptors/visual-acuity

Chapter 8
Lasers

8.1 General Comments

The word "laser" is the acronym for "light amplification by stimulated emission of radiation." Since the invention of the first laser in 1960, lasers have rapidly gained wide application in industry, communications, scientific research, medical, military, and entertainment. The current global laser market is about ten billion dollars. There are several types of lasing mechanisms and lasing materials. The emission type of lasers ranges from an extremely narrow pulse of femtoseconds to continuous wave. The power range is from submilliwatts to megawatts. The wavelength range is from tens of nanometers in the UV to tens of microns in the IR. Among all of these lasers, laser diodes are the most widely used lasers and account for about half of the laser market. Laser diodes and fiber lasers are the two fastest-growing sectors.

Laser types include gas, solid state, semiconductor (laser diodes), fiber, metal vapor, chemical, dye, Raman, free electron, x-ray, etc. Hundreds of different types of lasers are commercially available. References 1 and 2 provide detailed studies of laser principles.

8.2 Working Principle

8.2.1 Planck relation

To study lasers, the Planck relation[3] (also called the Planck–Einstein relation) and Maxwell–Boltzmann distribution must be introduced. The Planck relation states that

$$E = h\nu = hc/\lambda, \tag{8.1}$$

where $h \approx 6.626 \times 10^{-34}$ joules·second is the Planck constant, ν is the frequency of light, c and λ is the light velocity and wavelength in a vacuum, respectively, and E is the energy that one photon with a frequency ν carries.

8.2.2 Maxwell–Boltzmann distribution

Consider a system that has several energy levels and contains a total of N particles; the Maxwell–Boltzmann distribution[4] can be written in a simple form as

$$\frac{N_i}{N} \sim \exp\left(-\frac{E_i}{kT}\right), \tag{8.2}$$

where N_i is the average number of particles at energy level E_i, $k \approx 1.38 \times 10^{-23}$ joules/K is the Boltzmann constant, and T is the equilibrium temperature of the system. Although particles in the system collide and gain or lose energy, Eq. (8.2) states that the average number of particles at a certain energy level is a constant that decreases if the energy increases, and vice versa. This is similar to how water at a higher altitude naturally flows to a lower altitude.

8.2.3 Gain medium, three- and four-level systems, and population inversion

Every laser must have a gain medium to generate laser light. Gain media usually have a three- or four-energy level system, as illustrated in Figs. 8.1 and 8.2, respectively, where the vertical axis is the energy level, and higher horizontal lines have higher energies.

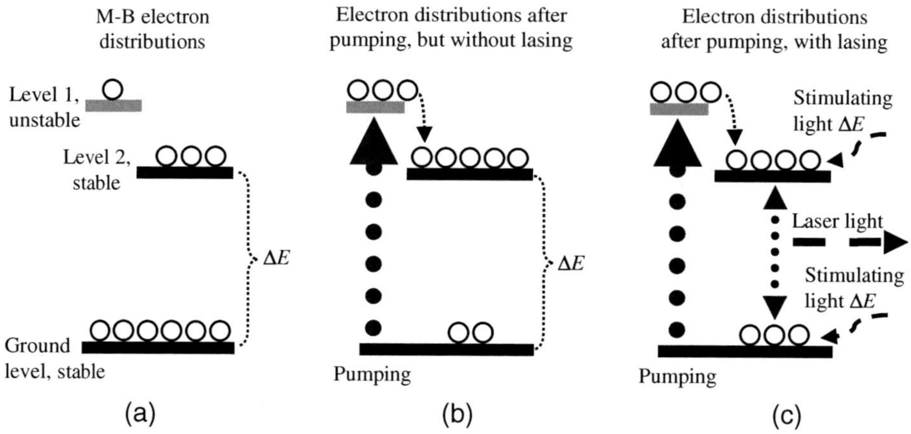

Figure 8.1 (a) A three-level system. The electrons obey the Maxwell–Boltzmann (M-B) distribution. Higher energy levels naturally have relatively fewer electrons. ΔE is the energy difference between level 2 and ground level. (b) Pumping electrons from the ground level to level 1. Because level 1 is unstable, the pumped electrons will fall to the stable level 2 and stay there. Then, level 2 has more electrons than the ground level. Electron populations at these two levels are inversed. (c) Light with energy ΔE stimulates electrons at level 2 falling to the ground level and emitting photons, and at the same time light stimulates electrons at the ground level jumping to level 2 and absorbing photons. Because of the electron population inversion, more electrons at level 2 than the electrons at the ground level are stimulated. A net optical gain is thereby achieved.

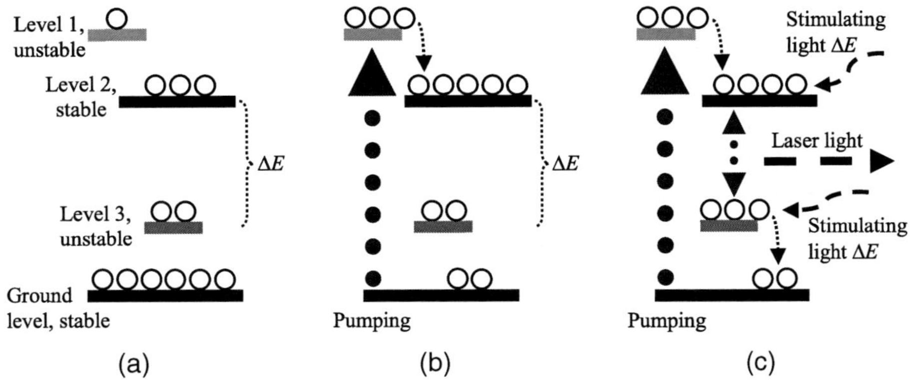

Figure 8.2 (a) A four-level system. The electrons obey the M-B distribution. Higher energy levels naturally have relatively fewer electrons. ΔE is the energy difference between levels 2 and 3. Levels 1 and 3 are unstable. The other two levels are stable. (b) Pumping electrons from the ground level to top level. Because the top level is unstable, the pumped electrons will fall to the stable level 2 and stay there. Then, level 2 has more electrons than level 3. The electron population is inverted between levels 2 and 3. Because level 3 is unstable, electrons tend to fall from this level to ground level. It is relatively easier to invert the population between levels 2 and 3. (c) Light with ΔE energy stimulates electrons at level 2 falling to level 3 and emitting photons, and at the same time stimulates electrons at level 3 jumping to level 2 and absorbing photons. Because of the electron population inversion, more electrons at level 2 than the electrons at level 3 are stimulated. A net optical gain is achieved.

In a three-level system, level 1 is not stable, but the other two energy levels are. The Maxwell–Boltzmann distribution states that higher energy levels naturally have relatively fewer electrons, as illustrated in Figs. 8.1(a) and Fig. 8.2(a). Electrons at the unstable level 1 tend to fall to the stable level 2 and stay there. A pumping mechanism can be applied to the gain medium to pump the electrons from ground level to level 1. These electrons then fall to level 2 by themselves, and the electron number at level 2 is larger than the electron number at ground level, as illustrated in Fig. 8.1(b). This phenomenon is called "population inversion"[5] and is a precondition for lasing.

If light with energy equal to ΔE is incident on the gain medium, where ΔE is the energy difference between level 2 and the ground level, the light will stimulate the electrons at level 2 to fall to the ground level and emit photons with energy ΔE, so that one electron emits one photon and increases the light intensity. At the same time, the light will stimulate the electrons at the ground level to jump to level 2 and absorb photons with energy ΔE, so that one electron absorbs one photon and decreases the light intensity. Since the electron population is inverted between level 2 and the ground level, more electrons at level 2 are stimulated than at the ground level. Thereby, a net optical gain is achieved, as illustrated in Fig. 8.1(c).

In a four-level system, levels 1 and 3 are not stable, and level 2 and the ground level are stable, as illustrated in Fig. 8.2(a). Electrons at level 1 and 3 tend to fall to level 2 and the ground level, respectively, and stay at level 2 and the ground level, as illustrated in Fig. 8.2(b). Therefore, it is relatively easier to achieve population inversion between levels 2 and 3 in a four-level system than in a three-level system. That is the main advantage of four-level systems over three-level systems. The lasing mechanism for the latter is the same as for the former. A population inversion between two energy levels is required.

Any materials that have a three- or four-level system and are nearly transparent at $\nu = \Delta E/h$ are potential laser media. A relatively small number of electrons may be thermally or otherwise disturbed, fall to the lower level, and emit light with energy ΔE. This light is called spontaneous emission and serves as the stimulating light to trigger the lasing process.

8.2.4 Pumping sources

Every laser needs a pumping source to provide the lasing energy by pumping electrons to higher levels. Some lasers have electrical current as pumping sources, for example, laser diodes. Some lasers have electrical discharge as the pumping sources, e.g., He-Ne lasers and other gas lasers. Some lasers have optical light from lamps as pumping sources, e.g., ruby lasers and some other solid state lasers. Some laser have laser light from laser diodes as the optical pumping source, e.g., fiber lasers, Nd-YAG lasers, and some other solid state lasers.

The efficiency of converting pumping energy to laser energy is low, from below 1% for the widely used He-Ne lasers to about 30% for other widely used laser diodes. Therefore, pumping sources must output power significantly greater than the desired laser power, and the pumping power should be concentrated on the laser medium.

8.2.5 Resonant cavity and threshold condition

Besides a gain medium and a pumping source to achieve population inversion, a resonant cavity is required to maintain lasing oscillation. There are various types of laser cavities. Figure 8.3(a) shows the schematic of a planar-planar laser cavity consisting of two glass plates with an optical distance nL in between them. A gain medium is inside the cavity. n is the refractive index of the gain medium. The right-side glass plate is coated with reflectance $R_1 < 1$. When a laser beam hit this reflector from inside the cavity, a portion of the beam transmits through the plate as the output laser beam. Another portion of the laser beam is reflected back. The left-side glass plate is high-reflection coated with a reflectance $R_2 \approx 1$. The medium has a linear gain and loss coefficient g and α, respectively. The gain is provided by the pumping source, and the loss is mostly the absorption of the gain medium.

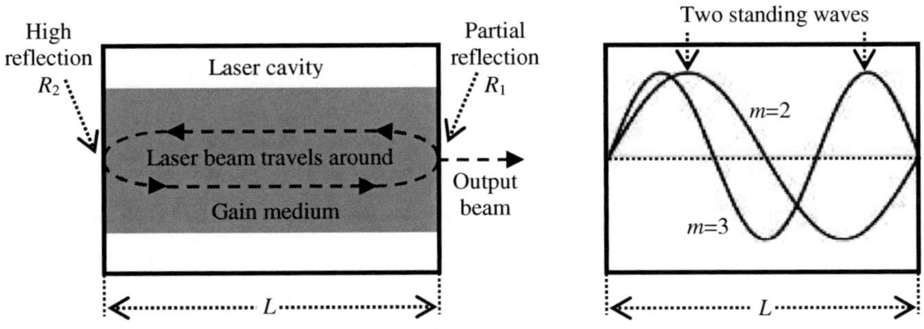

(a) (b)

Figure 8.3 (a) Schematics of a planar-planar laser cavity consisting of two glass plates with an optical distance nL between them, n is the refractive index of the gain medium, and L is the physical length of the cavity. The two plates are coated with a reflectance of $R_1 < 1$ and $R_2 \approx 1$, respectively. The medium has a gain and a loss coefficient g and α, respectively. The lasing condition requires $R_1R_2\exp[2L(g - \alpha)] \geq 1$. (b) Two standing waves with $m = 2$ and 3, respectively, in a cavity.

Spontaneous emission light bounces back and forth inside the cavity. For every round trip inside the cavity, the light experiences a gain and loss given by $\exp(2Lg)$ and $\exp(-2L\alpha)$, respectively. For every round trip inside the cavity, the light is reflected once by the two glass plates, respectively, and the fraction of light energy remaining inside the cavity is given by R_1R_2. For every round trip inside the cavity, the total gain and loss of the light is given by

$$G = R_1R_2\exp[2L(g - \alpha)]. \tag{8.3}$$

$G = 1$ is the threshold condition. If $G > 1$, the light has more gain than loss and is amplified as it travels back and forth. The newly gained light has the same phase as the existing light. Such a light is laser light. Since $R_1 < 1$, a portion of the laser beam incident on the right-side plate from inside the cavity exits the cavity and becomes the output of the laser device. As the laser light is amplified, the output laser power continues to increase while the gain provided by the pumping is limited. Eventually, the laser power will stop increasing and remain at a certain level, which means that the value of G must change from >1 to $=1$, or the gain coefficient g in Eq. (8.3) must decrease as the laser power increases. Some types of laser, e.g., laser diodes, allow easy adjustment of laser power by adjusting the pumping magnitude (electrical current intensity).

8.2.6 Standing waves

When traveling back and forth inside the cavity, the laser light must repeat itself and add constructively to form a coherent light, which is the key property of laser light. The lasing wavelength must thus meet the standing wave condition given by

$$\frac{m\lambda_m}{2} = nL, \tag{8.4}$$

where m is any integer, λ_m is the corresponding wavelength, n is the refractive index of the gain medium, and L is the physical length of the cavity, as illustrated in Fig. 8.3(b). One cavity can theoretically support a large number of standing waves, but only the wavelengths of one or a few of the standing waves are inside the spectral band of the gain medium and can be amplified.

8.2.7 Cavity stability

To sustain stable lasing, a laser cavity must meet the stability condition, which means that laser light must be completely contained within the cavity and retrace its path after one round trip through the cavity. The mathematical form of the stability condition is given by[6]

$$0 \leq g_1 g_2 \leq 1, \tag{8.5}$$

where $g_1 = 1 - L/r_1$, $g_2 = 1 - L/r_2$, L is the spacing between the two reflectors that form the cavity (assuming the medium inside the cavity is air), and r_1 and r_2 are the surface curvature radii of the two reflectors, respectively. Equation (8.5) is plotted in Fig. 8.4.

Figure 8.5 shows the schematics of four types of stable cavity. The stability of these four cavities are examined using Eq. (8.5):

1. For the plane-plane cavity shown in Fig. 8.5(a), $r_1 = r_2 = \infty$, and $g_1 \times g_2 = 1 \times 1 = 1$ as marked in Fig. 8.4. This cavity is stable.

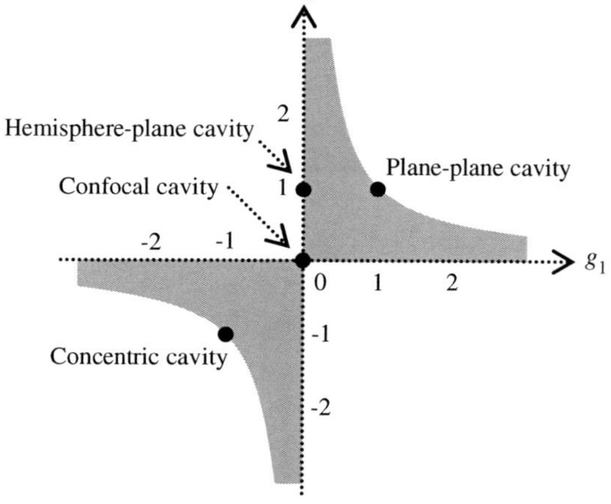

Figure 8.4 Graphic explanation of the cavity stability criterion Eq. (8.5). The shaded area meets Eq. (8.5) and is the stable area for laser cavities. The position of the four cavities shown in Fig. 8.5 are marked by black dots.

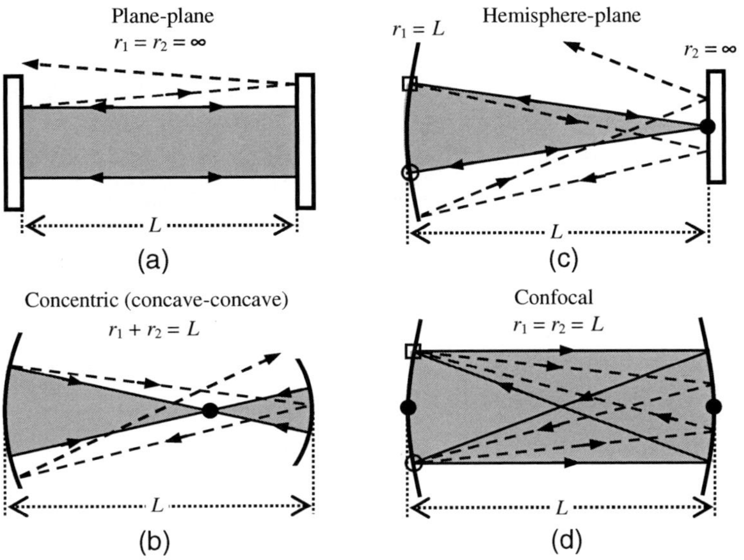

Figure 8.5 Four types of stable laser cavities. For each cavity, L is the spacing between the two reflectors (assuming the medium inside the cavity is air), r_1 and r_2 denote the surface curvature radius of the two reflectors, respectively, and the standing wave condition is assumed being met. (a) A planar-planar cavity. The two reflectors are planar and parallel. Rays denoted by the solid lines travel in the normal direction of the plates back and forth along the same path inside the cavity. Such a cavity is stable. However, once the ray deviates from the normal direction due to misalignment of the ray or the reflectors, as denoted by the dashed ray, the ray will have cumulative transverse displacement after every reflection and eventually leave the cavity after a certain number of round trips inside the cavity. Therefore, the parallelism of the two reflectors must be well aligned and maintained. (b) Hemispherical cavity. One spherical reflector with a radius L and one planar reflector form the cavity. The spherical curvature center denoted by the solid dot is on the planar reflector. Considering a ray denoted by the solid line originated from the open square on the spherical reflector, this ray travels along the radius to the solid dot position, and then is reflected by the planar reflector and travels along the other radius to hit the spherical reflector at the open circle position. This ray will trace back to the open square position along the same path and complete one round trip inside the cavity. This cavity is stable. Once the ray deviates from its original path due to misalignment of the ray or the reflectors, as denoted by the dashed line, the ray will no longer travel along any radius of the spherical reflector and will eventually escape the cavity. This cavity is also sensitive to alignment. (c) Concentric (concave-concave) cavity. The two spherical reflectors have a surface curvature center at the same spot marked by the solid dot, so $r_1 + r_2 = L$. Two rays denoted by solid lines travel along two radii of the cavity, respectively. This is a stable cavity. Again, once there is a deviation in the ray direction due to misalignment of the ray or the reflectors, as denoted by the dashed line, the deviation is increased after each reflection, and the ray will eventually escape the cavity. Such a cavity is still sensitive to alignment. (d) Confocal cavity. Two identical spherical reflectors with radius L and a distance of L in between them form the cavity. Each reflector has its curvature center on the other reflector, as marked by the two solid dots. Considering a horizontal ray denoted by the solid line starts from the open square, the ray will hit the open circle after being reflected once before returning to the open square after being reflected once more. This cavity is stable. If the ray from the open square is deviated as denoted by the dashed line, the ray will still return to the open square position after being reflected three times. Any rays will at least theoretically trace their way back to their original location. This cavity is the least sensitive to misalignment and is the most widely used.

2. For the hemisphere-plane cavity shown in Fig. 8.5(b), $r_1 = L$, $r_2 = \infty$, $g_1 \times g_2 = 0 \times 1 = 0$, as marked in Fig. 8.4. This cavity is also stable.
3. For the concentric (concave-concave) cavity shown in Fig. 8.5(c), $r_1 + r_2 = L$. A general situation is assumed that $r_1 = aL/(a+b)$ and $r_2 = bL/(a+b)$, where a and b are two arbitrary coefficients, then $g_1 \times g_2 = (-b/a) \times (-a/b) = 1$. Figure 8.4 illustrates the case of $a = b = 1$. This cavity is stable.
4. For the confocal cavity shown in Fig. 8.5(d), $r_1 = r_2 = L$, and $g_1 \times g_2 = 0 \times 0 = 0$, as shown in Fig. 8.4. This cavity is stable. More detailed analysis performed in Fig. 8.5 shows that the confocal cavity is the most stable cavity and is widely used in gas laser cavities and confocal Fabry–Pérot interferometer cavities.

The captions of Fig. 8.5 explain in more detail about the stability features of these four cavities. The analysis technique described in Fig. 8.5 and its caption can be used to analyze the stability of other cavity types.

All of the cavities discussed here do not have a waveguide mechanism associated with them. The cavity stability is therefore of vital importance. Some lasers have a waveguide-type cavity, e.g., fiber lasers and laser diodes, whereby the laser light is confined inside the cavity by the waveguide. In this case, cavity stability is not an issue.

8.3 Laser Specifications

Several parameters are used to describe a laser. The commonly used parameters are wavelength, laser power level, continuous wave or pulsed, and longitudinal mode and transverse mode. Some other parameters, such as linewidth, wavelength tunability, power modulation, Q-switched, mode-locking, etc., are also used sometimes.

8.3.1 Wavelength and tunability

Most lasers emit only a single wavelength. Based on the type of gain medium, the wavelength can cover from the UV to the mid IR. A few lasers can be tuned to emit a few wavelengths. For example, a multicolor He-Ne laser can alternatively emit green (543 nm), yellow (594 nm), orange (612 nm), and red (633 nm) by carefully tuning the specially designed laser cavity. A few lasers can be continuously tuned over a wavelength range of tens of nanometers, e.g., external cavity laser diodes. The following sections list many types of lasers according to their wavelengths to provide a complete view of wavelength selections.

8.3.2 Power and energy: continuous, pulsed, or tunable

The output optical power of lasers can be continuous wave (CW), or pulsed or tunable (modulatable), depending on the type of laser. CW output lasers are the most commonly seen, although pulsed output lasers are frequently used. The proper unit for pulsed output is energy per pulse. Tunable output power lasers are the least seen. Once the output power of a laser is tunable, it may be modulatable too.

In principle, all CW lasers can be pulsed or modulated by placing a modulatable absorber inside the laser cavity and then pulsing or modulating this absorber. In reality, most gas laser cavities are sealed glass tubes filled with a special gas. It is rare to place an absorber inside such a glass tube. Gas lasers can be pulsed by pulsing the electrical discharge pumping sources, but such pumping sources cannot be tuned or modulated. Laser diodes are pumped by DC electrical current, so it is easy to pulse, tune, or modulate the output power of laser diodes.

Most solid state laser cavities are open for easy access and can be pulsed or modulated using intracavity absorbers. Some solid state lasers use a flash lamp as the pumping source. Such lasers can be pulsed by pulsing the flash lamp. Some solid state lasers and all fiber lasers are pumped by laser diodes. It is easy to pulse, tune, or modulate these solid state lasers by applying the effect to the pumping laser diodes.

8.3.3 Longitudinal modes

A laser cavity with an optical length nL can support a large number of standing waves that meet the condition set by Eq. (8.4). After differentiating Eq. (8.4) and rearranging the result, the result is

$$\frac{\Delta\lambda}{\lambda_m} = -\frac{\Delta m}{m}, \tag{8.6}$$

where $\Delta\lambda = \lambda_m - \lambda_{m-1}$ is the cavity longitudinal mode spacing,, and $\Delta m = 1$ is considered here. Combining Eqs. (8.4) and (8.6) leads to

$$\Delta\lambda = -\frac{\lambda_m^2}{2nL}. \tag{8.7}$$

Equation (8.7) indicates that longitudinal mode spacing is inversely proportional to the cavity optical length. Figure 8.6 illustrates Eq. (8.7) for two cavities with different lengths nL_S and nL_L, respectively. The short cavity has much larger mode spacing than the longer cavity.

A given laser gain medium has a certain spectral bandwidth. Such a gain medium can support either a single or multilongitudinal mode to lase, depending on the laser cavity length, as illustrated in Fig. 8.7.

8.3.4 Transverse modes

Most laser gain media have a circular cross-section. A small cross-section can support only one transverse mode, but a large cross-section can support

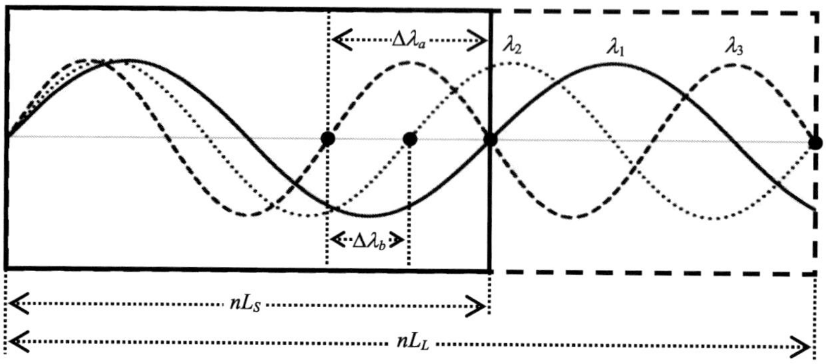

Figure 8.6 Three waves denoted by the solid, dot and dash curves with wavelength $\lambda_1 > \lambda_2 > \lambda_3$, respectively. For a short cavity with length nL_S, the solid and dashed waves are standing waves with a wavelength difference $\Delta\lambda_a = \lambda_1 - \lambda_3$, The dot line wave is not a standing wave. For a long cavity with length nL_L, the dotted and dashed waves are standing waves with a wavelength difference $\Delta\lambda_b = \lambda_2 - \lambda_3$. The solid wave is no longer a standing wave. It is obvious that $\Delta\lambda_b < \Delta\lambda_a$, so a longer cavity has smaller longitudinal mode spacing.

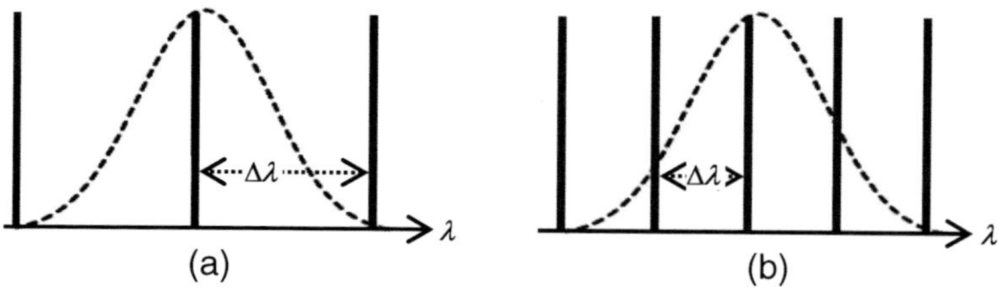

Figure 8.7 The dashed curve is the spectrum of a laser gain medium. The solid straight lines denote longitudinal modes supported by a laser cavity. (a) The larger mode spacing leads to single-mode operation. (b) The smaller mode spacing leads to multi- (three) mode operation.

multitransverse modes. This is similar to the transverse modes in an optical fiber. The transverse mode structures shown in Fig. 6.37(a) for a multimode optical fiber also well describe the transverse mode structure in lasers with a circular cross-section medium.

For semiconductor lasers, the gain medium is a P-N junction, also called an "active layer." The cross-section of active layers is rectangular. The two facets of the active layer form the resonant cavity. A large rectangular waveguide can also support a multitransverse mode, which is usually undesired.

The transverse mode structure in a rectangular waveguide is analyzed here using geometrical optics and with the help of Fig. 8.8. Note that the constructive interference condition requires that the wave repeats itself after being reflected twice by the interface of the waveguide. The mathematical form of this condition is[7]

$$\frac{2\pi n_1}{\lambda}(AC - AB) + 2\varphi_R = 2m\pi, \qquad (8.8)$$

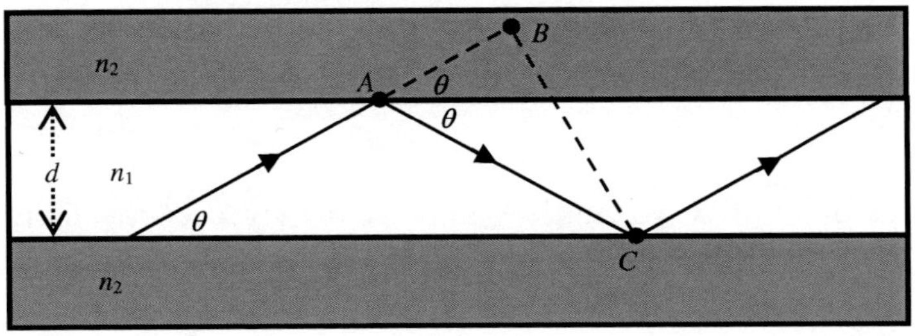

Figure 8.8 2D schematics of a rectangular dielectric waveguide has thickness d and refractive index n_1, and is surrounded by another material with index $n_2 < n_1$. A ray is bouncing in between the two interface of the waveguide. The constructive interference condition requires $AC - AB + 2\varphi_R = 2m\pi$, where AC and AB denote the distances between points A and C and points A and B, respectively, m is an integer, and φ_R is the reflection phase change occurring at the interface. Only the rays that meet this constructive interference condition can exist in the waveguide.

where λ is the wavelength in air, n_1 is the index of the waveguide, AC and AB are the distances between points A and C and points A and B, respectively, φ_R is the phase change caused by the reflection at the inner interface of the waveguide, and m is an integer. The details about φ_R are shown in Fig. 2.4. $\varphi_R = 0$ is assumed here for simplicity. With some triangular geometry, the two relations can be found to be

$$\frac{d}{AC} = \sin(\theta), \tag{8.9}$$

$$\begin{aligned} \frac{AB}{AC} &= \cos(2\theta) \\ &= \cos^2(\theta) - \sin^2(\theta). \end{aligned} \tag{8.10}$$

After combining Eqs. (8.9) and (8.10) to solve for AC and AB, inserting AC and AB found into Eq. (8.8), and rearranging the result, the final result is

$$\sin(\theta_m) = m\frac{\lambda}{2dn_1}. \tag{8.11}$$

Equation (8.11) indicates that only a series of discrete θ_m is allowed. $m < 1$ is the single transverse mode condition, which is $(2dn_1/\lambda)\sin(\theta_m) < 1$.

In Fig. 8.8, Snell's law and the total reflection condition take the form

$$n_1 \sin\left(\frac{\pi}{2} - \theta_m\right) = n_1 \cos(\theta_m) < n_2. \tag{8.12}$$

The maximum allowed reflection angle is then found to be $\theta_m \leq \cos^{-1}(n_2/n_1)$. In real laser didoes $n_1 - n_2 \ll 1$, and thus $\theta_m \approx 0$. In a dielectric waveguide where $\varphi_R \neq 0$, the mathematics involved is more complex. Reference 7 provides a detailed discussion about this topic.

8.3.5 Linewidth

The product of light wavelength λ and frequency ν is the light velocity c, which is a constant:

$$\lambda\nu = c. \tag{8.13}$$

Differentiating Eq. (8.13) and rearranging the result leads to

$$-\frac{\Delta\nu}{\nu} = \frac{\Delta\lambda}{\lambda}, \tag{8.14}$$

where $\Delta\nu$ is the linewidth in term of frequency with a unit Hz or so, chemists are used to use $\Delta\nu$, $\Delta\lambda$ is the laser linewidth in term of length with unit nm or so, physicists are used to use $\Delta\lambda$. In this book, both $\Delta\lambda$ and $\Delta\nu$ are used to denote linewidth. Any real laser light has a certain wavelength linewidth $\Delta\lambda$ or frequency linewidth $\Delta\nu$. The negative sign in Eq. (8.14) only means that λ is inversely proportional to ν, as shown in Eq. (8.13). The absolute value of $\Delta\lambda$ or $\Delta\nu$ is the value of concern.

For stable lasing, $G = 1$ in Eq. (8.3), which can be written as

$$2gL_G = \ln(R_1R_2) + 2\alpha L_G, \tag{8.15}$$

where L_G is the physical length of the section of the cavity that is filled with a gain medium with a refractive index n. The left- and right-side of Eq. (8.15) are the total gain and total loss, respectively, that occur in one round trip of light inside the cavity. The round-trip time of light inside a laser cavity is

$$\tau_R = \frac{2(nL_G + L_A)}{c}, \tag{8.16}$$

where L_A is the physical length of the other section of the cavity that is filled by air. Note that the inverse of the round-trip time is called "mode spacing." For every round trip of light inside a laser cavity, a fraction of the laser power exits the cavity and becomes the output power. An equation of photon lifetime τ_L can be written as

$$\tau_L = \frac{\tau_R}{\ln(R_1R_2) + 2\alpha L_G}. \tag{8.17}$$

Note that the inverse of the photon lifetime is called "cavity bandwidth."

The theoretically narrowest possible laser linewidth is given by Schawlow–Townes and the Melvin Lax equation[8]

$$\Delta\nu = \frac{n_{sp}h\nu}{2\pi P_{Out}\tau_L^2}, \tag{8.18}$$

where $\Delta\nu$ is the FWHM linewidth, n_{sp} is the "spontaneous factor" that can be neglected for the discussion here, h is the Planck constant, ν is the light frequency, $h\nu$ is the energy of one photon, P_{Out} is the output laser power, and τ_L is the photon lifetime given by Eq. (8.17).

In reality, several other types of noise increase the laser linewdith from the basic Schawlow–Townes linewdith, such as thermal noise, mechanical noise, pumping power noise, etc. The actual laser linewidth is often much larger than the basic linewidth given in Eq. (8.18). However, the equation still provides some important results, such as $\Delta\nu \sim 1/P_{Out} \sim 1/\tau_L{}^2 \sim 1/\tau_R{}^2 \sim 1/(nL_G + L_A)^2$. That means the laser linewidth decreases as the lasing power increases or the cavity length increases.

8.3.6 Coherent length

Any real laser light has a certain linewidth $\Delta\lambda > 0$. For illustration purposes, a simplified situation is considered that the laser line only contains two wavelengths λ_1 and λ_2, and $\lambda_2 > \lambda_1$, as shown in Fig. 8.9(a). The two wavelengths start with the same phase. After traveling a distance of

$$L_C = m\lambda_2 = (m+1)\lambda_1, \tag{8.19}$$

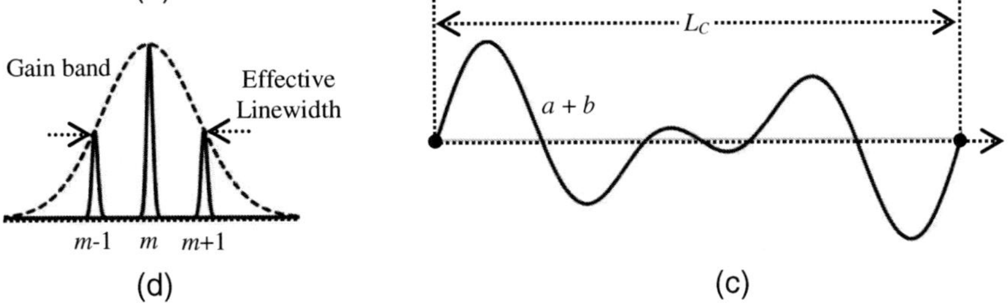

Figure 8.9 (a) A simplified laser line model that contains only two wavelengths λ_1 and λ_2. (b) Two waves *a* and *b* with wavelength λ_1 and λ_2, respectively. These two waves start with the same phase. After traveling a distance of L_C, they have the same phase again. (c) The amplitude sum of waves *a* and *b*. (d) Three longitudinal modes $m-1$, m, and $m+1$ are inside the band of a gain medium.

the two wavelengths have the same phase again, as illustrated in Fig. 8.9(b). L_C is the "coherent length" of this laser light. Equation (8.19) can be rearranged to take the form

$$m\Delta\lambda = \lambda_1, \tag{8.20}$$

where $\Delta\lambda = \lambda_2 - \lambda_1$. Multiplying both sides of Eq. (8.20) by λ_2 and rearranging the result leads to

$$\begin{aligned} L_C &= \frac{\lambda^2}{\Delta\lambda} \\ &= \frac{c}{\Delta\nu}, \end{aligned} \tag{8.21}$$

where $\lambda^2 = \lambda_1\lambda_2$ is the average of λ_1 and λ_2, and relations $\Delta\lambda/\lambda = \Delta\nu/\nu$ (neglect the minus sign here) and $\lambda\nu = c$ are used. Equation (8.21) is the widely used expression of coherent length, which is proportional to the inverse of the linewidth.

The amplitude sum of the two waves shown in Fig. 8.9(b) is plotted in Fig. 8.9(c), the wave shape is no longer sine or cosine. Figure 8.9(c) states that a laser light with linewidth $\Delta\lambda$ can be treated as single-wavelength light only if the light travel distance is much smaller than L_C. In applications involving interference, the optical path length difference between the two waves that interfere each other should be much smaller than L_C. Otherwise, the interference results can be erroneous.

The longitudinal modes of many lasers have a very narrow linewidth, but there are multi longitudinal modes, as illustrated in Fig. 8.9(d). The bandwidth of all these modes combined can be considered as effective linewidth. For example, He-Ne lasers usually have a linewidth of $\sim 10^3$ Hz and a bandwidth of about 1 GHz. The coherent length should be $\sim c/\Delta\nu \approx 3 \times 10^8$ m·s^{-1}/10^9 s^{-1} = 0.3 m. For comparison, Ng-YAG lasers have a typical linewidth of about 100 GHz, and the coherent length is $\sim$1 mm.

8.3.7 *Q*-factor and *Q*-switching

The widely used quality factor, or Q-factor, for a laser is defined as[9]

$$\begin{aligned} Q &= 2\pi \frac{\text{Laser energy stored in the cavity}}{\text{Laser energy lost in one round trip inside the cavity}} \\ &= 2\pi\nu\tau_L. \end{aligned} \tag{8.22}$$

The result $Q \sim \tau_L$ is understandable because a longer photon lifetime means less energy loss in one round trip of light inside the cavity. Note that a laser with a large Q-value does not necessarily mean a better-quality laser; it is just an indication of one characteristic of the laser.

Q-switching is a widely used technique to push a laser producing a pulsed output. The process is described below. A switchable attenuator at its high-attenuation state is placed inside the cavity of the laser to intentionally block the light traveling inside the cavity. That means that the Q-value is increased according to Eq. (8.22). The effect of continuous pumping is cumulative, and the consequence is a large population inversion, i.e., a large amount of energy stored in the gain medium. Then the attenuator is switched to low attenuation state, and an avalanche-type lasing process starts. The laser light intensity will dramatically increase because of the large amount of stored energy and its rapid depletion. Thereby, a laser pulse is generated.

8.3.8 Mode-locking

Many lasers have several longitudinal modes that run independently. The phase differences among them randomly vary. The phase, even the intensity, of the output laser beam, which is the sum of all these modes, can vary randomly. Such a laser beam is not very useful.

A technique called "mode-locking" can lock the phases of all these modes. The sum of these modes is a pulse. The GIF shown in Reference 10 provides an excellent illustration. In the picture, the phases of eight modes are locked, and the time evolution of the sum of these eight modes is a traveling pulse. Mode-locking can create pulses as short as femtoseconds. There are a few mode-locking methods. Among them, the active and passive versions are frequently used.

Active mode-locking is realized by placing an electro-optic modulator inside the laser cavity in the laser path. The modulator modulates the amplitude of the main mode with a frequency of $\Delta\nu$, which is selected to equal the mode spacing between two adjacent modes. This modulation creates two side modes $f + \Delta\nu$ and $f - \Delta\nu$, where f is the frequency of the main mode being modulated. The two side modes have the same phase as the phase of the main mode, have frequencies same as the frequencies of the two side modes adjacent to the main mode, respectively, and have an influence on these two side modes. Eventually, these two side modes will oscillate in phase with the main mode and thereby lock the phases of the three modes. Further operation of the modulator on the next two adjacent modes will phase lock those modes.

Passive mode-locking is realized by placing a saturable absorber inside the laser cavity in the laser path. The absorber has higher transmission to higher-intensity modes and acts as an amplitude modulator to create side modes. The following process is similar to active mode-locking.

8.4 Gas Lasers

8.4.1 General description

As the name suggests, gas lasers use certain types of gas or mixtures of gases as the gain media.[11] Gas lasers are pumped by high voltage electrical discharges, have low electrical to optical power conversion efficiency and are mostly designed to be operated by electrical power from wall sockets.

Gas laser cavities are usually a long and thin sealed glass tube filled with certain type(s) of gas. The two ends of the tubes are concave reflectors that form a confocal type cavity. The tube length can range from ten centimeters to a couple of meters. The long cavities often support more multilongitudinal modes but with small mode spacing and a narrower linewidth. A thin open-ended tube is often placed inside the sealed gas tube to reduce the cross-section size of the laser cavity and control the transverse mode numbers.

Gas lasers have a high beam spatial quality with a M^2 factor close to 1. Gas laser power can range from below 1 mW for a He-Ne laser to kW for a CO_2 laser. Most gas lasers output CW power, but some gas lasers can be pulsed. Most gas lasers output a fixed wavelength. But some gas lasers can output a few wavelengths, not simultaneously.

Among the many types of gas lasers, He-Ne lasers were the first invented and are the most widely used. Figure 8.10 shows the schematics of a He-Ne laser. A glass tube is filled with a mix of helium and neon gases and sealed. A DC power supply with ~900–2000-V pumps the He-Ne gas by discharging through an anode and a cathode.

Most He-Ne lasers output multilongitudinal modes at 0.6328 μm of red color with a bandwidth of about 1 GHz (0.001 nm). Some He-Ne lasers output yellow, green, or orange color lasers, or they can be switched among red, orange, yellow, and green. Commercial He-Ne laser power ranges from below 1 mW to tens of mW. The laser structure shown in Fig. 8.10 is common for other types of gas lasers.

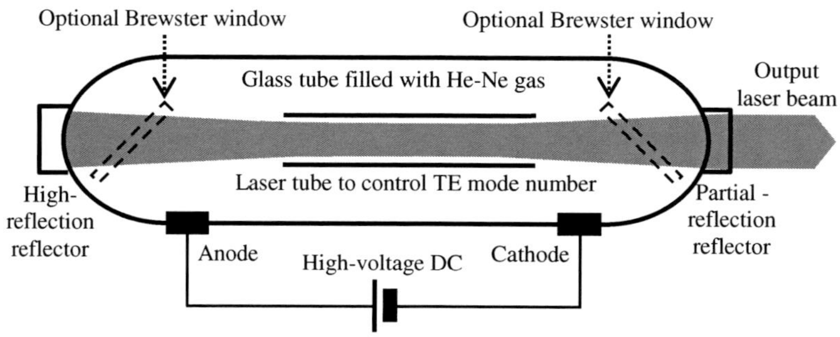

Figure 8.10 Schematics of a He-Ne laser. The optional Brewster window polarizes the laser beam.

8.4.2 Summary

Table 8.1 lists most gas lasers by their wavelength and briefly describes their applications.

8.5 Solid State Lasers

8.5.1 General description

The gain medium of solid state lasers is a piece of certain type of crystal or glass[12], usually has a rod shape. Solid state lasers are pumped by optical sources, such as flash lamp, laser diode or other lasers beams. The electrical to optical power conversion efficiency of solid state lasers is higher than that of gas lasers. Some solid state lasers can be operated by battery. For example, a green color laser pointer is a solid state laser, pumped by a laser diode which is operated by a battery.

Some solid state lasers have two concave reflectors to form the laser cavity. Some other solid state lasers have one flat facet of the gain medium

Table 8.1 Summary of various types of gas lasers. A: Wavelengths 0.3510 μm, 0.3638 μm, 0.4579 μm, 0.4658 μm, 0.4765 μm, 0.4727 μm, 0.5287 μm, and frequency-doubled 0.244 μm and 0.257 μm. B: Wavelengths 0.4067 μm, 0.4131 μm, 0.4154 μm, 0.4680 μm, 0.4762 μm, 0.4825 μm, 0.5208 μm, 0.5309 μm, 0.5682 μm, 0.6471 μm, and 0.6764 μm. Often mixed with argon to produce other wavelengths. C: The most commonly used laser of all kinds. Available wavelengths include 0.5435 μm, 0.5939 μm, and 0.6118 μm, also includes the less-seen 1.1523 μm, 1.52 μm, and 3.3913 μm.

Gain Medium	Wavelength (μm)	Applications
ArF	0.193	These are excimer lasers. Ultraviolet lithography for semiconductor manufacturing and laser surgery
KrF	0.248	
XeCl	0.308	
XeF	0.353	
Nitrogen	0.3371	Scientific research, measuring air pollution and pumping dye lasers
Argon[A]	0.4546, 0.4880, 0.5145	Retinal phototherapy, lithography, confocal microscopy, spectroscopy, and pumping other lasers.
Krypton[B]	Several wavelengths through visible range	Scientific research and mixing with argon to create other wavelengths.
Xenon ion	Many lines throughout visible spectrum and extending into UV and IR	Scientific research
Helium-neon[C]	0.6326	Interferometry, holography, spectroscopy, barcode scanning, alignment and optical demonstrations
Carbon monoxide	2.6–4.0 and 4.8–8.3	Material processing (engraving, welding, etc.) and photoacoustic spectroscopy
Carbon dioxide	9.4 and 10.6	Material processing (cutting, welding, etc.), surgery, dental lasers, and military.

and a separate concave reflector to form the laser cavity. Since the gain medium has a refractive index much higher than the index of the surrounding air, the medium acts like a waveguide and can confine the laser light inside the medium. Therefore, the laser cavity is stable and not very sensitive to alignment.

Solid state laser cavities can be as short as a few millimeters. Such cavities can support either single or multi longitudinal modes with large mode spacing and broad linewidth. The cross-section size of the gain medium determines the number of transverse modes. Solid state laser beam spatial quality is usually lower than that of gas lasers, can have a M^2 factor even larger than 2 and non-circular shape beams. Solid state laser power can be from below 1 mW for a Nd:YVO_4 green laser pointer to terawatt for a Nd:Glass laser. Solid state lasers can be operated in either CW or pulsed mode. Some solid state lasers can be modulated.

Ng-YAG lasers is the most widely solid state lasers. For example, green laser pointers are laser diode pumped Ng-YAG lasers emitting 0.532 μm wavelength beam. Ng-YAG lasers can also be pumped by flash lamp.

8.5.2 Summary

Table 8.2 lists most solid state lasers by their wavelengths and briefly describes their applications.

8.5.3 Solid laser cavities and pumping methods

Nowadays, many types of solid state lasers can be pumped by laser diodes. Laser diodes have an electrical-to-optical energy conversion efficiency as high as 50%. Laser-diode-pumped solid state lasers also have a relatively high electrical-to-optical energy conversion efficiency.

Figure 8.11(a) shows the schematics of a ruby laser. The cavity consists of two concave mirrors. The pumping source is a flash lamp tube placed along the ruby rod. Figure 8.11(b) shows a laser-diode-pumped Ng-YAG laser. The pumping wavelength is around 0.8 μm. The lasing wavelength is either 1.064 μm or 0.532 μm. A coupling lens focuses the laser diode beam onto the back facet of the laser rod. A spectral filter coated at the back facet of the laser rod lets the pumping light pass through and reflects the laser wavelength. The back facet and the concave reflector form the laser cavity.

Figure 8.12(a) shows the schematic of a ruby laser pumped by a coil shape flash lamp for higher pumping efficiency. To increase the pumping efficiency, an elliptical pumping cavity can be used along a tube-shaped flash lamp, as shown in Fig. 8.12(b). The ruby rod and the flash lamp tube are placed at the two foci of the pumping cavity, respectively. Every ellipse has two foci with a focal length given by $f = (a^2 + b^2)^{0.5}$, where a and b are the major and minor axis of the ellipse, respectively. The unique feature of an elliptical cavity is that

Table 8.2 Summary of various types of solid state lasers. A: The first type of visible-light laser invented in May 1960. B: One of the most commonly used high-power lasers, usually pulsed down to fractions of a nanosecond.

Gain Medium	Wavelength (μm)	Applications
Cerium-doped lithium strontium (or calcium) aluminum fluoride (Ce:LiSAF, Ce:LiCAF)	~0.280 to 0.316	Remote atmospheric sensing, LIDAR and scientific research
Titanium sapphire (Ti:sapphire)	0.650 to 1.100	Spectroscopy, LIDAR and scientific research. Often used to produce ultrashort pulses
Ruby[A]	0.6943	Holography and tattoo removal
Chromium-doped chrysoberyl (alexandrite)	0.70 to 0.82	Dermatological uses, LIDAR, laser machining
Ytterbium-doped glass (rod, chip, and fiber)	1	Material processing: cutting, welding and marking. Pump and amplification for fiber lasers and telecommunications.
Ytterbium YAG (Yb:YAG)	1.03	Laser cooling, materials processing, ultrashort pulse research, multiphoton microscopy, and LIDAR
Ytterbium:$_2O_3$	1.03	Ultrashort pulse research
Neodymium YLF (Nd:YLF)	1.047 and 1.053	Pulsed pumping of other types lasers
Neodymium-doped yttrium calcium oxoborate $Nd{:}YCa_4O(BO_3)_3$ or Nd:YCOB	~1.060 (~0.530 2nd harmonic)	Self-frequency doubling material for second harmonic generation.
Neodymium glass (Nd:Glass)	~1.062 and ~1.054	Extremely high power (terawatt scale), high energy (megajoules) multiple beam systems for inertial confinement fusion. Usually frequency tripled to the 3rd harmonic at 351 nm in laser fusion devices.
Neodymium-doped yttrium orthovanadate ($Nd{:}YVO_4$)	1.064	Pumping of other lasers, pulsed for marking and micromachining and making green laser pointers.
Nd:YAG[B]	1.064 and 1.320	Material processing, rangefinding, target designation, dental and surgery, tattoo and hair removal, scientific research, and pumping other lasers
NdCrYAG	1.064 and 1.320	Experimental production of nanopowders
Erbium-doped and erbium–ytterbium-co-doped glass (rod, chip, and fiber)	1.53 to 1.56	Optical amplifiers for telecommunications
Thulium YAG (Tm:YAG)	2.0	LIDAR
Holmium YAG (Ho:YAG)	2.1	Tissue ablation, kidney stone removal and dentistry.
Chromium ZnSe (Cr:ZnSe)	2.2 to 2.8	MWIR laser radar and military
F-center	2.3 to 3.3	Spectroscopy
Er:YAG	2.94	Dental and skin resurfacing

light emitted from one focus will hit another focus after being reflected by the wall of the cavity, as illustrated in Fig. 8.12(b). The light from the flash lamp can either directly hit the ruby rod or be mostly focused on the rod for efficient pumping.

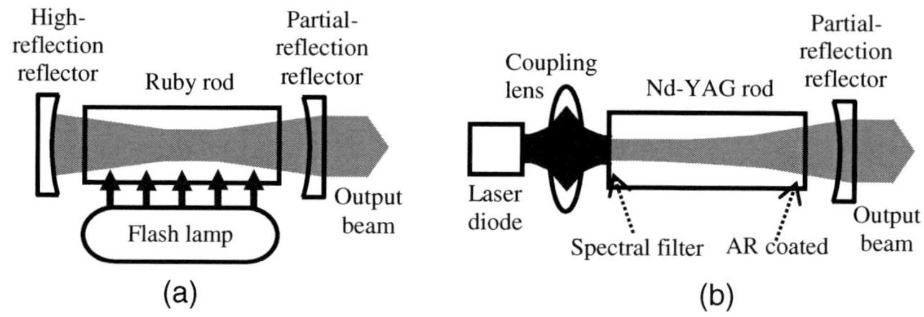

Figure 8.11 (a) Schematics of a ruby laser pumped by a flash lamp. (b) Schematics of a Nd-YAG laser pumped by a laser diode.

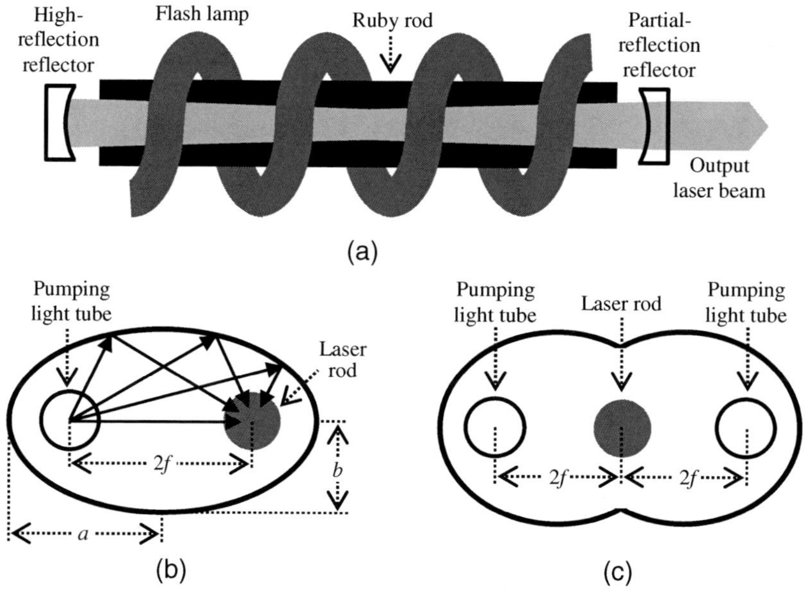

Figure 8.12 (a) Schematics of a ruby laser that is pumped by a coil-shaped flash lamp surrounding the ruby rod. (b) Schematics of the cross-section of an elliptic pumping cavity. One laser rod and one pumping flash lamp tube are placed at the two foci of the cavity, respectively. Most of the light from the flash lamp will hit the laser rod either directly or after being reflected by the elliptical cavity. (c) Schematics of the cross-section of a twin elliptical pumping cavity. The two cavity have one foci coincide. Two flash lamp tubes are placed at the other two foci of the twin cavity, respectively. The laser rod is placed at the common focus of the twin cavity. This structure allows the use of two pumping lamps.

To further increase the pumping light power, a twin-ellipse cavity can be used, as illustrated in Fig. 8.12(c). The two ellipses have one focus coincide where the ruby rod is placed. Two flash lamp tubes are placed at the other two foci of the twin cavity, respectively, to pump the ruby rod. Although the net pumping efficiency of the twin-ellipse cavity is lower than the pumping efficiency of one elliptical cavity, the net pumping light power can still be much higher. Note that the pumping cavity has nothing to do with the laser cavity.

Both elliptical pumping cavity and the coil shape flash lamp are used in various types of solid state lasers.

8.5.4 Frequency doubling

When a laser is pumped by a light source, such as a flash lump, the process converts one pumping photon to one laser photon. The energy that one pumping photon carries must be larger than the energy that one laser photon carries. Equation (8.1) states that the energy that one photon carries is inversely proportional to the wavelength, which means that the wavelength of the pumping light must be shorter than the laser wavelength. For example, a Ng-YAG laser with a 1.064-μm wavelength is pumped by a light source with a ~0.8-μm wavelength. The remaining energy of the pumping photon is converted to heat.

Several solid state laser media can convert two or even three pumping photons to one laser photon. Such a process is called frequency doubling[13] (second-harmonic generation) or frequency tripling.[14] For example, a Ng-YAG laser pumped by a ~0.8-μm-wavelength light source can emit 0.532 μm wavelength laser light. Frequency doubling and tripling processes create new laser wavelengths that cannot be created by other methods.

8.6 Laser Diodes

8.6.1 General description

Laser diodes (more formally called semiconductor lasers) are the most widely used lasers, accounting for about half of global laser sales. Applications include using them as a light source for fiber optic telecom to carry signals, for data reading and storage, and for material processing. The gain medium of laser diodes is apparently solid. However, laser diodes are traditionally not categorized as one type of solid state laser, but treated as a standalone type of lasers.

The advantages of laser diodes:

1. All laser diodes are pumped by an electrical current with the highest electrical-to-optical power conversion efficiency up to ~50% among all types of lasers. Laser diodes can be comfortably operated by battery.
2. There are wide selections of wavelength, from ~375 nm to ~20 μm.
3. The wavelength is tunable up to 100 nm or so.
4. Output power can be either CW or pulsed or modulated to GHz.
5. A wide power-range selection from 1 mW for a single transverse-mode laser diode to kW for a laser diode pile.
6. They feature a small size and light weight.

The disadvantages of laser diodes:

1. The tiny size and rectangular cavity cross-section results in elliptical beams with large divergence and astigmatism. Such beams are difficult to manipulate.
2. The spatial quality of a high-power laser diode is particularly low because of the many transverse modes in the linear active layer.
3. It is very easy to cause damage by static discharge or electrical surge.
4. Large thermal wavelength shifts of about 0.1–0.3 nm/°C.
5. Large linewidth of up to about 0.1 nm.
6. Large manufacturing tolerance on every specification, such as power, wavelength, threshold current, boresight, etc.

Reference 15 provides detailed studies on laser diode physics. Reference 16 provides detailed studies on laser diode beam manipulations.

8.6.2 Single-TE-mode cavity structure

The basic structure of a single-TE-mode laser diode is illustrated in Fig. 8.13. A P-N junction or "active layer" is sandwiched between two other layers and acts as a waveguide. The active layer has a rectangular cross-section. The active layer of most modern laser diodes consists of several parallel quantum wells, which are intended to increase the electron and hole densities inside the quantum wells to increase the lasing efficiency. Laser diode users do not need

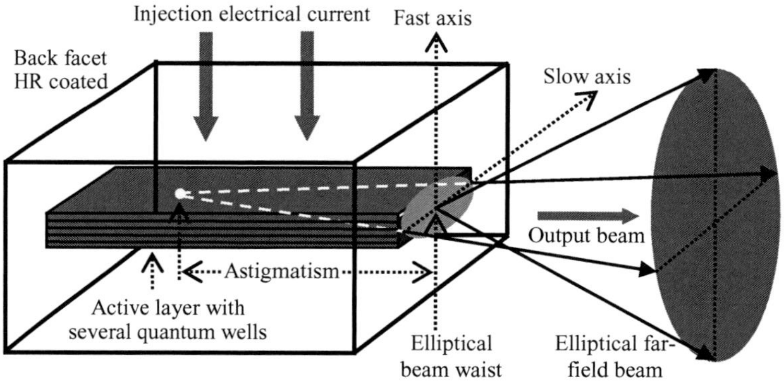

Figure 8.13 Schematics of an "edge-emitting" laser diode. An active layer is sandwiched between two other materials. The active layer is submicron thick, a fraction of a millimeter long, and has a rectangular cross-section. For single-TE-mode laser diodes, the width of the active layer is a few microns. The two facets of the active layer form the laser cavity. A forward-biased DC electrical current injects electrons and holes into the active layer, where they recombine and emit photons. The tiny size and rectangular cross-section of the active layer leads to an elliptical beam with large divergence and astigmatism. The astigmatism magnitude is a few microns, exaggerated here for illustration purposes. The terms "fast axis" and "slow axis" are commonly used to denote the two transverse directions.

to care about such details. The two facets of the active layer form a planar-planar cavity. The back facet is high-reflection coated. The front facet is usually uncoated with a natural reflectance of about 0.3. Since laser light is confined inside the active layer, the cavity is stable and not sensitive to alignment. The active layers are a fraction of millimeter long and can support either single or multilongitudinal mode with large mode spacing and a broad linewidth. The active layer is submicron thick in the fast axis direction. In this direction, the laser always has a single TE mode. For laser power below tens of mW or so, the active layer is a few microns wide and only supports a single TE mode. The width of the active layer is usually increased in order to produce more laser power. The wider active layer supports multi-TE mode lasing in the slow axis direction. Laser diodes with light emitted from the edge of the active layer are also called "edge-emitting laser diodes."

As shown by Eq. (3.6), the far-field divergence of a Gaussian beam is inversely proportional to its waist size. Because of the tiny rectangular cross-section of the active layers, laser diode beams have large divergence, are elliptical, and have astigmatism of a few microns, as illustrated in Fig. 8.13. The spatial quality of singe TM semiconductor lasers is comparable to that of solid state lasers with a M^2 factor value from about 1.1 to 1.3.

8.6.3 Multi-TE-mode cavity structure

In order to obtain more laser power, the active layer width can be increased to tens of microns or even up to a couple of hundred microns, while the thickness is usually not increased to maintain the high carrier density that leads to high lasing efficiency. Such active-layer cross-sections virtually have linear shapes, support many TE modes in the line direction, and are often called "wide-stripe laser diodes" or "broad-area laser diodes."

In the fast axis direction, wide-stripe laser diode beams still behave like a laser beam. In the slow axis or the line direction, the many transverse modes make the beam somehow like a mixture of laser beam and flashlight.

In order to further increase the laser power, several wide-stripe laser diodes are arranged to form a laser diode stack, as illustrated in Fig. 8.14(a). The laser beams then behave like a mixture of laser beam and flashlight in both the fast and slow axis directions.

Several wide-stripe laser diodes can be mounted on a metal base to form a laser diode "bar" several millimeters long, as illustrated in Fig. 8.15(a). Beams from laser diode bars behave like a laser beam in the fast axis direction and like a mixture of laser beam and flashlight in the slow axis direction. To further increase the laser power, several laser diode bars can be arranged to form a laser diode "pile," as illustrated in Fig. 8.15(b). Beams from laser diode piles behave like a mixture of laser beam and flashlight in both the line direction and the direction perpendicular to the line.

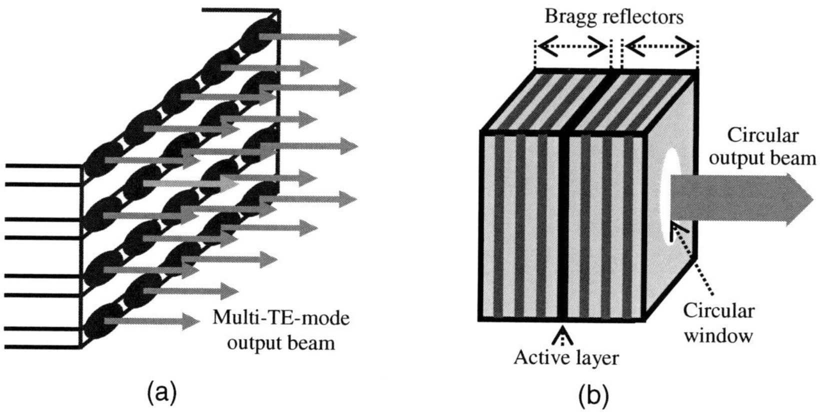

Figure 8.14 (a) Schematics of a laser diode pile consisting of four wide-stripe laser diodes. Each diode emits a multi-TE-mode beam. The total output beam is the combination of many TE modes and no longer behaves like a typical laser beam. (b) Schematics of a vertical cavity surface emitting laser (VCSEL) that has a circular output window and outputs a circular laser beam.

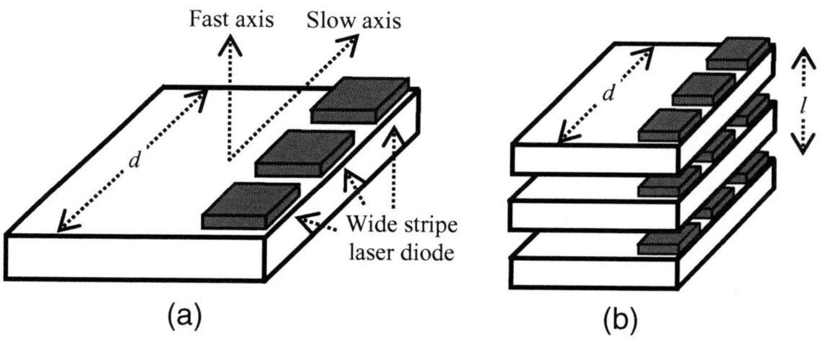

Figure 8.15 (a) A laser diode bar has three wide-stripe laser diodes, each of which emits a multi-TE-mode beam. This bar is a linear light source with length *d*. (b) A laser diode pile consists of nine wide-stripe laser diodes. This pile is a rectangular light source with size $d \times l$.

8.6.4 Laser diode packages and modules

Most low-power, single-TE-mode laser diodes available on the market are mounted in a cap, as shown in Fig. 8.16(a). Bare, low-power laser diodes are rarely seen. Because of the tiny size of the active layer, laser diodes will be very hot during operation, so a heat sink is often required in addition to the metal base of the laser diode cap.

Most commercially available, high-power laser diodes are mounted in an open metal base/heat sink. The metal base has no standard shape and size. High-power laser diodes often need air or water cooling.

Laser diodes are easy to damage by static discharge or overdriving. Hands that directly touch laser diodes must be grounded. To avoid accidently damaging laser diodes and to facilitate the use of lasers, most

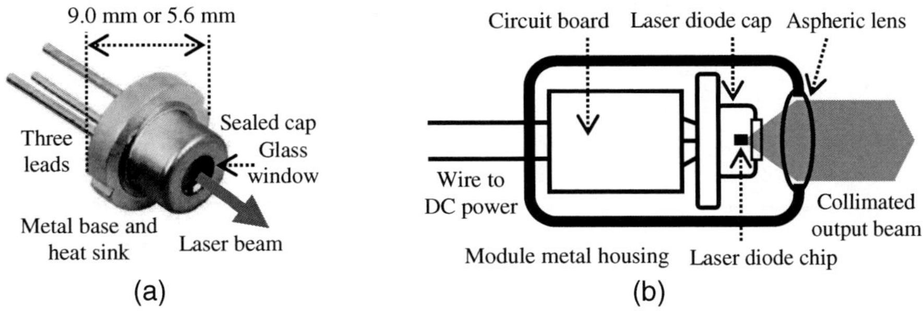

Figure 8.16 (a) A typical low-power, single-TE-mode laser diode package consists of a sealed metal cap filled with inert gas to slow down the oxidizing process, a metal base plate to serve as a mounting support and heat sink, a submillimeter-thick glass window to let the laser beam output, and three metal leads to DC power and grounding. The metal base come with two standard sizes, 9 mm and 5.6 mm. (b) Schematics of a typical laser diode module that consists of a laser diode, a circuit board, a collimating lens, and a shielded metal housing tube.

low-power, single-TE-mode laser diodes are sold as a "module," as illustrated in Fig. 8.16(b). An aspheric lens collimates the laser beam for easy use. The circuit board controls the current to avoid overdriving the laser diode or to stabilize the laser power and suppress the possible electrical surge from the DC power source. The metal housing shields the laser diode from static discharge. Inexperienced users should avoid directly handling laser diodes and instead use laser diode modules.

8.6.5 Three special types of laser diodes and devices

There are a few special types of laser diodes, three of which are briefly introduced here:

1. Vertical cavity surface emitting laser (VCSEL). The active layer is sandwiched between two Bragg reflectors that form the laser cavity (as illustrated in Fig. 8.14(b)), reflect only one wavelength, and force the laser to lase at this wavelength. A circular output window is etched on one Bragg reflector so that the output beams are circular with small divergence and no astigmatism. Compared with edge-emitting laser diodes, a VCSEL is surface emitting and has a very short cavity, which limits the gain the laser can obtain during one round trip inside the cavity and limits the maximum laser power to a value lower than edge-emitting laser diodes (although VCSELs with a laser power of a few hundred mW have been reported).
2. Distributed feedback (DFB) laser diodes. In such laser diodes, a refractive index grating is built along the active layer, as shown in Fig. 8.17. As the laser light travels inside the cavity, only the wavelength that equals the grating period is reflected along the way. The theory

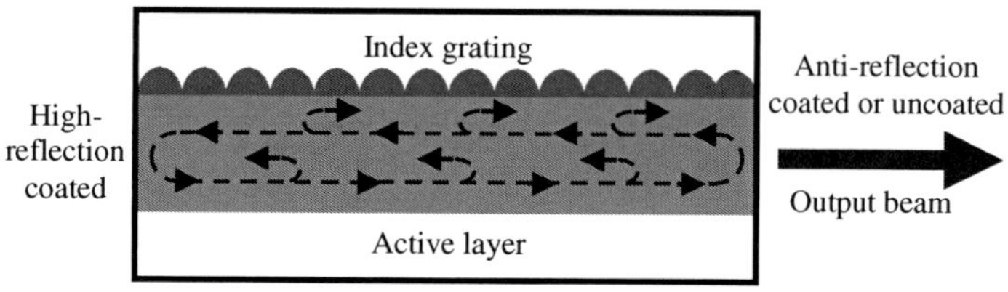

Figure 8.17 Schematics of a DFB laser. The built-in distributed index grating selects the lasing wavelength.

related to this mechanism, i.e., "coupled wave theory," is beyond the scope of this book. Because of the grating, DFB laser diodes have a single longitudinal mode with a very narrow linewidth of about 10^{-5} nm. As a comparison, conventional laser diodes have a broad linewidth up to about 0.1 nm. DFB laser diodes are widely used in fiber optic telecom systems as signal carriers.

3. External cavity wavelength tunable laser diode system (EWLS). There are a couple of slightly different structures; the one shown in Fig. 8.18 is a Littrow-type EWLS. The wavelength tuning is realized by rotating the mirror. The wavelength tuning range is limited by the bandwidth of the gain medium. The tuning range is about 100 nm around 1.55 μm and is about 10 nm in the red-color range. Because the external cavity is much longer than the laser diode active layer, the mode spacing is much smaller and the laser linewidth is much narrower.

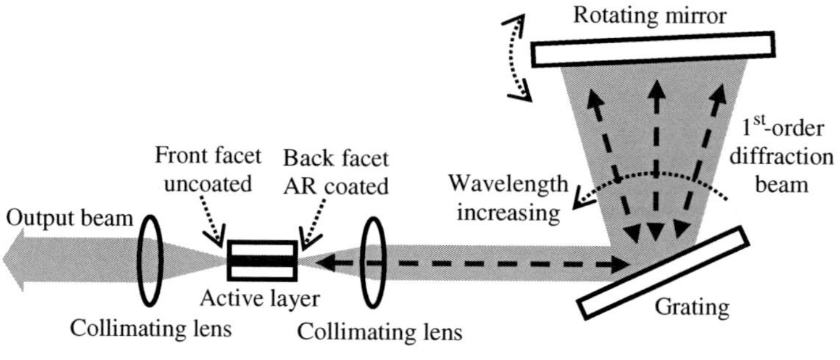

Figure 8.18 Schematics of an EWLS. The back facet of the active layer is AR coated. The uncoated front facet and the rotating mirror form the laser cavity. The laser light exiting the back facet is collimated by a lens and incident on a grating, which disperses the different wavelength components of the light. Different wavelength components propagate in different directions. The first-order diffraction light is incident on a mirror, which reflects a certain wavelength in the first-order diffraction based on the mirror orientation. The reflected light traces the same way back to the grating, and then to the active layer, before forcing the laser to oscillate at this wavelength. The reflected wavelength can be tuned by rotating the mirror, and the laser wavelength is thereby tuned. The beam from the front facet is the output beam, which must be collimated for easy use.

Single-longitudinal-mode operation can still be achieved by using a powerful grating and adjusting the external cavity length while tuning the wavelength so that the lasing wavelength is always one of the standing waves of the external cavity.

8.7 Optical Fiber Lasers

Optical fiber lasers are also becoming rapidly more popular.[17] Although fiber lasers are a type of solid state lasers, they are often treated as a standalone type.

Figure 8.19(a) shows the schematics of a simple structure fiber laser that has only one pumping laser diode. To increase the laser power, multiple pumping laser diodes can be coupled into one fiber laser, as illustrated in Fig. 8.19(b). The most widely used fiber laser is erbium doped with a 1.53–1.62-μm wavelength, pumped by a 0.98-μm laser diode source and used as an optical amplifier in fiber optic communication systems. The other main application of fiber laser is material processing.

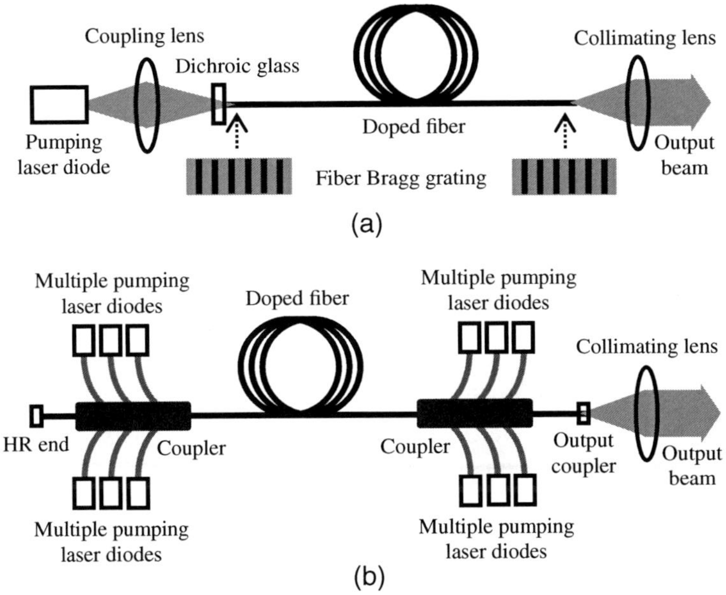

Figure 8.19 (a) Schematics of a fiber laser. The pumping light from a laser diode is focused at one end of a doped single-mode optical fiber mainly into the cladding of the fiber. The pumping wavelength is shorter than the lasing wavelength. The dichroic glass lets the pumping light pass through and reflects the fiber laser light. The fiber lasing wavelength range is determined by the type of doping in the fiber. Two Bragg gratings are built at the two ends of the fiber to provide wavelength-selected reflection and thus achieve a single wavelength and narrow linewidth lasing. The output beam has very good spatial quality but must be collimated for easy use because of its large divergence. (b) The light from multiple pumping laser diodes can be coupled into the fiber to increase the pumping power and thereby increase the laser power.

The advantages of fiber lasers include

1. High energy-conversion efficiency, partially due to the efficiency of the pumping laser diodes.
2. The ability to output either CW or pulsed or modulated power.
3. High laser power, up to thousands of watts.
4. Large wavelength selections, from ~1 μm to the mid IR.
5. A large wavelength-tunable range, over 0.3 μm for some wavelengths.
6. A high-spatial-quality, single-TE-mode beam with $M^2 < 1.1$.
7. A beam that can be delivered through a zigzag path.

8.8 Other Types of Lasers

There are some other types of lasers that are used less often:[18]

1. Metal vapor lasers, several types of which have been developed. They are pumped by electrical discharge, output single-line laser power in the UV and visible range, and are used in scientific research, printing and typesetting, fluorescence excitation, medicine, etc.
2. Dye lasers, which are pumped by flash lamp or other lasers, famous for their very broad wavelength tuning range in the visible range, and are used for scientific research, medicine, spectroscopy, and isotope separation.
3. Chemical lasers, which are pumped by chemical reaction, have a wavelength in the near-IR range, and are mainly used for weaponry.
4. Raman lasers, which are mostly optical fiber lasers. They are pumped by other lasers, have a wavelength from 1–2 μm, and are mainly used as optical signal amplification for telecommunications.
5. Free electron lasers, which have a large size and are expensive. Laser beams are generated by high-speed electron beams passing through a periodic magnetic field, have a very broad wavelength tuning range from 0.1 nm to several mm, and are used in scientific research and medical applications.
6. X-ray lasers, which are generated by terawatt-irradiation fluence pulses, have a sub-10-nm wavelength, and may be used in high-resolution microscopy and holography.

8.9 Laser Safety

Medium-to-high-power/energy laser beams are hazardous to the human eye or even skin and must be handled with cautions. Various specifications, classifications, and revisions have been developed to control the risk of injury.

Lasers are divided into four classes and some subclasses based on their power, energy, pulsed rate and repetition, time of exposure to laser radiation, human reaction speed, and wavelength. The contents of these classifications are rigorous with various damaging mechanisms considered. There is no simple analytical way to classify lasers, although the following list provides a brief summary:

Class 1	Lasers are safe under all conditions of normal use.
Class 2	Lasers are to be considered safe because of the blink reflex.
Class 3R	Lasers are to be considered safe if handled carefully and with restricted beam viewing.
Class 3B	Lasers are hazardous if the eye is exposed directly.
Class 4	Lasers are to be considered dangerous.

Labs equipped with lasers should have a warning sign on the door that clearly states which classes of lasers are inside the lab. Most lasers also have a warning sign on them. Laser operators often need to wear protection glasses. Reference 19 is a good guide to laser safety.

References

1. A. E. Siegman, *Lasers*, University Science Books, Sausalito, CA (1986).
2. P. W. Milonni and J. H. Eberly, *Laser Physics*, Wiley-Interscience, Hoboken, NJ (1988).
3. Wikipedia, "Planck–Einstein relation," wikipedia.org/wiki/Planck%E2%80%93Einstein_relation
4. Wikipedia, "Maxwell-Boltzmann distribution," wikipedia.org/wiki/Maxwell%E2%80%93Boltzmann_distribution
5. Wikipedia, "Population inversion," wikipedia.org/wiki/Population_inversion
6. B. E. A. Saleh and M. C. Teich, "10.2 Spherical-mirror resonant," *Fundamentals of Photonics*, 2nd ed., John Wiley & Sons, Inc., Hoboken, NJ, pp. 378–381 (2007).
7. B. E. A. Saleh and M. C. Teich, "8.2 Planar dielectric waveguides," *Fundamentals of Photonics*, 2nd ed., John Wiley & Sons, Inc., Hoboken, NJ, pp. 299–303 (2007).
8. P. W. Milonni and J. H. Eberly, *Laser Physics*, Wiley-Interscience, Hoboken, NJ, p. 763, Eq. (15.4.8) (1988).
9. Wikipedia, "Q factor," en.wikipedia.org/wiki/Q_factor
10. Wikipedia, "Mode-Locking," wikipedia.org/wiki/Mode-locking
11. Wikipedia, "Gas Laser," wikipedia.org/wiki/Gas_laser
12. Wikipedia, "Solid-State Laser," wikipedia.org/wiki/Solid-state_laser
13. Wikipedia, "Second-Harmonic-Generation," en.wikipedia.org/wiki/Second-harmonic_generation
14. Wikipedia, "Optical Frequency Multiplier," en.wikipedia.org/wiki/Optical_frequency_multiplier
15. L. A. Coldren and S. W. Corzine, *Diode Lasers and Photonic Integrated Circuit*, 2nd ed., John Wiley & Sons, Hoboken, NJ (2012).

16. H. Sun, *A Practical Guide on Handling Laser Diode Beams*, Springer, New York (2015).
17. Wikipedia, "Fiber lasers," wikipedia.org/wiki/Fiber_laser
18. Wikipedia, "List of laser types," wikipedia.org/wiki/List_of_laser_types
19. Wikipedia, "Laser safety," wikipedia.org/wiki/Laser_safety

Chapter 9
Laser Optics and Devices, and Manipulable Laser Beams

9.1 Laser Mirrors

Laser beams are coherent, and they have high wavefront qualities. Laser optics, mainly mirrors and lenses, have quality and other requirements different than those of conventional optics. Optical component vendors often set up a separate category for laser optics.

9.1.1 Surface qualities and coatings

Tiny defects on a mirror surface can cause noticeable scattering because of the high power/energy densities of laser light. Laser mirrors usually have high surface qualities, such as a $<\lambda/8$ surface flatness and 10-5 scratch–dig.

Laser mirrors are coated with high-reflection, narrowband multilayer dielectric coatings. The band is around the specific laser wavelength. The reflectance is at least >99% and often >99.8% for the laser wavelength. (Such coatings were discussed in Section 5.7.) Besides a high surface quality, laser mirror substrates are the same as conventional mirror substrates (see Section 4.6).

9.1.2 Mirror laser-damage threshold

For laser mirrors, laser damage is an issue because laser lights often have high power/energy densities. In particular, pulsed laser lights have their energy concentrated in a narrow pulse, so the energy density can be very high.

Since most laser beams have Gaussian-shaped spatial intensity profiles, the on-axis intensity $I_0(z)$ is the peak intensity and determines the damage threshold. $I_0(z)$ is relatively more complex to measure, whereas the total power P_0 of a laser beam is easy to measure. Equation (3.9) relates $I_0(z)$ to the laser beam power P_0 and is rewritten here:

$$I_0(z) = \frac{2P_0}{\pi w(z)^2}, \tag{9.1}$$

where $w(z)$ is the $1/e^2$ intensity radius of the laser beam at distance z from the laser beam waist. In this case, $w(z)$ is the $1/e^2$ intensity radius of the laser beam on the surface of the optical component of interest.

For pulsed lasers, their outputs are specified by energy E per pulse. P_0 in Eq. (9.1) can be replaced by $P_0 = E/T$, where T is the pulse width. The result is

$$I_0(z) = \frac{2E}{\pi w(z)^2 T}. \tag{9.2}$$

The damage mechanism of pulsed lasers is avalanche ionization or dielectric breakdown.[1] The surfaces of the optical components show breaks or cracks. The laser pulse repetition rate and laser wavelength also affect the damage threshold. The following is a typical damage-threshold specification for a Nd:YAG laser mirror:

$$25\,\mathrm{J/cm^2}(1064\,\mathrm{nm},\ \varnothing 0.55\,\mathrm{mm},\ 10\,\mathrm{ns},\ 10\,\mathrm{Hz}),$$

where 1064 nm is the peak wavelength, ⌀0.55 mm is the specified diameter of the laser beam on the mirror, 10 ns is the specified pulse width, and 10 Hz is the specified pulse repetition rate.

If the laser in use has a different size and/or pulse width than the specifications, the corresponding laser damage threshold can be found using Eq. (9.2). If the laser in use has a different wavelength from the specifications, an approximation relation can be used to find the laser damage threshold:[1,2]

$$LDT(T_2,\lambda_2) \approx LDT(T_1,\lambda_1)\left(\frac{T_2}{T_1}\right)^{0.5}\left(\frac{\lambda_2}{\lambda_1}\right), \tag{9.3}$$

where T_1 and λ_1 are the specified pulse width and wavelength, $LDT(T_1, \lambda_1)$ is the specified laser damage threshold, T_2 and λ_2 are the pulse width and wavelength of the laser in use, and $LDT(T_2, \lambda_2)$ is the damage threshold for the laser in use. The accuracy of Eq. (9.3) will be sacrificed if the pulse width and wavelength in use are very different from the specified values, e.g., a difference of 10 times is too large.

The damage mechanism of a CW laser, long-pulse lasers with a pulse width >1 μs or so, and very-high-repetition lasers is thermal melting caused by absorbing the laser energy.[1] Because the damage mechanism of pulsed and CW lasers are very different, the damage threshold of these two types of lasers cannot replace each other; they must be specified separately.

In reality, most laser damage occurs at the surface of the optical components because the surfaces are more or less polluted. Dust and/or oil on the surface absorb laser energy. Dielectric coatings often have a lower laser damage threshold than bare glass surfaces.

Laser damage starts as a tiny defect on the optical surfaces or inside optical components and is cumulative. Once a tiny defect is created, the defect absorbs more energy. If the defective spot is repeatedly hit by the laser pulses, the size and depth of the defect will increase rapidly. Severe laser damage can cause breaks or cracks on optical surfaces or even inside the bulk of the components. Reference 3 reviews optical coating damage.

9.2 Laser Lenses

9.2.1 Surface qualities and coatings

Laser lenses are used to manipulate laser beams. Just like laser mirrors, laser lenses have high surface qualities, such as a $<\lambda/8$ surface flatness and 10-5 scratch–dig in order to minimize scattering because laser light has high power/energy densities. Laser lens surfaces are narrowband or V-shaped AR coated for the specified laser line.

9.2.2 Laser lens materials

Since laser beams have very narrow spectral bandwidths, color dispersion is never an issue, although transmission at the working wavelength can be an issue.

- For lasers with a wavelength in the visible to near-IR range, all optical glasses are suitable lens materials.
- For lasers with a wavelength in the UV range, fused silica is the most commonly used material. Fused silica is clear in the visible range and has >80% transmission down to about 180 nm. Fused silica is also used to fabricate lenses for the visible range. Calcium fluoride and sapphire are also used in the UV range. Calcium fluoride is clear in the visible with >80% transmission down to about 160 nm. Sapphire is clear in the visible and has >80% transmission down to about 240 nm. Quartz and calcite have high transmission in the UV range, but they are rarely used to make lenses because of their birefringence.
- For lasers with a wavelength in the mid-IR range, zinc selenide, germanium, sapphire, and calcium fluoride are the commonly used lens materials. Zinc selenide has a clear orange color in the visible range as shown in Fig. 9.1(a) and >80% transmission up to 18 μm. Germanium is opaque black in the visible range, as shown in Fig. 9.1(b), and has >80% transmission up to 15 μm. Sapphire and calcium fluoride have >80% transmission up to 5 μm and 8 μm, respectively.

IR materials also often have a large refractive index. Germanium has a refractive index of about 5 at a 5-μm wavelength. Zinc selenide has a refractive index of about 2.43 at a 5-μm wavelength. The reflectance of an uncoated surface of these two materials is about 36% and 20%, respectively. Antireflection coating on germanium and zinc selenide lenses is very important. The residual reflectivity of an AR coating can be <0.5% for a single laser line and a couple of percent for a broadband.

(a) (b)

Figure 9.1 (a) Zinc selenide has a clear orange color in the visible range. (b) Germanium is opaque black in the visible range.

9.2.3 Lens laser-damage threshold

Some laser lenses have their laser damage threshold specified, but some do not because AR coatings on laser lenses are less susceptible to damage than high-reflection coatings on laser mirrors. However, laser lenses can still be damaged by high-power/energy laser beams. For example, CO_2 lasers with a wavelength of 10.6 μm often have kW power, are often used for material processing, and can damage laser lenses in certain circumstances. Laser lens-material damage mechanisms are the same as the laser mirror-material damage mechanisms discussed earlier.

9.3 Focusing or Collimating Laser Beams

9.3.1 Geometrical optics modeling

When a laser beam is focused, color dispersion is not a problem. Spherical aberration may be an issue, depending on how quickly the beam is focused. The spherical aberration occurring for a laser beam is the same as the spherical aberration occurring for geometrical rays. Geometrical optics can be used with good accuracy to analyze spherical aberration in a laser beam. The ray bundle diameter can be chosen to equal the $1/e^2$ intensity diameter of the laser beam. The focused laser beam waist is about the same size as the Airy disk if spherical aberration is corrected. This approach can simplify the calculation. A detailed example that compares the focusing of a Gaussian beam and a planar wave is provided in Subsection 3.8.3.

9.3.2 Focusing a beam: three examples

Three cases of focusing geometrical ray bundles are analyzed with spherical aberration being considered and the previous explanation in mind. Figure 9.2(a) shows the focusing of a collimated geometrical ray bundle. A Zemax-generated Airy disk and focused RMS spot for $D = 1$ mm, 2 mm, and 3 mm are plotted in Figs. 9.2(b), (c), and (d), respectively. The actual

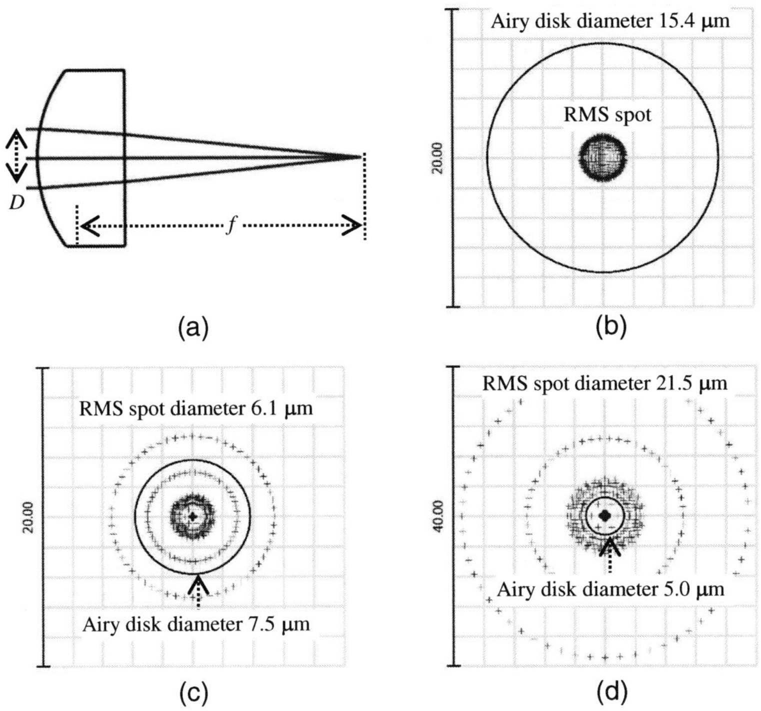

Figure 9.2 (a) A convex-planar N-BK7 lens with a 3-mm center thickness and 10-mm focal length focuses a collimated ray bundle with diameter D and a 0.633-μm wavelength. (b) For $D = 1$ mm, $F/\# = f/D = 10$, and the Zemax-generated RMS spot is much smaller than the 7.8-μm Airy disk diameter (the smallest possible spot). Spherical aberration is negligible. The real focused spot size is the Airy disk size. (c) For $D = 2$ mm, $F/\# = f/D = 5$, and the RMS spot is still smaller than the Airy disk. Spherical aberration is still negligible. The real focused spot size is still the Airy disk size. (d) For $D = 3$ mm, $F/\# = f/D = 3.33$, and the RMS spot is 21.5 μm larger than the Airy disk. Spherical aberration is not negligible.

focused spot size is the larger of the Airy disk size or the RMS spot size. Figure 9.2 shows that $D = 2$ mm ($F/5$) results in the most balanced focus with the smallest focused spot size of 7.5-μm diameter. For $D = 3$ mm ($F/3.3$), the relatively larger spherical aberration results in a larger RMS spot size of 21.5 μm, although the Airy disk size of 5.0 μm is the smallest.

9.3.3 Collimating laser beams

Collimating a laser beam can be considered as the reversal of focusing. Whether spherical aberration is an issue or not depends on the beam divergence or the value of the numerical aperture (NA), which is related to the $F/\#$ by $NA = 1/(2F/\#)$. The results obtained earlier for focusing laser beams can be applied to collimating laser beams. $NA = 0.1$ is about the dividing line, whereas $NA < 0.1$ can be considered as free of spherical aberration.

Most laser beams have a circular cross-section and a small divergence when they are emitted from the lasers. Manipulation of these beams is not complex.

9.3.4 Effects of lens shape and orientation

If the lens shown in Fig. 9.2 is used in a flipped orientation, as shown in Fig. 9.3(a), the spherical aberration will be larger. For a 3-mm input-ray-bundle diameter, the Zemax raytracing result shows that the geometrically focused RMS spot diameter is 94.9 μm, which is much larger than the 21.5-μm diameter shown in Fig. 9.2(d), which proves that the lens orientation matters.

If a lens has an equi-convex shape, although the size, thickness, and focal length remain the same, as shown in Fig. 9.3(b), for a 3-mm input-ray-bundle diameter the geometrically focused RMS spot diameter is 28.7 μm—slightly larger than the 21.5-μm diameter shown in Fig. 9.2(d).

If both surfaces of this lens are optimized to minimize the spherical aberration while the size, thickness, and focal length remain the same, the resultant lens shape is shown in Fig. 9.3(c). For a 3-mm input-ray-bundle diameter, the geometrically focused RMS spot diameter is 19.2 μm—slightly smaller than the 21.5-μm diameter shown in Fig. 9.2(d).

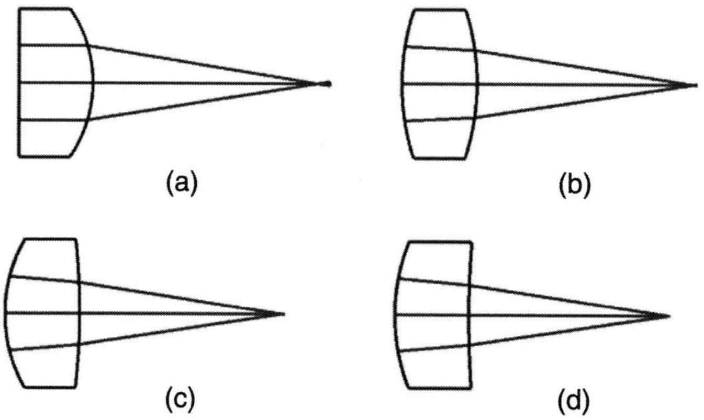

Figure 9.3 Comparison of various shapes of lenses. All of these lenses have a 3-mm center thickness, 10-mm focal length, and focus a 3-mm-diameter geometrical ray bundle with a 0.633-μm wavelength. The glass used in (a)-(c) is N-BK7 with an index value of 1.52, and in (d) the glass used is S-LAH79 with an index value of 2.00. (a) The wrong lens orientation results in significantly larger spherical aberration and a large focused spot of 94.9-μm RMS diameter as compared with the 21.5-μm focused spot shown in Fig. 9.2(d). (b) Equi-convex shape is neither the best nor the worst shape for focusing. The RMS focused spot diameter is 28.7 μm. (c) The best shape for N-BK7 glass with 19.2 μm RMS spot diameter. The two convex surfaces have different radius of curvature. (d) Meniscus-shaped lens (convex-concave) is the best shape for Ohara S-LAH79 glass with a large index. The RMS focused diameter is only 8 μm.

For the same focal length, lenses made of glass with a larger index will have a larger surface curvature radius and smaller spherical aberration. The lens shown in Fig. 9.3(d) has the same size, thickness, and focal length as the other lenses shown in Figs. 9.2 or 9.3(a)–(c) but uses Ohara S-LAH79 glass with a large index of 2 (N-BK7 glass has an index of 1.52). The two optimized surface shapes are convex and concave. Lenses with such a shape are called "meniscus lenses." For a 3-mm input-ray-bundle diameter, the geometrically focused RMS spot diameter is only 8 μm.

9.3.5 Effects of aperture truncation on laser beams

A laser beam with large divergence can sometimes be truncated by a circular aperture, such as the edge of a lens that is smaller than the laser beam. The truncation will cause complex diffractions because of the coherent nature of laser beams. As a comparison, the truncation of an incoherent ray bundle can be ignored.

Figure 9.4 shows the simulation results of a laser beam truncated by an aperture. The laser beam has a Gaussian intensity profile with a 4-mm $1/e^2$ intensity diameter and 0.633-μm wavelength. The aperture has a 2-mm diameter. The intensity profiles at four different distances after the aperture are plotted using Kirchhoff's diffraction formula[4] and the mathematics software Mathcad.

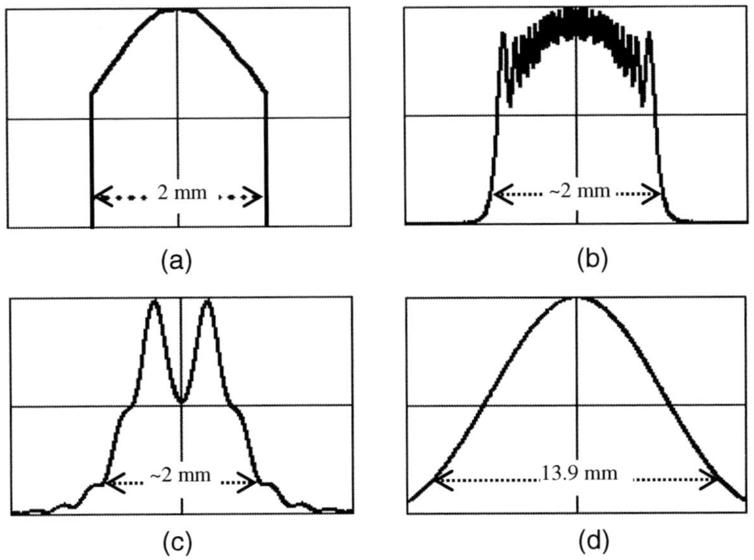

Figure 9.4 Simulated normalized intensity profiles of a truncated laser beam at four different distances. The laser beam has a wavelength of 0.633 μm and a Gaussian intensity profile with a $1/e^2$ intensity diameter of 4 mm before being truncated. The truncation aperture is circular with a diameter of 2 mm. The beam profiles at four distances after the aperture are simulated. The four distances are (a) 1 mm, (b) 50 mm, (c) 1 m, and (d) 30 m.

At a 1-mm distance after the aperture, the beam has two clean-cut edges, as shown in Fig. 9.4(a), the $1/e^2$ intensity diameter is determined by the aperture diameter. This result is quite understandable.

At a 50-mm distance after the aperture, the beam edges are no longer cleanly cut, as shown in Fig. 9.4(b). The edge intensities gradually vary but are still steeper than a Gaussian profile. The $1/e^2$ intensity diameter approximately equals the aperture diameter. Complex diffraction rings appear at the central portion of the beam. As the beam travels forward, the intensity profile continues to change. The beam edge profiles will gradually approach the Gaussian profile, and the diffraction ring number will reduce.

At a 1-m distance after the aperture, the intensity profile of the beam is plotted in Fig. 9.4(c). The diffraction ring at the central portion of the beam reduces to two. The maximum intensity is no longer at the center of the beam, which may look weird but is normal. The beam edge shows some ups and downs (rings) caused by diffraction. For such a beam intensity profile, the definition of the $1/e^2$ intensity diameter does not make much sense, but it can still be calculated or measured. As the beam travels forward, the beam intensity profile continuously changes, and the diffraction ring patterns at the beam edge can change quickly. As a result, the $1/e^2$ intensity diameter can suddenly become either larger or smaller.

A laser beam that has almost any intensity profile (except perhaps a Bessel intensity profile) will eventually regain a Gaussian intensity profile as the beam travels forward because of diffraction, which is also the case here. At a 30-m distance after the aperture, the beam regains a Gaussian intensity profile, as shown in Fig. 9.4(d). The beam is also much larger than the aperture.

As explained earlier, the intensity profile of a truncated laser beam undergoes complex changes shortly after being truncated because of diffraction. The beam size is difficult to define. If the goal is to visualize the beam size change as the beam travels forward, Fig. 9.5 provides an approximate

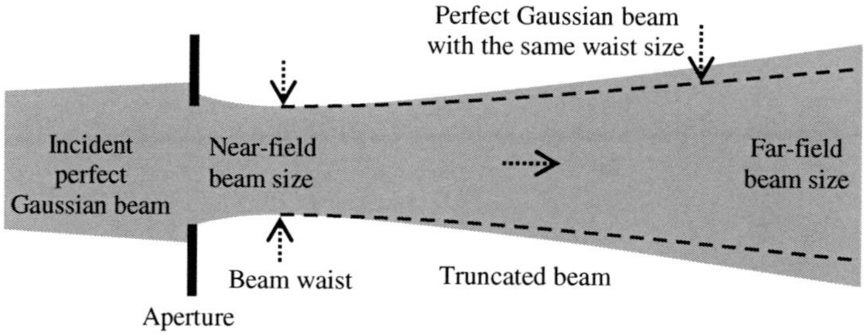

Figure 9.5 Indication of size changes to a truncated laser beam as it travels forward. Shortly after being truncated, the beam size is smaller than the truncation aperture. The far-field size of the truncated beam is larger than the far-field size of a perfect Gaussian beam with the $1/e^2$ intensity diameter that equals the truncated beam.

indication. The $1/e^2$ intensity diameter of the truncated beam shortly after being truncated is slightly smaller than the diameter of the aperture. The far-field size of the truncated beam is a little larger than the far-field size of a perfect Gaussian beam that has a $1/e^2$ intensity diameter that equals the $1/e^2$ intensity diameter of the truncated beam, as illustrated in Fig. 9.5.

Equation (3.6) states that the far-field divergence θ of a beam is proportional to M^2/w_0, where M^2 is the M-squared factor that describes the beam quality, and w_0 is the beam waist size. Compared with the perfect Gaussian beam of the same waist size, the field divergence of a truncated beam is larger. The only explanation of this phenomenon is that the value of the M-squared factor of the beam is increased by the truncation. The deeper the truncation is, the lower the truncated beam quality. See Ref. 5 for a detailed analysis of truncated-laser-beam size changes.

Aperture truncation will also affect a focused laser beam in a similar way. The intensity profile of the focused spot of a truncated beam will undergo changes similar to those shown in Fig. 9.4. In addition, the truncation of a focused beam will shift the focal point of the beam closer to the aperture and increase the focused spot size, as illustrated in Fig. 9.6 with an exaggerated scale. Aperture truncation will reduce the beam quality by increasing the value of the M-squared factor.

9.4 Low-Power, Single-TE-Mode Laser Diode Optics

Laser diodes are the most widely used lasers, accounting for about half of all laser sales. Laser diode beams have elliptical or even linear shapes, large divergence, and astigmatism; they are much more difficult to manipulate than other kinds of lasers. In addition, laser diodes have notoriously large tolerances. Different laser diodes of the same type can behave differently.

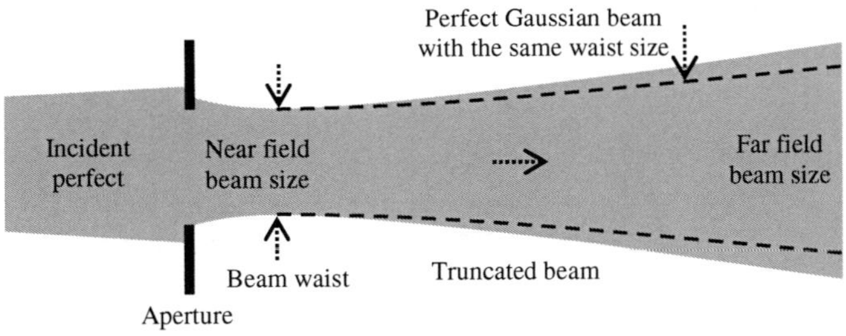

Figure 9.6 Illustration of the truncation effect on a focused beam. If not truncated, the focal plane and focused spot size of a focused beam is shown by the dashed lines. After being truncated, the beam focal plane moves closer to the aperture, and the focused spot (beam waist) size is increased, but the far-field divergence is not reduced. Thus, the quality of the truncated beam is reduced or the truncation increases the M^2 factor value.

9.4.1 Beam characteristics

Low-power, single-TE-mode laser diode beams have two large and differently divergent angles in the fast axis and slow axis directions, respectively. The beams are elliptical (see Fig. 8.13 for the schematic structure of such a laser diode). Laser diode manufacturers traditionally specify the beam divergence at FWHM, whereas the optical community traditionally specifies the beam divergence at the $1/e^2$ (13.5%) level, which is about 1.7 times larger than the FWHM divergence. There are many types of laser diodes, and their beam FWHM divergences in the fast axis direction range from about 20° to about 40°. In the slow axis direction, the beam FWHM divergences range from about 6° to 12°.

Laser diode beams are astigmatic. For single-TE-mode beams, the magnitude of astigmatism ranges from ~3–7 μm. For high-power, wide-stripe, multi-TE-mode beams, the magnitude of astigmatism is tens of microns.

9.4.2 Lenses for collimating or focusing

Because of the very large divergence, laser diode beams must be collimated for effective use. The collimating lenses must have a large numerical aperture, ideally $NA > 0.5$. If a single lens is used to collimate the beam, the lens must be aspheric. LightPath Inc. offers a series of molded aspheric lenses specially designed for collimating or focusing laser diode beams. These lenses maintain their original trademark GelTech™ and are the most widely used laser diode lenses. Spherical lens groups are used less often to collimate laser diode beams because of their large size and weight.

Figure 9.7 shows a Zemax-generated raytracing diagram for a laser diode collimating lens made by LightPath Technologies (part number 354453) as an example. This lens has a focal length of 4.6 mm, a NA of 0.5, a clear aperture of 5.1 mm, an aspheric convex surface, and a planar surface. When the lens is designed, a parallel ray bundle is incident on the lens, and the shape of one surface of the lens is optimized to minimize the focused spot. This lens has a

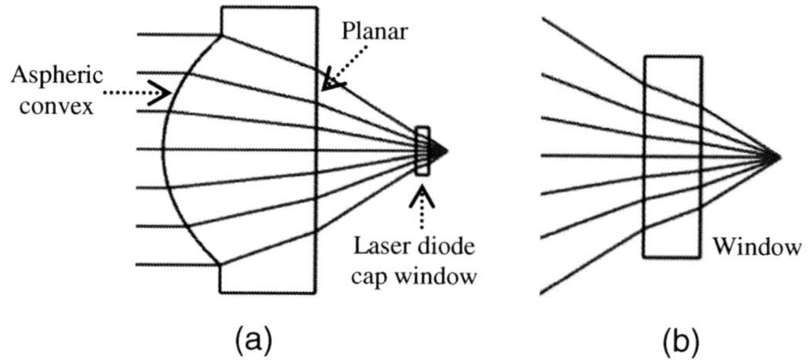

Figure 9.7 (a) Zemax-generated raytracing diagram for a laser diode collimating lens made by LightPath Technologies, part number 354453. (b) Detailed view of the focusing area.

1.5-μm Airy disk radius and a near-perfect RMS focused spot size of 0.011 μm. As a comparison, if no aspheric surface is used, even with a glass with index value of 2, the RMS focused spot size will be about 200 μm.

When the lens is used, the rays are emitted by the laser diode, which can be considered as a point source, travel in the direction opposite that shown in Fig. 9.7 through the lens, and are collimated.

Lenses that can well collimate beams from laser diodes can also well focus collimated beams of any type of laser.

9.4.3 Spatial evolution of collimated laser diode beams

The spatial evolution of a single-TE-mode laser diode beam is illustrated in Fig. 9.8. The beam shape changes as it travels. The far-field beam shape is always similar to the beam waist shape at the laser diode facet. Note that the focused spot is optically at the far field. Therefore, the focused spot shape is also similar to the beam waist shape at the laser diode facet.

9.4.4 Far-field pattern of collimated beams

Because laser diode beams have large divergence in the fast axis direction, most single-element, aspheric collimating lenses do not have a numerical aperture large enough for the beams. As a consequence, the beam will be truncated in the fast axis direction by the collimating lens, as illustrated in Fig. 9.9(a). The far-field pattern of a collimated beam will have truncation-generated diffraction

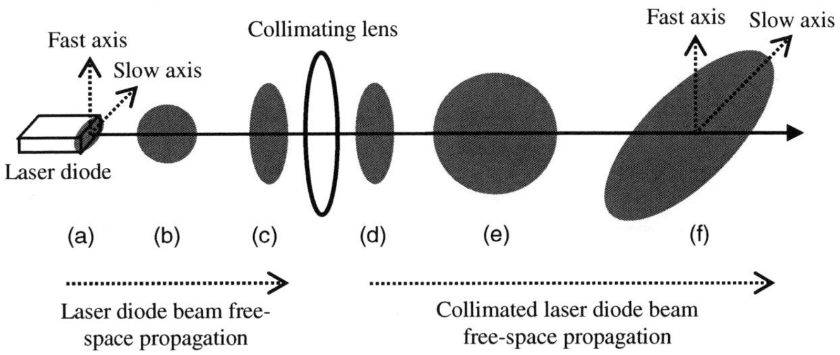

Figure 9.8 Illustration of the evolution of a laser diode beam as it propagates. (a) Elliptical beam waist at the laser-diode active layer facet with the minor axis of the beam waist and larger divergence in the fast axis direction. (b) After propagating tens of microns, the beam shape becomes circular at a certain point because the divergence in the fast axis direction is larger. (c) As the beam continues propagating, the beam shape becomes elliptical with the major axis in the fast axis direction. (d) If there is a lens to collimate the beam, the beam shape is about the same shortly before and shortly after the lens. The collimated beam has larger divergence in the slow axis direction. (e) After propagating several meters, the collimated beam shape becomes circular at a certain point because now the divergence in the slow axis direction is larger. (f) As the beam propagates further, the beam shape becomes elliptical again with the major axis in the slow axis direction again.

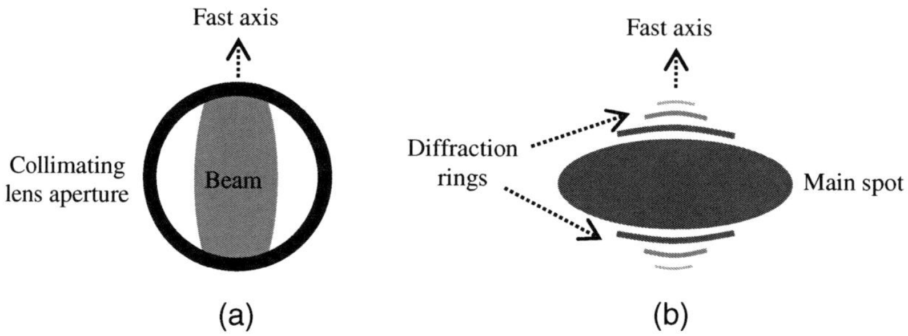

Figure 9.9 (a) Most single-element aspheric collimating lenses will truncate a laser diode beam in the fast axis direction. (b) The far-field beam pattern, including the focused spot, will have diffraction rings in the fast axis direction caused by lens aperture truncation.

rings in the fast axis direction, as illustrated in Fig. 9.9(b). The main beam spot and the diffraction rings should have a symmetric shape; if not, the collimating lens is not well centered with the laser beam. Since the focused spot is optically at the far field, the focused spot will have diffraction rings similar to that shown in Fig. 9.9(b).

9.4.5 Circularization and astigmatism correction

Elliptical cross-sections and astigmatism are the two major shortcomings of laser diode beams. Several techniques have been developed to circularize the elliptical beams and correct the astigmatism.

The most common technique uses a pair of anamorphic prisms to expand the elliptical beam in the minor axis direction, or in a reversed way compress the elliptically shaped beam in the minor axis direction, as shown in Fig. 9.10. The beam expanded/reduced by the prism pair does not have an ideal circular shape. A circular aperture is often used to further circularize the beam, as illustrated in Fig. 9.11(a).

However, the circularized beam still has astigmatism, which means that the beam cannot be truly collimated simultaneously in both the slow and fast axis directions. If the beam is truly collimated in the slow axis direction, the beam will be slightly divergent in the fast axis direction, as shown in Fig. 9.11(b). A weak cylindrical lens, e.g., a positive one, can be used to further collimate the beam in the fast axis direction. The focal length of this cylindrical lens is about 1 m or so. Since every laser diode of the same type can have a slightly different magnitude of astigmatism, the power of the cylindrical lens cannot match every laser diode. The cylindrical lens can be rotated about the beam axis to move part of the focusing power from the fast axis direction to the slow axis direction, but the rotation angle must be chosen so that the resultant divergences in both the slow and fast axis directions are the same. Thus, the astigmatism of the circularized beam is corrected. Visual inspection of the

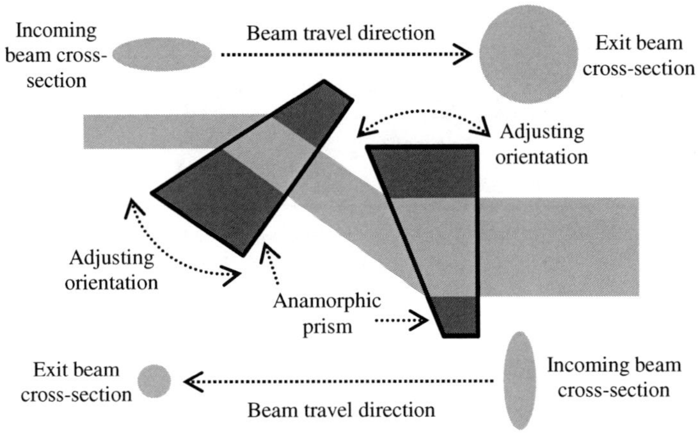

Figure 9.10 An anamorphic prism pair is used to circularize an elliptical beam by expanding (compressing) the beam size in the minor (major) axis direction. The beam travel direction determines whether the beam is expanded or compressed. The expansion/ compression ratio can be adjusted by rotating the two prisms.

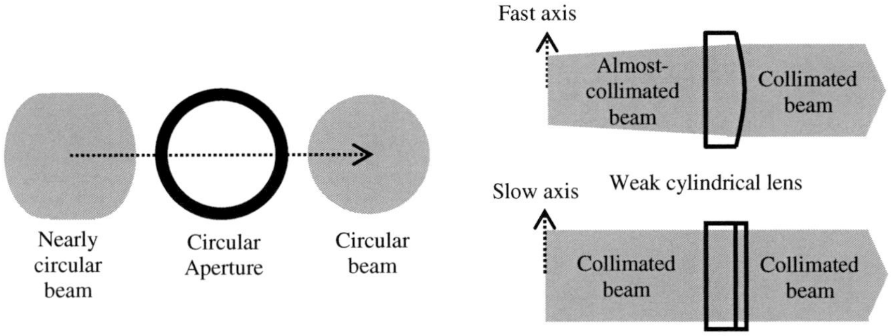

(a) (b)

Figure 9.11 (a) The shape of a beam circularized by an anamorphic prism pair is nearly circular. After the beam passes through a circular aperture, the beam shape becomes circular. (b) Even when the beam is circularized, the astigmatism in the beam will prevent the beam from being truly collimated in both the slow and fast axis directions. If the beam is truly collimated in the slow axis direction, it will be slightly divergent in the fast axis direction. A weak positive cylindrical lens can further collimate the beam in the fast axis direction.

beam is not accurate enough to evaluate the result of the astigmatism correction. A wavefront sensor can be used to monitor the process.

Another way of circularizing an elliptically shaped beam and correcting astigmatism couples the laser diode beam into a single-mode optical fiber. Regardless of the input beam characteristics, the output beam will have a suitable circular shape and no astigmatism. The output beam from the fiber has too large divergence, and must be collimated by a lens for effective use, as

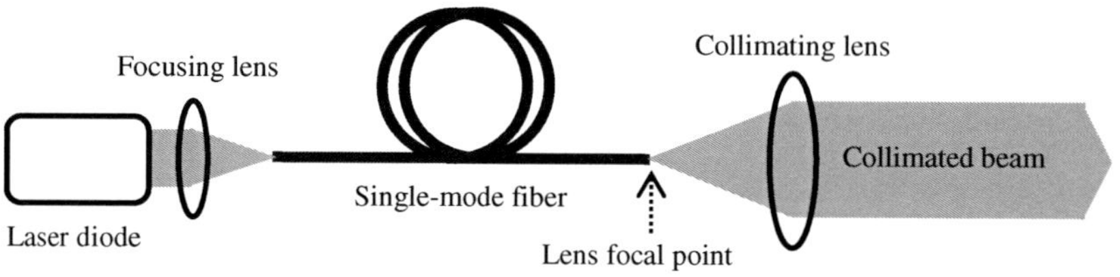

Figure 9.12 Schematic of coupling a laser diode beam into a single-mode optical fiber. The beam output from the fiber is circular and astigmatism free, regardless of the input beam characteristics.

shown in Fig. 9.12. Coupling a laser diode beam into a single-mode fiber with high coupling efficiency is not easy, as will be discussed in Section 9.7.1.

The second method of circularizing an elliptical beam and correcting astigmatism uses a special micro-acylindrical lens, as illustrated in Fig. 9.13. The microlens only works on the beam in the fast axis direction. The front surface of the microlens collimates the beam in the fast axis direction. The lens thickness is so chosen that at the back surface of the lens, the beam sizes are the same in the fast and slow axis directions. The back surface of the lens then diverts the beam in the fast axis direction to a divergent angle that is the same as the divergent angle in the slow axis direction. Thus, the elliptical beam is circularized, and the astigmatism is corrected. A collimating lens must be used to collimate the circularized and astigmatism-corrected beam.

The microlens has a focal length of a fraction of millimeter. If the laser diode is sealed inside a cap, the cap must be cut open so that the microlens can be placed close enough to the laser diode. Both the cutting of the cap and the

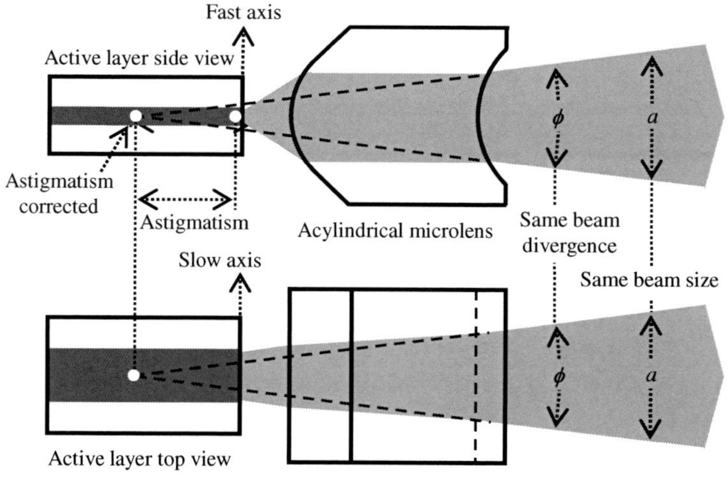

Figure 9.13 Schematic of an acylindrical microlens that works on the beam in the fast axis direction. The beams output from the lens have the same sizes and same divergences in the fast and slow axis. Thus, the elliptically shaped and astigmatic beam is circularized, and the astigmatism is corrected.

mounting of the lens are not easy. The microlens will cause a lot of scattering because it is placed very close to the laser diode, where the beam size is small and the light intensity is high. The alignment of the microlens to the laser diode is also sensitive. A wavefront sensor can be used to monitor the alignment. After the microlens is mounted, the cap must be filled with inert gas and then sealed.

The third method to circularize the elliptical beam and correct astigmatism uses two cylindrical lenses in an orthogonal way, as shown in Fig. 9.14. The first cylindrical lens is acylindrical to minimize the "spherical" aberration, has a large NA to reduce beam truncation, and collimates the beam in the fast axis direction. The second cylindrical lens collimates the beam in the slow axis direction. The focal length and position of the second cylindrical lens are chosen so that the beam size after being collimated equals the beam size in the fast axis direction, as illustrated in Fig. 9.14. The astigmatism is also corrected. If the beam comes from a wide-stripe, multi-TE-mode laser diode, the collimated beam is still divergent in the slow axis direction, as will be illustrated later in Fig. 9.15(b). The two cylindrical lenses can be combined to form one very thick cylindrical lens with two toroidal surfaces.

9.5 High-Power, Wide-Stripe Laser Diode Optics

In general, beams from wide-stripe laser diodes, bars, or piles can never be collimated or focused as well as single-TE-mode beams.

9.5.1 Beam characteristics

High-power laser diodes have one wide, active layer with a width from tens of microns to a couple of hundred microns. Such laser diodes are linear light sources. Some high-power laser diodes have several wide-stripe laser diodes

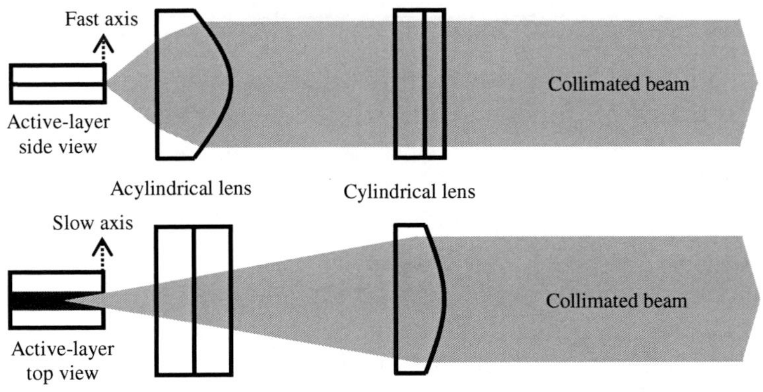

Figure 9.14 Schematic of two cylindrical lenses used to collimate a laser diode beam and to circularize and correct astigmatism.

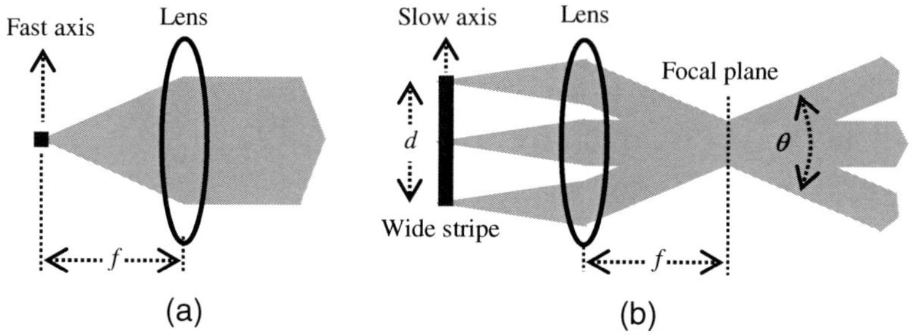

Figure 9.15 Illustration of collimating a wide stripe laser diode beam. (a) The beam can be well collimated in the fast axis direction by using a proper aspheric lens. (b) There are many TE modes along the active layer stripe in the slow axis direction. Three representative TE modes are considered here in an exaggerated way. Although the beam from every TE mode can be collimated, the combination of all the collimated beams still has a divergence angle of $\theta = d/f$, where f is the focal length of the collimating lens, d is the wide stripe length.

stacked together, as illustrated in Fig. 8.14(a). Such laser diodes are rectangular light sources. Some high-power laser diodes have several wide-stripe laser diodes mounted side by side on a bar, as illustrated in Fig. 8.15(a). Such laser diodes are linear light sources with several line sections and up to several millimeters of total length. A very high laser power can be obtained by piling up several laser bars, as illustrated in Fig. 8.15(b). Such laser diodes are rectangular light sources with a size of several millimeters or up.

Because high-power laser diodes are either long and linear or rectangular light sources, true collimation of the beams from wide-stripe laser diodes is impossible, although the term "collimating" is still used. Some optical devices have been invented to reduce the divergence of wide-stripe laser diode beams.

9.5.2 Collimating a wide-stripe laser diode beam

As illustrated in Fig. 9.15(a), the beam at the active layer facet has a tiny size of about 1 μm in the fast axis direction and can be well collimated by using a proper aspheric lens. In the slow axis direction, a wide-stripe light source is a line and has many TE modes. Although the beam from each TE mode can be well collimated, not all of the collimated beams are parallel, as illustrated in Fig. 9.15(b). The combination of all of the collimated beams have a divergent angle of $\theta = d/f$, where d is the wide-stripe length, and f is the lens focal length.

Considering a typical case of $d = 100$ μm and $f = 5$ mm, $\theta = 0.02$, which is not negligible. For example, if the beam needs to work at a 1-m distance, the beam size in the slow axis direction will be 0.02×1 m $= 20$ mm, which is not a "focused spot." This explains why the setup shown in Fig. 9.15 cannot well collimate the beam in the slow axis direction if the beam comes from a wide-stripe laser diode.

If the beam is from a laser diode stack like that shown in Figs. 8.14(a) or 8.15(b), the beam cannot be well collimated in both the slow and fast axis directions.

9.5.3 Slow axis collimator

To further increase the laser power, several wide stripe laser diodes can be mounted in a bar with a total length up to several millimeters, as illustrated in Fig. 9.16. If a single lens is used to collimate the beams from the bar, the divergence of the combination of all of the collimated beams is $\phi = L/f$, as shown in Fig. 9.16(a), where L is the length of the bar.

A "slow axis collimator" consists of a micro-cylindrical lens array and can be used to collimate the beams from the bar, as shown in Fig. 9.16(b). The collimated beams are parallel to each other, and the divergence of the combination of all collimated beams is $\theta = d/f$, where d is the width of individual laser diode active layer. Since $d \ll L$, the slow axis collimator can collimate the beams from a bar better than a single lens can. However, a divergence of $\theta = d/f$ is not considered to be "well collimated."

9.5.4 Fast axis collimator

A "fast axis collimator" is an acylindrical microlens bar that is longer than the wide-stripe laser diode or the laser diode bar and can simultaneously collimate in the fast axis direction the beam(s) from the wide stripes or the bar, as shown in Fig. 9.17. A fast axis collimator usually has a small focal length <1 mm, a large numerical aperture of 0.8 or so to avoid truncating the beam, and an

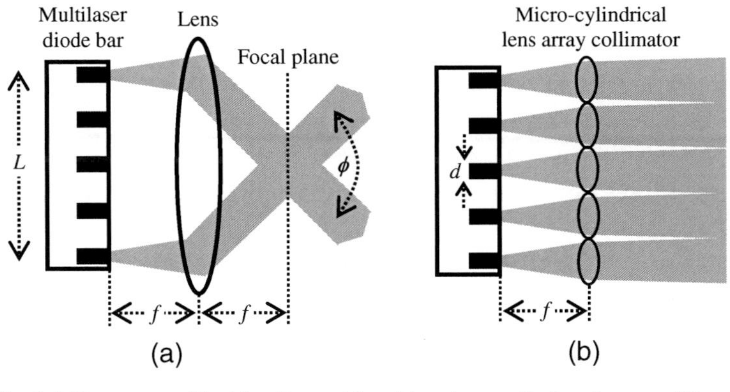

Figure 9.16 (a) Beams emitted by five wide-stripe laser diodes in a multilaser diode bar with length L are collimated by a single lens with focal length f. Only the beams from the two side bars are drawn here for clarity. The combination of all of the collimated beams has a large divergence of $\phi = L/f$. (b) A slow axis collimator consists of a micro-cylindrical lens array with focal length f. Each microlens collimates the beam from one stripe laser diode. Each collimated beam still has a divergence θ, as shown in Fig. 9.15(b). All of the collimated beams are parallel.

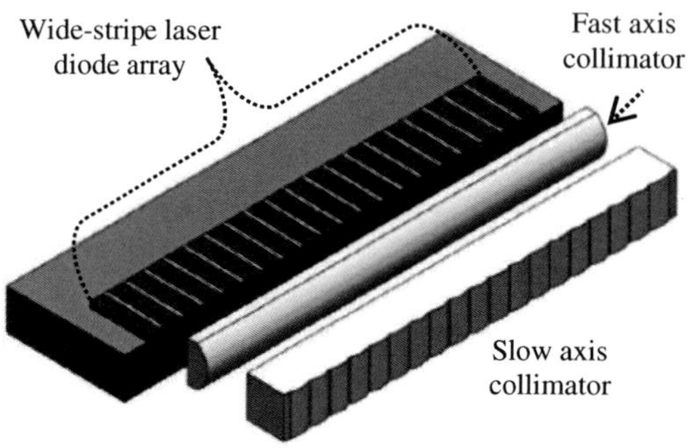

Image courtesy of PowerPhotonic

Figure 9.17 A fast axis collimator and a slow axis collimator can be used together to collimate the beams from a wide-stripe laser diode bar.

acylindrical surface to minimize the "spherical" aberration caused by the large divergence of the beam(s) in the fast axis direction.

Fast axis collimators have a very short focal length and are placed very close to the laser diode(s), where the intensity of the beam(s) is high, and the scattering caused by the collimator will be strong.

A fast axis collimator and a slow axis collimator can be used together to collimate in both fast and slow axis directions the beams from a multilaser diode bar, as shown in Fig. 9.17.

9.6 Laser Beam Shaping Optics

The term "shaping" can refer to either the beam shape or beam intensity profile.

9.6.1 Laser line generator

Laser lines are used in construction, industrial, and medical alignment. The simplest laser line generator is a glass rod. A cylindrical lens, either positive or negative, can also be used to generate a laser line.

Figure 9.18(a) shows a Zemax-simulated raytracing diagram and an intensity pattern of a laser line generated by a glass rod. The incident laser beam has a Gaussian intensity pattern. The intensity pattern of the laser line is near Gaussian with tails longer and weaker than those of Gaussian profile. However, in most applications, the desired intensity patterns are flat.

Axicon lenses are developed to improve the flatness of the laser line intensity. An axicon lens is an acylindrical lens with strong optical power around the tip to spread the high intensive central portion of the laser beam incident on the lens. The edge of an axicon lens does not have much optical power, and so the laser line generated is flatter. Figure 9.18(b) shows a Zemax-simulated raytracing diagram and an intensity pattern of a laser line generated by an axicon lens. The intensity profile of the incident laser beam is still Gaussian, whereas the intensity profile of most of a laser line is much flatter.

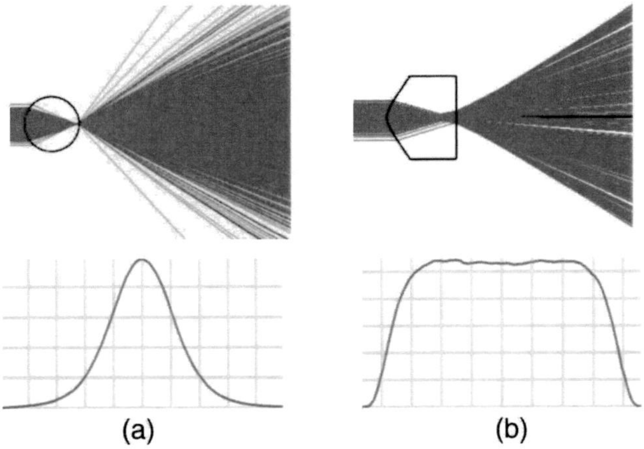

Figure 9.18 Two laser line generators. (a) A glass rod. The intensity pattern of the laser line generated by a glass rod is not flat. (b) An axicon lens. The intensity pattern of the laser line generated by an axicon lens is much flatter.

9.6.2 Flat-top beam shaper

The natural intensity pattern of a laser beam is Gaussian, but flat intensity patterns are often desired. Flat-top laser beam shapers have been developed to transform Gaussian intensity patterns to flat-top patterns.

Flat-top beam shapers utilize aspheric surfaces to transform the intensity pattern. Figure 9.19(a) shows a Zemax-generated raytracing diagram for a simple flat-top beam shaper that consists of two lenses. The back surface of the first lens is conic and has more power at the center of the lens than at the edge. Thus, the central portion of the beam incident on the lens is more spread. All three other surfaces of the two lenses are spherical. The input beam has a 1-mm diameter and a Gaussian intensity pattern.

The flat-top or nearly-flat-top intensity pattern can only be maintained within a certain range. For this simple flat-top beam shaper, the Zemax-simulated intensity patterns and profiles at three different distances are plotted in Figs. 9.19(c)–(e). For any intensity patterns at the near field, the intensity pattern at a distance much larger than the beam size will become nearly Gaussian because of diffraction. To maintain a flatter and/or larger flat-top range, more lenses are needed.

9.6.3 Beam expanders

Beam expanders can have two types: two positive lenses, as shown in Fig. 9.20(a), or one negative and one positive lens, as shown in Fig. 9.20(b). Both beam expanders shown here use simple N-BK7-glass spherical lenses, have the same magnification of 3X, the same diameter, and lens surfaces that are optimized to minimize spherical aberration.

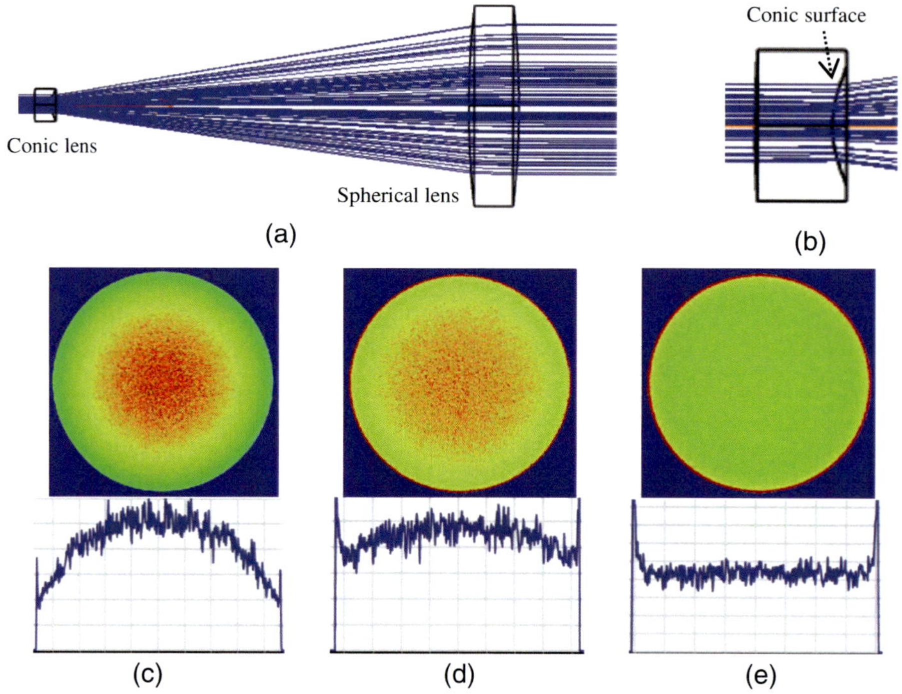

Figure 9.19 (a) Zemax-generated raytracing diagram of a simple two-lens flat-top beam shaper with one conic surface. (b) The detailed view of the conic lens. (c)–(e) The Zemax-simulated, 2D false-color beam intensity patterns and intensity profiles for this beam shaper at a distance of 250 mm, 350 mm, and 450 mm, respectively.

The second type of beam expander performs much better than the first type, as proven by the focused spot size shown in Figs. 9.20(c) and (d), because the two lenses in the second type have negative and positive types of spherical aberrations that offset each other. Furthermore, the second type can accept a larger-angle incident beam without clipping the beam and is only half the length of the first type of beam expander.

Note that if a beam is expanded N times, the field angle of the expanded beam will simultaneously be decreased by about N times, where N can be any positive number. The product of the beam size and divergence is a constant. This is the etendue of geometrical optics; it cannot be changed. If a beam expander is used in reverse, it becomes a beam reducer, and so $N < 1$.

9.6.4 Making a linear light source rectangular

In some applications, such as a laser rangefinder, a beam shaper is used to transform the linear beam of a wide-stripe laser diode into a rectangular shape to minimize the beam size at a long distance. This type of beam shaper neither collimates nor focuses the beams; therefore, it is included in Section 9.6 rather than Section 9.5.

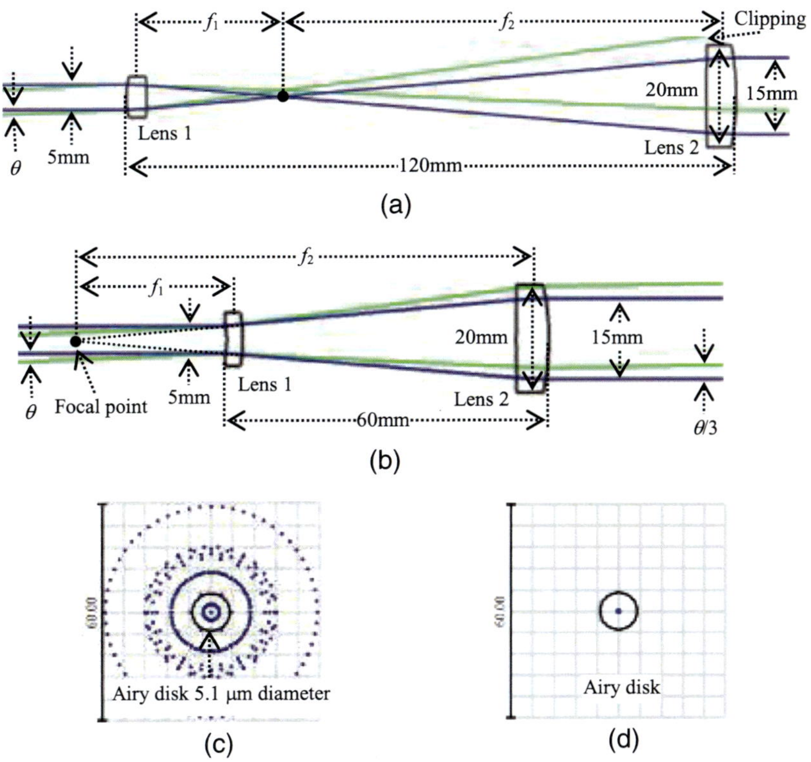

Figure 9.20 (a) Zemax-generated raytracing diagram of a 3X beam expander consisting of two N-BK7-glass spherical positive lenses and an aperture of 20 mm. If the incident beam has an angle $\theta = 2.4°$ to the optical axis, part of the beam will be clipped. (b) Zemax-generated raytracing diagram of a 3X beam expander consisting of two N-BK7-glass spherical lenses, one positive and one negative, and an aperture of 20 mm. If the incident beam has an angle $\theta = 2.4°$ to the optical axis, no beam clipping occurs. (c) A 100-mm-focal-length paraxial (perfect) lens is used to focus the expanded beam in (a); the RMS focused spot radius is 17.0 μm, much larger than the Airy disk radius of 5.1 μm because the residual spherical aberration is still large. (d) A 100-mm-focal-length paraxial (perfect) lens is used to focus the expanded beam in (b); the RMS focused spot radius is much smaller than the Airy disk radius of 5.1 μm because the two lenses have negative and positive types of spherical aberration that well cancel each other.

As explained in Fig. 9.15(b), even when collimated, the slow-axis-direction divergent angle of a wide-stripe laser diode beam is still large. For example, when the stripe length is 100 μm and the collimating lens focal length is 10 mm, the collimated beam still has a divergence of 100 μm / 10 mm = 0.01 in the slow axis direction. If delivering such a beam to a target 1000 m away, the beam in the slow axis direction will have a size of 0.01×1000 m = 10 m and low energy density.

Various beam shapers were invented to make a linear beam rectangular. Figure 9.21 shows the schematics of such a beam shaper. The linear light with

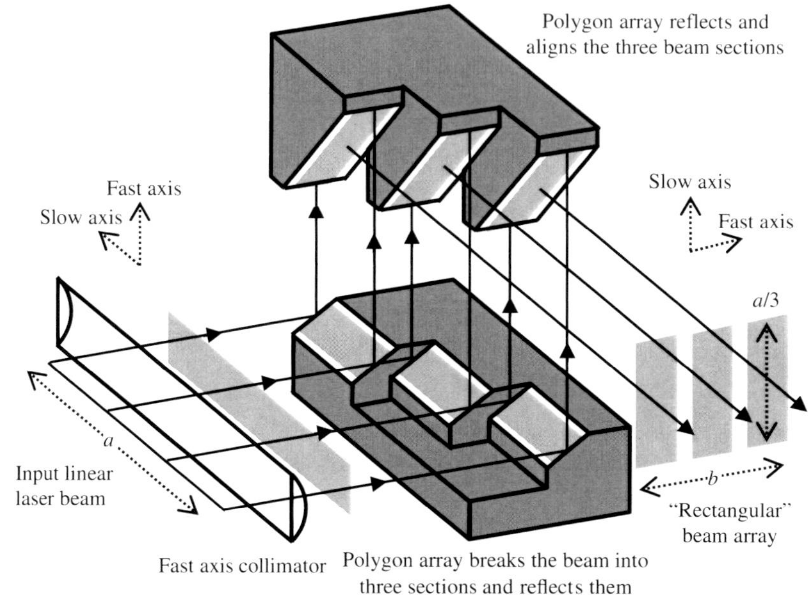

Figure 9.21 Schematics of a beam shaper that consists of two polygon arrays. Each array consists of three reflectors. A linear incident beam with length *a* from a wide-stripe laser diode is first collimated by a fast axis collimator and then transformed to three sections of a beam by the beam shaper. The three sections of the beam form a rectangular beam with size $a/3 \times b$.

length a is first collimated by a fast axis collimator and has a long, rectangular shape that is then incident on the lower polygon array. The array breaks the incident light into three sections and reflects them upward toward the upper polygon array, which reflects and realigns the three section of light. Three such sections of light form a rectangular light array with size $a/3 \times b$. A long-focal-length cylindrical lens must be used to collimate the rectangular light array in the slow axis direction. The slow axis divergence is reduced to 1/3 of the original divergence by the beam shaper. Some beam shapers can break the beam into five sections or more to further reduce the slow axis divergence.

9.7 Laser Devices

9.7.1 Pigtail laser modules: coupling a single-TE-mode laser beam into a single-mode optical fiber

"Pigtail" laser modules are the nicknames of lasers with an output beam coupled into a single-mode optical fiber. The vast majority of pigtail laser modules use laser diodes, although other types of lasers, such as He-Ne lasers, are sometimes used. Once a laser beam is coupled into an optical fiber, it is confined inside the fiber and guided by the fiber through a zigzag path to the working spot.

The coupling lens (group) is a positive lens (group) that focuses the laser beam onto the fiber core facet. To minimize the coupling loss, the focused spot diameter should be smaller than the fiber core diameter, and the focused cone angle should be smaller than the fiber acceptance angle given by Eq. (6.15).

Equation (2.18), which is rewritten here for convenience, relates the diffraction-limited focused spot radius a to the wavelength λ, lens focal length f, and beam size D:

$$a = \frac{1.22 f \lambda}{D}. \tag{9.4}$$

If the fiber core diameter is 7 μm ($2a \leq 7$ μm), the laser wavelength is $\lambda = 0.633$ μm, and the coupling lens focal length is $f = 3$ mm, then D can be found from Eq. (9.4) to be $D \geq 0.66$ mm. This result means that in order to produce a diffraction-limited spot diameter of <7 μm, the beam diameter must be >0.66 mm. Many laser beams are larger than 0.66 mm.

For such a beam, the focusing cone angle in terms of numerical aperture is $NA = 0.66\ \text{mm} / (2f) = 0.11$, which is smaller than the NA of most types of single-mode fibers. To achieve a diffraction-limited focused spot size with $F/\# = 1/(2NA) \approx 4.5$, the coupling lens needs to have one mildly aspheric surface.

Aligning a micron-size focused spot to a micron-size fiber core is somewhat challenging. Three translation stages are required. One stage controls the movement in the optical axis direction, and the other two stages control the movements in the two transverse directions, respectively. The transverse alignment is very sensitive. If the coupling lens has good quality and the alignment is good, a coupling efficiency of ~90% is achievable for He-Ne laser beams. In this case, fiber-core facet reflection is the primary loss factor.

Pigtail laser diode modules are commonly seen because laser diodes are the most widely used lasers and can be battery operated; furthermore, a single-mode fiber can circularize the elliptical shape and correct the astigmatism of a laser diode beam. However, coupling single-TE-mode laser diode beams into a single-mode fiber is more complex than coupling circular He-Ne laser beams because laser diode beams are elliptical. If a laser diode beam is focused with a spot size in the fast axis direction that matches the fiber core size, then the focused spot size in the slow axis direction will be a few times larger than the fiber core size that will cause a large coupling loss. If a short-focal-length lens is used to focus the laser diode beam with a spot size in the slow axis that matches the fiber core size, then the focusing cone angle in the fast axis direction will be a couple of times larger than the fiber acceptance angle that will cause another large coupling loss. A "balanced" approach is the best, as illustrated in Fig. 9.22. The focused spot in the slow axis direction is slightly

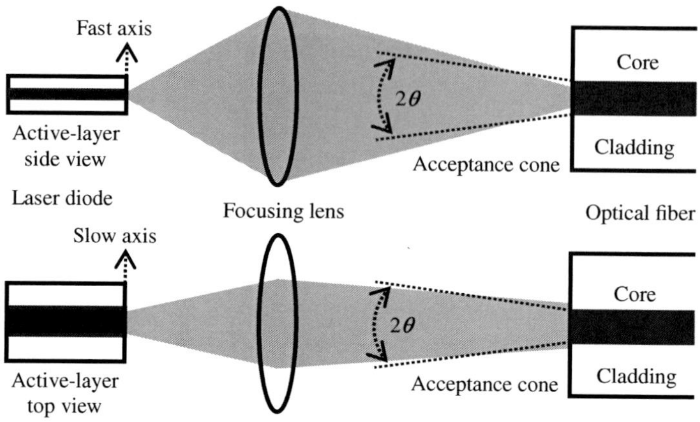

Figure 9.22 Illustration of a "balanced" approach to coupling a single-TE-mode laser diode beam into a single-mode optical fiber. In the fast axis direction, the focused spot at the fiber facet is slightly smaller than the fiber core size, but the focusing cone angle is slightly larger than the fiber acceptance angle. In the slow axis direction, the focused spot at the fiber facet is slightly larger than the fiber core size, but the focusing cone angle is slightly smaller than the fiber acceptance angle.

larger than the fiber core size. The focusing cone angle in the fast axis direction is also slightly larger than the fiber acceptance angle. The realistic coupling efficiency depends on the ellipticity of the laser diode beam involved. A coupling efficiency of 60% can be used for estimates.

Pigtail laser modules with multimode optical fiber are rare, but coupling a laser beam into a multimode fiber is much easier than into a single-mode fiber.

9.7.2 Laser diode power stabilization

Laser diodes have a long life of tens of thousands of hours if they are properly operated. However, laser diodes are easily damaged with electrical static, surges in the driven electrical current, or by overdriving. Laser diodes are usually built as modules for protection, as shown in Fig. 8.16(b).

The two most basic operational modes of laser diode modules are constant current and constant power. The former means that the electronics in the module stabilizes the driven current of the laser diode. The latter is more complicated. The back facet of the active layers of laser diodes are high-reflection coated. A photodiode is often mounted behind the back facet of laser diode active layers to pick up the laser power leaking through the high-reflection coating. The output electrical current of the photodiode is proportional to the output laser power. This current is used in a feedback loop as an indicator to automatically adjust the driven current of laser diodes to stabilize the laser power.

In constant-current mode, the laser power can vary because temperature variations change the lasing efficiency and threshold. In constant-power mode, the driven current also varies due to temperature variations. Therefore, a laser diode module cannot simultaneously operate in both modes unless the temperature of the laser diode is accurately stabilized.

9.7.3 Laser diode temperature stabilization

Some laser diode modules have their temperature and their power/current stabilized to stabilize the wavelength. Because the laser diode active layer where lasing occurs has a needle-tip size, its temperature cannot be directly measured by touching it with a thermistor. The so-called "temperature stabilization" actually only stabilizes the temperature of the metal heat sink of the laser diode. There is a thermal resistance, a temperature gradient, and a temperature-change time delay between the laser diode active layer and the heat sink. A stable environment and some time are necessary for the laser diode active layer to reach thermal equilibrium with the heat sink. If the driven current of laser diodes is tuned quickly and frequently, the active layer temperature cannot be effectively stabilized.

9.7.4 Laser rangefinders

Before the invention of lasers, it required significant time and effort to find the distance to an object. The object is viewed from two locations A and B with known distance l between them. The object and the locations A and B form a triangle. At location A, the angle θ_A between the viewing direction and the line connecting A and B can be measured. At location B, the angle θ_B between the viewing direction and the line connecting A and B can also be measured. The distance to the object from either location A or B can then be calculated utilizing triangular geometry and data l, θ_A, and θ_A. To reduce the measurement error, l should not be much smaller than the distance under measurement.

Modern laser rangefinders emit laser pulses to the object under measurement. The object scatters the laser pulses. The scattering back signal is detected by the laser rangefinder. The round-trip time t a laser pulse takes from the laser rangefinder to the object and back is measured. The distance d to the object can be calculated as $d = ct/2$, where c is the velocity of light in air. Since laser rangefinders are often used in open areas, the laser should be battery operable. Laser diodes are the best choice.

To measure the distance to an object that is kilometers away, the laser diode must have high pulse energy. Wide-stripe laser diodes are the optimal choice. On the other hand, the laser beam should have the smallest possible divergence so that the laser spot projected onto the object can be small with a high energy density. Wide-stripe laser diodes have linear light sources with a length up to hundreds of microns, and such laser beams collimated by conventional collimating lens will have large divergences in the light-source line direction. For example, a collimating lens with a

100-mm focal length collimates the beam from a wide-stripe laser diode with a source line length of 100 μm. The collimated beam will have a divergence of 100 μm / 100 mm = 1 mR in the source line direction. Such a "collimated" laser beam will make a 1-m-long linear spot on an object 1 km away. (This topic was discussed in Section 9.5.2.) To reduce the collimated beam divergence, some type of beam-shaping optics, similar to the prism array shown in Fig. 9.21, must be used to optically convert the linear light source to a square or rectangular light source.

Various commercial laser rangefinders can provide a ranging distance from about half a meter to over ten kilometers. The ranging accuracy can be up to ±0.5 m for an object distance of 1 km.

9.7.5 Laser detection cards and viewing scopes

UV and IR laser beams are invisible to human eyes. Laser detection cards and viewing scopes have been developed to convert invisible radiation to visible, not only for the convenience of handling lasers but also for laser safety. Detection cards in the visible range are also used to avoid directly viewing laser beams.

Detection cards are made by coating a phosphor layer on substrates. This layer absorbs light in certain spectral ranges and emits visible light. There are several selections of detection cards from the UV to the MIR. UV photons have higher energy than visible photons. The conversion of a UV photon to a visible photon is a relatively simple process. Although IR photons have lower energy than visible photons, an IR photon must get energy from some other energy source so that it can be converted to a visible photon. Room light is usually the other energy source, and the energy from room light is stored in the phosphor layer. After being illuminated by an IR laser beam and emitting visible light for a while, the stored energy is depleted, after which the card must be charged. This phenomenon becomes more apparent as the IR wavelength increases. In real applications, the user can move the card somewhat so that the laser beam will hit different portions of the card where the stored energy is not yet depleted. The charging or recovery time is several seconds.

The minimum power to stimulate visible emission is about 1 nW/cm^2 in the UV range, a few μW/com^2 in the visible, and about 100 μW/com^2 in the NIR. Laser detection cards cost only tens of dollars per piece.

Laser detection cards can approximately display the intensity pattern of a laser beam incident on the cards, but they cannot show the details of an object that emits UV or IR radiation. Viewing scopes are one step further forward. They see not only a laser spot on an object but also the details of an object in the UV or IR. Commonly used viewing scopes are IR viewing scopes, but they actually cover a spectral range from 350 nm to 1550 nm. They are battery operated, and the current price is over $2000 per unit.

9.7.6 Spatial filters

A spatial filter is an optical device used to "clean up" an aberrated laser beam. A spatial filer consists of a high-quality spherical-aberration-free lens and an opaque mask with a pinhole at the center, as shown in Fig. 9.23. The pinhole diameter almost equals the Airy disk size of the lens. An incident laser beam with an aberrated wavefront is focused by the lens onto the mask. The nonspherical components of the wave, which are the aberration in the wave, can only be focused to a spot larger than the pinhole and is blocked by the opaque mask. Only the spherical component of the incident beam can be focused to a spot small enough to pass through the pinhole. Thereby, the beam passing through the pinhole is aberration free and clean.

9.7.7 Laser speckle

When a laser beam is reflected by a surface with random micro-structures and then illuminates an object, the illumination will have a random granular pattern, as shown in Fig. 9.24(a). Such a pattern is called laser speckle. Because the reflection surface is not optically smooth, it is an assembly of a large number

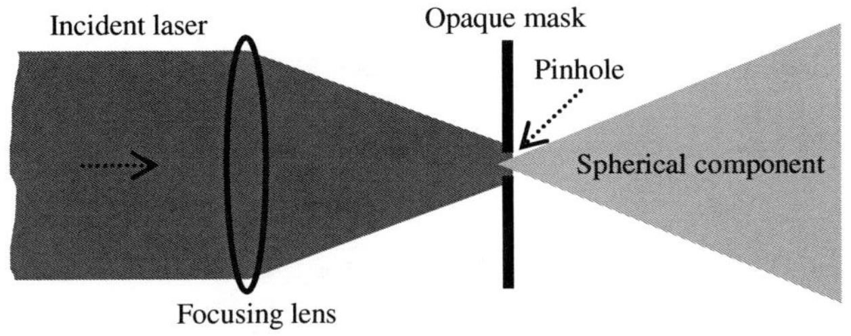

Figure 9.23 Schematics of a spatial pinhole filter that consists of a high-quality lens and an opaque mask with a pinhole at the center. The pinhole size nearly equals the Airy disk size of the lens. A laser beam is focused onto the mask. Only the spherical component of the laser beam wavefront can be focused to a spot small enough to pass through the pinhole.

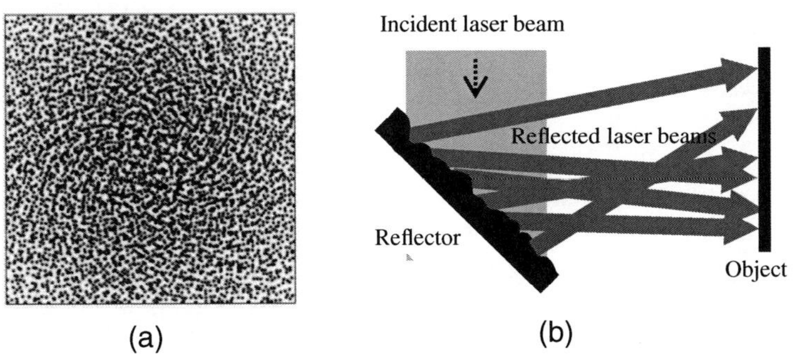

Figure 9.24 (a) A typical laser speckle pattern. (b) A reflector with random micro-structures decomposes an incident laser beam into a large number of micro-beams that meet at an object and form a random interference pattern called laser speckle.

of micro-reflectors with random sizes and orientations. Each micro-reflector reflects the laser beam incident on it in a different direction with a different intensity and phase, as illustrated in Fig. 9.24(b) in a dramatically simplified and exaggerated way. These micro-beams meet each other at the object and form an interference pattern, which is laser speckle. A laser beam passing through a diffusive medium can also form speckle on an object. Laser speckle is very sensitive to physical disturbance. A subwavelength magnitude movement of the reflector can completely change the speckle pattern.

Laser speckle can significantly reduce the image quality. If laser beams are used for illumination, the possible speckle must be considered. One way to get rid of speckle is to let the laser beam pass through a jittering diffuser. The speckle is also jittered and evens out. The speed of physical movement is often much slower than the speed of electronics. If high-speed images are taken of the object being illuminated by speckle, the jittering of the diffuser will be too slow.

If laser diodes are used for illumination, there is a way to get rid of speckle. The wavelength of laser diodes is affected by the driven current magnitude and can be modulated by modulating the driven current. When the wavelength changes, the speckle pattern also changes. The modulation speed of laser diodes can easily be megahertz or higher. As long as the modulation speed is faster than the detector speed, the speckle pattern will even out.

The wavelength of most other types of lasers, such as He-Ne lasers, cannot be modulated, and laser speckle created by these lasers cannot be easily eliminated.

9.8 Opto-electrical Devices and Other Devices

The working principles of some frequently used opto-electrical devices are briefly described in this section. These devices can be used with or without lasers.

9.8.1 Basic principles

In general, the refractive index n of any material is a complex number and can be written as

$$n = n_r + in_i, \tag{9.5}$$

where i is the imaginary number, and n_r and n_i are the real and imaginary parts of the index, respectively. In most literature, including most of this book, only the real part of the refractive index is considered, whereas the imaginary part of the index can be important in certain circumstances.

When an optical field $E(z,t)$ propagates through a piece of material, the electrical field can be written as

$$E(z,t) = E_0 \exp(inz + i\omega t), \tag{9.6}$$

where E_0 is the field amplitude, z is the coordinator in the field propagation direction, ω is the angular frequency of light, and t is the time. Inserting Eq. (9.5) into Eq. (9.6) produces

$$E(z,t) = E_0 \exp(in_r z + i\omega t) \times \exp(-n_i z), \tag{9.7}$$

where the term before the multiplication symbol is a standard wave propagation term that contains the commonly considered refractive index n_r, and the term after the multiplication symbol represents the attenuation of the material. If the n_r of a material is a function of the electrical field, the value of n_r can be modulated by modulating the electrical field. Such a material can be used to make a phase modulator or focusing adjustable lens. Similarly, if the n_i of a material is a function of the electrical field, the value of n_i can be modulated by modulating the electrical field. Such a material can be used to make an amplitude/intensity modulator or switch.

The refractive index of all materials is somehow affected by the electrical field applied to the materials. However, in most materials the effects are weak and negligible. Only in a small number of materials are the effects apparent. Various opto-electrical devices have been developed utilizing the apparent dependence of the refractive index on the applied electrical field.

Nonlinear optical effects are also used to make opto-electrical devices. The two widely used nonlinear optical effects are the Kerr effect[6] and the Pockels effect.[7] In the former, a voltage applied to a piece of certain material makes the material birefringent. The indices in the voltage direction and perpendicular to the voltage direction have a difference $\Delta n = \lambda K E^2$, where λ is the wavelength of light incident on the material, K is the Kerr constant, and E^2 is the intensity of the electrical field applied to the material. The Pockels effect is similar, except $\Delta n \sim E$ instead of E^2. Both effects generate or enhance birefringence in a material and can be used to make amplitude/intensity modulators, phase modulators, tunable retarders, or tunable polarization rotators.

9.8.2 Electro-optic amplitude modulators

Modern optical modulators utilize the nonlinear property of some crystals, such as the Pockels effect in lithium niobate crystal. In a lithium niobate electro-optic amplitude modulator,[8] the crystal is placed between two linear polarizers, which have their polarizations in parallel and at an angle to the crystal ordinary axis. The optical beam being modulated passes through the first polarizer to become linearly polarized. The birefringence of the crystal is modulated by applying an electrical field to the crystal. The polarization direction of the beam passing through the crystal is rotated, and the second polarizer filters the polarizing rotating beam. Thus, the amplitude of the final output beam is sinusoidally modulated.

The modulation depth can be 1, which means that the amplitude of the electrical field applied rotates the polarization direction of the beam by 90° so that the second polarizer completely blocks the beam. The modulation frequency can be from DC to RF, determined by the frequency of the electrical field applied to the crystal. The wavelength range is usually from the visible to the near IR.

9.8.3 Electro-optic phase modulators

The commonly used electro-optic phase modulator also utilizes the Pockels effect in a lithium niobate crystal.[9] The optical beam being modulated is linearly polarized in the extraordinary direction of the crystal. An electrical field is applied to the crystal and changes the refractive index of the crystal. Thus, the phase of the optical beam passing through the crystal is modulated. This application does not involve the birefringent property of the crystal.

The modulation frequency can be from DC to RF, determined by the frequency of the electrical field applied to the crystal. The wavelength range is usually from the visible to near IR.

Note that when an AC electrical field with a frequency of f_m is applied to the crystal, the phase of the optical beam passing through the crystal is modulated at a frequency of f_m, which is effectively frequency modulation of the optical beam. As a consequence of the frequency modulation, sidebands are created with frequencies $f_0 \pm f_m$, $f_0 \pm 2f_m$, $f_0 \pm 3f_m$, ..., where f_0 is the frequency of the optical beam. A certain amount of energy will be converted from the optical beam to these sidebands; the specific amount is determined by the modulation depth (the phase-varying magnitude) and can be significant.[10] However, in most applications, the modulation frequency is below or around MHz, which is much smaller than the optical frequency of around $\sim 10^{14}$ Hz. The sidebands can be considered to be merged with the main band.

9.8.4 Acousto-optic modulators

An acousto-optic modulator is also called a Bragg cell. In this type of device, a piezoelectric transducer is attached to a material, such as quartz. An electrical AC signal drives the transducer to vibrate and create an acoustic wave in the material. The acoustic wave causes periodic expansion and compression in the material, as well as periodic refractive index variations. This piece of material behaves like a volume diffraction grating and diffracts light incident on it.

The refractive-index variation depth and period can be modulated by modulating the amplitude and frequency, usually in the RF range, of the electrical AC signal. The diffraction of light is also modulated. The process is an interaction of a light wave with an acoustic wave, which results in the creation of a sum-frequency wave and a difference-frequency wave.

Figure 9.25(a) shows the schematic of an acousto-optic modulator, and Fig. 9.25(b) shows the detail of the diffraction. Diffraction can occur in those directions at which the Bragg condition:

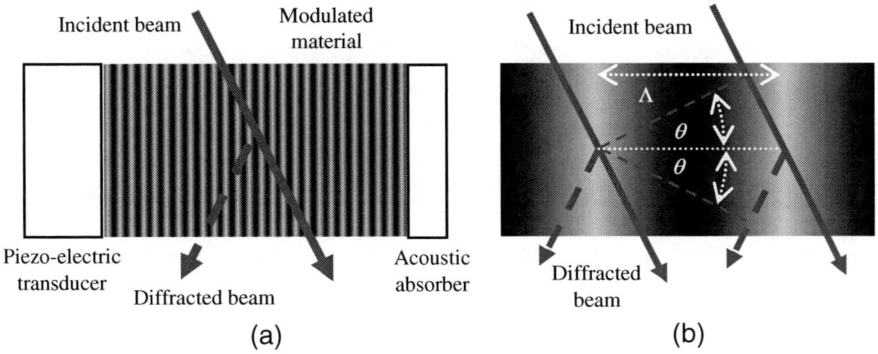

Figure 9.25 (a) Schematic of an acousto-optic modulator, where only one diffracted beam is shown. (b) Details of the Bragg condition, where Λ is the wavelength of acoustic wave.

$$2\Lambda \sin(\theta) = m\lambda \tag{9.8}$$

is met, where Λ and λ are the wavelengths of the acoustic wave and incident light, respectively, and $m = 0, \pm 1, \pm 2 \ldots$ is the diffraction order. Note that the Bragg diffraction condition is similar to the grating equation [Eq. (6.11)]. The difference is that Eq. (9.8) has a factor of 2 on the left side because it describes a volume grating, as illustrated in Fig. 9.25(b).

9.8.5 Frequency mixing

Frequency mixing is a nonlinear optical phenomenon that occurs in certain crystals, such as BBO (β-barium borate), KDP (potassium dihydrogen phosphate), KTP (potassium titanyl phosphate), lithium niobate, Nd:YAG, Nd:YVO_4, etc. Frequency mixing includes frequency doubling, tripling, etc. Among these, frequency doubling has already been widely used. For example, a green laser pointer has a KDP crystal in it. A laser diode placed behind the crystal emits a laser beam with a 1064-nm wavelength. The laser beam travels through the crystal, and its frequency is doubled or the wavelength is converted to 532 nm. A piece of Nd:YAG or Nd:YVO_4 crystal can convert a 946-nm laser beam emitted by a laser diode and passing through the crystal to 473 nm.

Note that frequency doubling, tripling, etc. are not widely occurring phenomena; they can occur only inside certain crystals for certain frequencies (wavelengths). For example, a piece of KDP crystal cannot convert 946 nm to 473 nm, whereas a piece of Nd:YAG crystal can. Conversely, a piece of Nd:YAG crystal cannot convert 1064 nm to 532 nm, whereas a piece of KDP crystal can.

9.8.6 Liquid crystal devices

Liquid crystals are used to make almost all the display screens for computers, TVs, cellphones, etc. They can also be used to make other useful devices, such as tunable bandpass filters (LCTBFs)[11] for multispectral and hyperspectral imaging applications. The bandpass width is tens of nanometers, and the tuning range covers the visible or near IR. A LCTBF can be incorporated in the optical path of a camera. When the camera takes an image of an object, the LCTBF is tuned, and the spectrum of the object is taken. The chemical or biological substances of the object can often be found based on the spectrum. Therefore, LCTBFs have found useful applications in the security and medical fields. The shortcomings of LCTBFs are their low transmission of ~10% and a slow tuning speed of tens of seconds to cover the entire tuning range. Liquid crystals can also be used to make tunable retarder/waveplates.[12] The spectral range covers the visible or near/mid IR, respectively.

9.8.7 Chopper modulators

The classic optical modulator is a dish-shaped mechanical chopper with several blades attached to it.[13] An electronic motor system spins the dish, and the blades chop the optical (mostly laser) beam propagating through the dish. The modulation frequency can be adjusted by altering the dish spin rate up to only 10 kHz or so. The modulation depth has to be 1 (the beam completely passes through) and 0 (the beam is completely blocked). Dishes with different blade widths can offer different duty cycles. Such a device does not cost much. Figure 9.26 draws the schematic of two eight-blade dishes with different blade widths.

9.8.8 Beam-steering devices

Laser beams can be steered by opto-mechanical or opto-electrical methods. Figure 9.27 shows the schematic of two opto-mechanical laser-beam-steering methods. In Fig. 9.27(a), a glass plate with index n and thickness d is placed with an angle θ to the incident-laser-beam propagation direction. The beam passing through the glass plate has a displacement Δ relative to the incident beam. The displaced beam is parallel to the incident beam. Δ can be calculated by

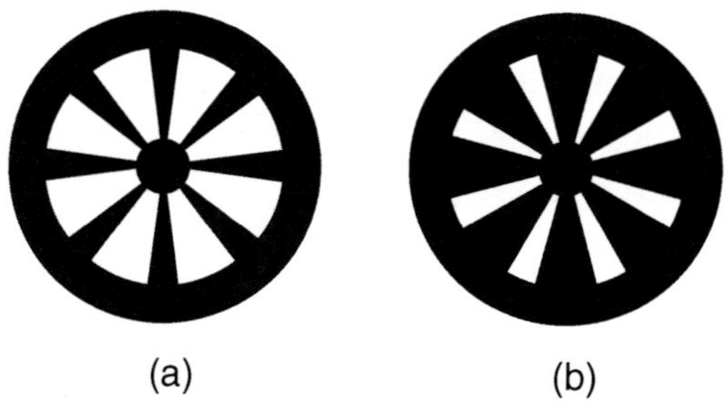

Figure 9.26 Eight-blade dishes with (a) a narrow blade width and (b) a wide blade width.

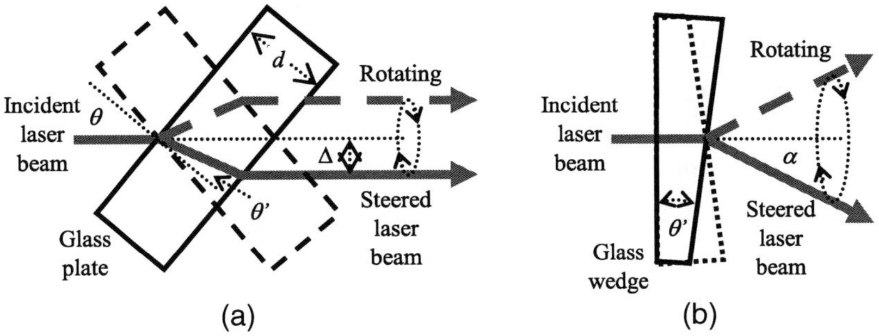

Figure 9.27 (a) A glass plate with angle θ to the propagation direction of an incident laser beam can displace the beam. The laser beam propagation direction is not changed. (b) A glass wedge with a wedge angle θ′ can deflect the beam. The direction of displacement or deflection can be changed by rotating the glass plate or glass wedge about the incident laser beam, respectively.

$$\begin{aligned}\Delta &= d[\tan(\theta) - \tan(\theta')]\\ &= d\left[\frac{1}{\cos(\theta)} - \frac{1}{\left[n^2 - \sin^2(\theta)\right]^{0.5}}\right],\end{aligned} \tag{9.9}$$

where θ' is the refraction angle, which is related to θ by Snell's law $\sin(\theta) = n\sin(\theta')$. The glass plate can be rotated to change the displacement direction.

In Fig. 9.27(b) a glass wedge with index n and wedge angle θ' is placed with the front surface of the wedge perpendicular to the incident beam. The beam passing through the wedge will be deflected with an angle α to the incident laser beam. The deflection angle α is given by

$$\begin{aligned}\alpha &= \theta - \theta'\\ &= \theta - \arcsin[n\sin(\theta)],\end{aligned} \tag{9.10}$$

where θ' is related to θ by Snell's law $\sin(\theta) = n\sin(\theta')$. The glass wedge can be rotated to change the deflection direction.

A laser beam can also be steered by a few opto-electrical methods. For example, a laser beam is incident on a liquid crystal grating and is diffracted.[14] The grating period of the liquid crystal can be changed by changing the electrical voltage applied on the crystal. The diffraction angle of the laser beam is then changed.

References

1. *Laser Induced Damage Threshold Tutorial*, www.thorlabs.com/tutorials.cfm?tabID=762473b5-84ee-49eb-8e93-375e0aa803fa
2. *Laser Damage Threshold Testing*, www.edmundoptics.com/resources/application-notes/lasers/laser-damage-threshold-testing

3. *The Complexities of High-Power Optical Coatings*, www.edmundoptics.com/resources/application-notes/optics/the-complexities-of-high-power-optical-coatings
4. *Kirchhoff's Diffraction Formula*, en.wikipedia.org/wiki/Kirchhoff%27s_diffraction_formula
5. H. Sun, "3.5 Aperture beam truncation effects," *A Practical Guide to Handling Laser Diode Beams*, Springer, Berlin, pp. 91–92 (2015).
6. *Kerr effect*," wikipedia.org/wiki/Kerr_effect
7. *Pockelseffect*, en.wikipedia.org/wiki/Pockels_effect
8. *EO amplitude modulators*, www.thorlabs.com/newgrouppage9.cfm?objectgroup_id=2729
9. *EO phase modulators*, www.thorlabs.com/newgrouppage9.cfm?objectgroup_id=2729
10. *Acousto-optic modulator*," wikipedia.org/wiki/Acousto-optic_modulator.
11. *Liquid Crystal Tunable Bandpass Filters*," www.thorlabs.com/newgrouppage9.cfm?objectgroup_id=3488
12. *Half-Wave Liquid Crystal Variable Retarders/Wave Plates*, www.thorlabs.com/newgrouppage9.cfm?objectgroup_id=6179
13. *Optical Chopper System and Chopper Wheels* www.thorlabs.com/newgrouppage9.cfm?objectgroup_id=287
14. J. Kim et al, "Wide-angle, nonmechanical beam steering using thin liquid crystal polarization gratings," pdfs.semanticscholar.org/d1ed/72df608973c3b23a7ddd10aa8dd45d140644.pdf

Chapter 10
Instruments to Characterize Laser Beams

The characterization of laser beams and optical measurements partially overlap. Some instruments introduced in this chapter can be used for other optical measurements, e.g., laser power meters can also be used to measure the optical power of other light sources. Some instruments that can be used to characterize laser beams are introduced in the next chapter, e.g., a laser beam wavefront can be characterized using the instruments introduced in Section 11.2.

The following four aspects apply to the characterization of laser beams:

1. Characterizing the laser beam size, shape, and intensity profile;
2. Measuring the laser beam power or energy;
3. Measuring the laser beam wavelength or characterizing the spectrum (see Section 11.3);
4. Characterizing the laser beam wavefront (see Section 11.2).

10.1 Laser Beam Profilers

Laser beam profilers are instruments used to characterize the size, shape, and intensity pattern of laser beams. Commonly used profilers are based on either cameras or scanners.

10.1.1 Camera-based beam profilers

A camera-based beam profiler consists of a 2D sensor array that takes the image of the beam and a computer installed with special software for data processing and display. The advantage of using a 2D sensor array is that it can provide a true 2D image of the beam. The software and computer can display the whole beam intensity profile and various beam parameters, such as the $1/e^2$ intensity diameter, $1/e^2$ intensity encircled power, beam center position, etc.

One disadvantage of camera-based beam profilers is the limited wavelength range. Currently, commercial profilers cover from 190–1100 nm and 900–1700 nm using two types of camera. Sensor arrays below 190 nm are rare. A cooled InGaAs sensor array can cover up to 2000 nm or so.

Another disadvantage of camera-based beam profilers is the relatively low spatial resolution limited by the pixel size of at least a few microns or up to thirty microns or so in the IR range. However, the technology is advancing rapidly, and these numbers are prone to change.

10.1.2 Scanner-based beam profilers

A scanner-based beam profiler consists of a scanner and a computer installed with special software for data processing and display. Figure 10.1(a) shows the schematic of a widely used slit beam scanner. A drum has two slits on it

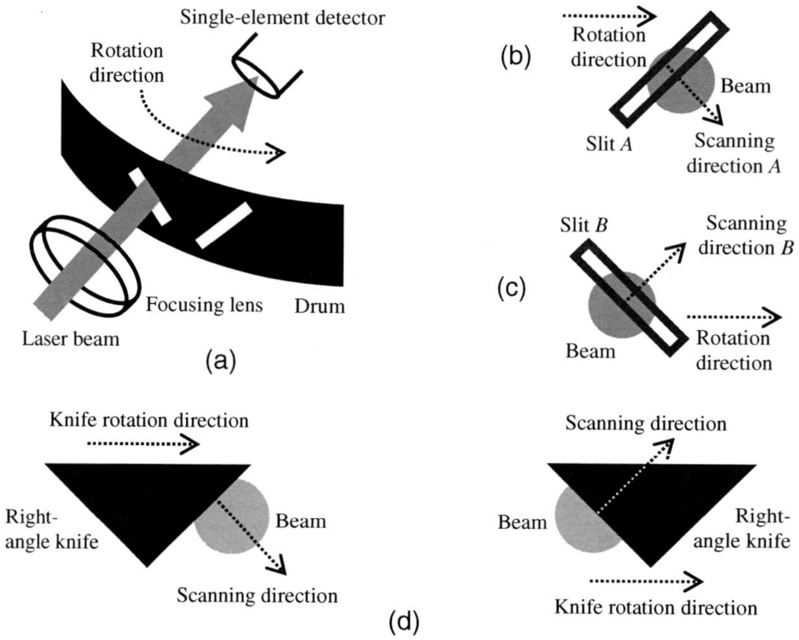

Figure 10.1 (a) Schematic of a slit laser beam profiler. The laser beam being profiled is focused onto a single-element sensor by a lens. A drum has two slits with orthogonal orientation. The drum rapidly rotates, and the slits scan the beam. The sensor records the beam intensity variation and the slit positions. Software calculates the beam intensity profile in the two orthogonal directions based on the recorded data. (b) When slit *A* scans through the beam, the beam is scanned in direction *A* pointed by the arrow. (c) When slit *B* scans through the beam, the beam is scanned in direction *B* pointed by the arrow. Since slits *A* and *B* are orthogonal, directions *A* and *B* are also orthogonal. (d) A right-angle knife-edge beam scanner. The knife is on a drum; as the drum rapidly rotates, the knife scans the beam. The front and back edges scan the beam in two orthogonal directions, respectively.

that are orthogonal. The laser beam being profiled is focused by a lens onto a single-element detector. When the drum rotates, the two slits scan the beam in two orthogonal directions, as illustrated in Figs. 10.1(b) and (c). The software calculates and displays the beam intensity profile in the two orthogonal directions based on the detector output and the slit position.

There are other shapes of scanner, e.g., a right-angle knife-edge scanner, as shown in Fig. 10.1(d). All scanner-based beam profilers have the same working principle and similar performance.

Because the slit width is only a few microns, the beam intensity profiles in these two orthogonal directions can be calculated to a submicron resolution. Such a spatial resolution is higher than the spatial resolution of pixel-based 2D detector arrays. It is also much easier to find different types of single-element detectors to cover wide wavelength ranges than sensor arrays can cover. This is the major advantage of scanner beam profilers over camera versions. Another advantage is that scanner profilers in the IR or UV range cost much less than camera profilers.

The disadvantage of a scanning beam profiler is that the beam intensity profiles in the directions other than the two orthogonal scanning directions cannot be directly profiled. The computer software will data fit the beam intensity profiles in these other directions and display the best-guess results. This problem can be partially solved by rotating the scanner head about the beam under profiling and taking several scans.

The computer and software can display various beam parameters, such as the $1/e^2$ intensity diameter, $1/e^2$ intensity encircled power, beam center position, etc.

10.2 Laser Power Meters and Energy Meters

10.2.1 Specifications

A laser power meter consists of three basic components: a photodetector, an electronics conditioner to process the output signal of the photodetector, and a display device to display the measurement result in terms of watts or joules. To a large extent, the characteristics of the photodetector determine the performance of a laser power meter. When selecting a laser power meter, the following four issues must be considered:

1. Spectral response range. A photodetector has a certain spectral range that limits the spectral range of a laser power meter. The response of many photodetectors depends on the wavelength. Calibration is necessary when a laser power meter with such a photodetector is used to measure the laser power with different wavelengths.
2. Power range. The detection threshold and damage threshold usually determine the power range of a laser power meter. Incident laser power

lower than the detection threshold cannot be accurately measured, whereas incident laser power higher than the damage threshold may permanently damage the photodetector and must be attenuated before being measured. An optical attenuator placed in front of the photodetector can raise the damage threshold to megawatts and simultaneously raise the detection threshold.

The power range specifically includes the CW power range, peak power range, spatial power density range, single-pulse energy range, and spatial energy density range.

3. Response time. A photodetector needs certain time to respond to an input laser beam. If the output of a pulsed laser is measured, the response time must be much shorter than the width of the laser pulse to avoid large measurement error.
4. Detector size. The size (either diameter for circular detectors or side for square detectors) of commonly used photodetectors is from a few millimeters to tens of millimeters. When the size of a laser beam under measurement is larger than the photodetector size, a focusing lens must be used to reduce the beam size. Power loss caused by the focusing lens must be excluded to avoid measurement error.

10.2.2 Photodetector properties and selection

Three types photodetector are commonly used in laser power meters: photodiodes, thermopiles, and pyroelectric sensors. Photodiodes were previously described in Section 6.13, so this section covers thermopiles and pyroelectric sensors.

A thermopile is usually a light-absorbing disk onto which a ring of thermopiles has been deposited. The absorber converts the laser power incident onto it into heat and generates between the absorber and a heat sink a temperature difference that is proportional to the incident laser power. The thermopiles generate and output an electrical response proportional to the temperature difference. Thermopiles usually have a long response time of a few seconds and a flat spectral response range from 200 nm to 20 μm, and a power range of 1 mW to 5 kW for CW lasers or 0.01–300 J for pulsed lasers. Thermopiles are primarily used to measure moderate- to high-power CW lasers, moderate- to high-energy single-pulse lasers, and the energy of pulsed lasers with a repetition rate higher than 10 Hz.

A pyroelectric probe uses a ferroelectric material that is electrically polarized at a certain temperature. The material is placed between two electrodes. Any change in temperature of the material caused by the absorption of laser power produces a response electrical current in the external circuit. Pyroelectric probes are primarily used to measure the energy of pulsed lasers because they only respond to the rate of temperature change. Pyroelectric probes usually have a spectral response range from 100 nm to

Table 10.1 Summary of appropriate applications for three types of photodetectors.

Photodetector	Measurement Type
Photodiode	Primarily used to measure low CW laser power and low pulse energy.
Thermopile	Measures CW lasers and integrates the energy of pulsed lasers to produce an average power measurement; measures the energy of millisecond and longer pulse widths.
Pyroelectric sensors	Only measures the energy of pulsed lasers. The average power can be calculated by measuring the laser repetition rate and multiplying it by the pulse energy.

Table 10.2 Summary of photodetector criteria for various lasers.

Laser Type	Measurement Needed	Power Range	Wavelength Range	Sensor
CW	Average power	10 nW–50 mW	0.20–1.80 μm	Photodiode
		200 μW–>5 kW	0.15–12.00 μm	Thermopile
Pulsed	Average power	200 μW to >5 kW	0.15–12.00 μm	Thermopile
Pulsed	Energy per pulse	100 nJ to >10 J	0.15–12.00 μm	Pyroelectric
Pulsed	Energy per pulse	10 pJ–800 nJ	0.32–1.70 μm	Photodiode
Long pulsed (>1 ms)	Single-pulse integrated energy	1 mJ to >300 J	0.15–12.00 μm	Thermopile

100 μm, a pulse energy range from 10 nJ to 20 J, and a response time as short as a few picoseconds.

Each type of detector has its own advantages and disadvantages. Tables 10.1 and 10.2 summarize their properties and their appropriate measurement tasks.

10.3 Wavelength Meters

10.3.1 General comments

Most lasers have a fixed wavelength. There is no need to measure their wavelength. For example, a red He-Ne laser has a fixed wavelength of 632.8 nm. Wavelength meters (wavemeters) were first developed to measure the wavelength of tunable lasers, such as dye lasers. The wavelength of widely used laser diodes is not stable, can be tuned, and often need be measured.

Most commercial laser wavemeters in market nowadays can be categorized into two types: Fizeau wavemeters and Michelson interferometer wavemeters. There are other types of wavemeters commercially available. In this chapter, the working principles of four types of wavemeters are described.

10.3.2 Fizeau wavemeters

This type of wavemeter uses a Fizeau wedge as illustrated in Fig. 10.2, either an air wedge or glass wedge, to generate interference fringes from the laser beam under measurement, then uses a linear sensor array to detect the fringes, a computer calculates the fringe period, and the wavelength under

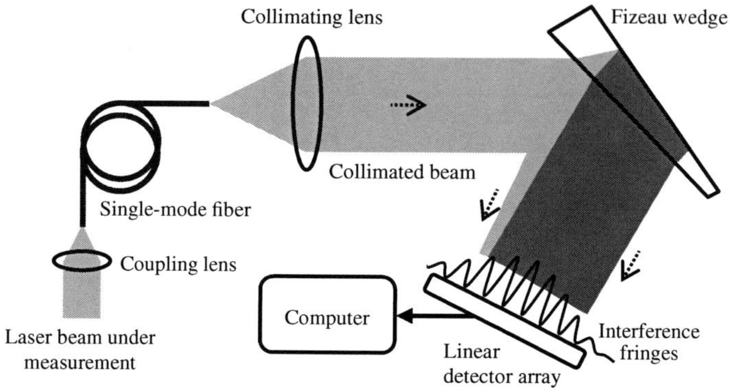

Figure 10.2 Schematics of a Fizeau wavemeter. The laser beam under measurement is coupled into a single-mode fiber. The beam output from the fiber is collimated and incident on a Fizeau wedge. The beams reflected by the two surfaces of the Fizeau wedge meet on a linear sensor array and interfere. A computer processes the interference data and calculates the wavelength.

measurement is proportional to the fringe period. If a glass wedge is used, the measurement results will be automatically adjusted by the data-processing software to compensate for the glass dispersion.

The wavefront curvature radius and aberrations of the laser beam under measurement will significantly affect the interference fringe period and thus the measurement accuracy. The beam under measurement must first be coupled into a single-mode fiber, and the beam output from the fiber must be collimated. Thereby, only a planar wavefront will be incident on the Fizeau wedge for measurement.

The wavelength under measurement λ_u is first calculated by

$$\lambda_u = \frac{P_u}{P_r}\lambda_r, \tag{10.1}$$

where λ_r is the wavelength of a calibration laser, usually a He-Ne laser, and P_u and P_r are the interference fringe periods of the laser under measurement and the calibration laser, respectively. P_r is already saved in the computer beforehand and need only be verified during a scheduled calibration process. P_u can be measured to an accuracy of 10^{-3} because the linear array used to detect the fringes usually has 1024 pixels. For example, the wavelength of a stabilized He-Ne laser with a wavelength of 632.8056 nm is measured. The measurement result at this stage is 632.XXXX nm, where the X symbols are unknown numbers.

The wavelength under measurement must be further calculated by

$$\lambda_u = \frac{d}{m}, \tag{10.2}$$

where m is an integer, and d is the optical thickness of the Fizeau wedge and is known to high accuracy beforehand by calibration. A properly chosen d value makes it possible to find the correct value of m. For example, if m_c is the correct value of m, plugging $m_c \pm 1$ into Eq. (10.2) will lead to a λ_u that contracts to the λ_u calculated from Eq. (10.1). With the correct m_c being plugged in Eq. (10.2), λ_u can be calculated to an accuracy of at least 10^{-6}, for example, 632.806 nm.

The "Relay" type of data processing is used in other wavelength measurement techniques.

A Fizeau wavemeter has no moving parts. The measurement can be completed at the same speed as the sensor array catches the signal. Therefore, Fizeau wavemeters can measure the wavelength of both continuous and pulsed lasers. The measurement error can be $<10^{-6}$. The shortcoming of Fizeau wavemeters is that the limited spectral response range of a linear sensor array limits the measurement range of Fizeau wavemeters. A wavemeter must be sealed in a box with temperature control. Details about aligning a Fizeau wavemeter can be found in Refs. 1–3, and the details about data processing can be found in Ref. 4.

The Fizeau wavemeter was invented by the late James J. Snyder[5] and is the most widely used wavemeter currently. Snyder had also invented other very successful optical devices and instruments, such as the micro-acylindrical lens, shown in Fig. 9.13, to circularize laser diode elliptical beams and correct astigmatism.

10.3.3 Michelson interferometer wavemeters

This type of wavemeter uses a moving-arm Michelson interferometer to generate interference fringes from the laser beam under measurement, as shown in Fig. 10.3. The laser beam under measurement is split in two by a cube beamsplitter. The two beams travel in two optical arms, are reflected by mirrors 1 and 2, respectively, recombined by the beamsplitter and focused onto a laser power meter, where the two beams interfere.

As the corner reflector moves, the optical path length difference between the two optical arms changes, and the intensity of the interference varies periodically from constructive to destructive. The laser power meter monitors the intensity of the interference and counts the periods.

The wavelength λ under measurement can be calculated from the corner reflector moving distance L and the number of the fringes, $m + \Delta m$, counted by the laser power meter by

$$\lambda = \frac{4L}{m + \Delta m}, \tag{10.3}$$

where m is an integer, and Δm is a fraction. Note that the beam makes a round trip in the optical arm, and the optical path length change is $2L$. Therefore, there is a factor of 4 in Eq. (10.3). For an arm moving a distance $L \sim 200$ mm,

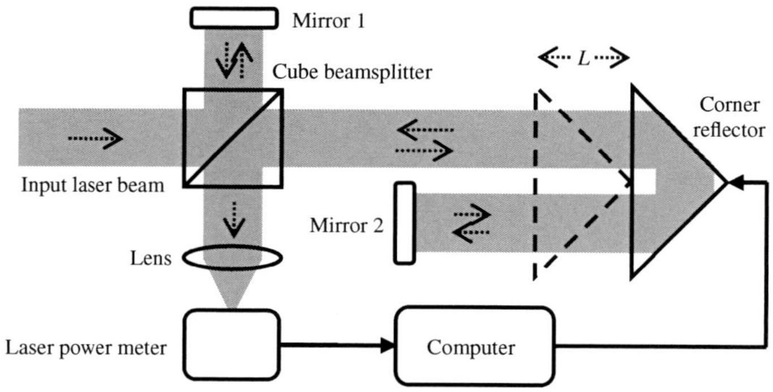

Figure 10.3 Schematics of a Michelson-interferometer laser wavemeter. The beam under measurement is split in two by a beamsplitter. The two subbeams are reflected by two mirrors, respectively, recombined by the beamsplitter, and interfere on a laser power meter. As the optical path length of one subbeam changes, the power meter counts the number of interference fringes. The wavelength of the beam can be calculated based on the fringe number and the magnitude of the optical path change.

$m \sim 10^6$, and Δm can be determined with an error <0.1. The wavelength calculated from Eq. (10.3) has an error of $\sim 10^{-7}$.

Michelson interferometer wavemeters have a moving arm. One measurement takes at least several seconds to complete and therefore cannot measure the wavelength of pulsed lasers, which is the disadvantage. The advantage is that various single-element detectors available on the market can cover a wider wavelength range than linear arrays can. Commercial Michelson interferometer wavemeters often have a temperature-control device inside the case to reduce the measurement error caused by thermal expansion and a calibration He-Ne laser inside the case for occasional self-calibration. Reference 6 describes in detail the design and construction of a Michelson wavemeter.

10.3.4 Other wavelength measurement methods

There are other types of wavemeters. One type uses three photodiodes of different types. The laser beam under measurement is incident on a strong diffuser that makes the three photodiodes receives the same amount of optical power. For any incident wavelength, the output signal magnitudes of the three photodiodes are different because the three photodiodes have different spectral responses. The wavelength can be calculated based on the difference among the three signal magnitudes. This type of wavemeter is small, inexpensive, and easy to use (almost no alignment requirement), but it has a low measurement accuracy of about 0.1 nm, and the outputs of the three photodiodes can be used after calibration to simultaneously measure the laser power. This type of wavemeter is best for measuring the laser diode wavelength since most laser diodes have a large linewidth; the high measurement accuracy of Fizeau wavemeters and Michelson interferometer wavemeters would be wasted.

Another type of wavemeter uses three gratings of different dispersion power and three photodiode linear arrays. The laser beam under measurement is split into three and incident on the three gratings. The diffraction beams of the three gratings are incident on the three photodiode linear arrays, respectively. Based on the positions of the diffraction beams on the linear arrays, the wavelength under measurement can be calculated again in a "relay."

The calculation details are as follows:

1. The first grating has the lowest dispersion power. Its ~1000 pixels cover a ~1000-nm range, and each pixel corresponds to ~1 nm. Thus, the wavelength can be measured to an accuracy of ~1 nm.
2. The second grating has medium dispersion power. Its ~1000 pixels cover a ~10-nm range, and each pixel corresponds to ~0.01 nm, which is the wavelength measurement accuracy.
3. The third grating has high dispersion power. Its ~1000 pixels cover a ~0.1-nm range, and each pixel corresponds to ~0.001 nm, which is the final measurement accuracy of the wavemeter.

For example, the wavelength of a stabilized He-Ne laser with a wavelength of 632.8056 nm is measured. The three grating measurement results are (the X symbol means an unknown number)

First grating measurement:	632.XXXX nm
Second grating measurement:	XX2.80XX nm
Third grating measurement:	XXX.X06 nm
Final combined measurement:	632.806 nm

Three Fabry–Pérot etalons with different free spectral ranges can be used in place of the three gratings to measure the wavelength using the same "relay" working principle.

References

1. J. J. Snyder, "Fizeau Wavemeter," *Proc. SPIE* **0288**, 258 (1981) [doi: 10.1117/12.932052].
2. B. Faust and L. Klynning, "Low-Cost Wavemeter with a Solid Fizeau Interferometer and Fiber-Optic Input," *Appl. Opt.* **30**, 5254–5259 (1991).
3. C. Reiser and R. B. Lopert, "Laser Wavemeter with Solid Fizeau Wedge Interferometer," *Appl. Opt.* **27**, 3656–3660 (1988).
4. J. J. Snyder, "Algorithm for Fast Digital Analysis of Interference Fringes," *Appl. Opt.* **19**, 1223–1225 (1980).
5. *In Memoriam: James J. Snyder*, www.osa.org/en-us/about_osa/newsroom/obituaries/earlier/snyder
6. J. P. Monchalin et al., "Accurate Laser Wavelength Measurement with a Precision Two-Beam Scanning Michelson Interferometer," *Appl. Opt.* **20**, 736–757 (1981).

Chapter 11
Instruments for Optical Measurements

Optical measurements partially overlap laser beam characterization, so some of the instruments introduced in this chapter can also be used to characterize laser beams. For example, Shack–Hartmann wavefront sensors can be used to characterize laser beam wavefronts. Scanning Fabry–Pérot interferometers can be used to analyze the laser beam spectrum.

11.1 Wavefront Modeling

11.1.1 General comments

Most optical measurements analyze a wavefront, and most analysis methods involve some kind of interference. Since wavefronts are 2D functions, a 2D mathematical model must be established to describe a wavefront. The aberration polynomials, Zernike polynomials, and wavefront polynomials are equivalent 2D mathematical models that are suitable to describe aberrated wavefronts. By analyzing these polynomials, it is easier to understand the wavefront, evaluate the performance of the optical systems that generate this wavefront, locate the origins of the aberrations, and modify the optical systems.

However, the mathematics involved with analyzing 2D wavefront is complex. Even though many optical engineering jobs do not require a mastery of wavefront theory, but it is still beneficial to know something about this topic. When other people discuss wavefront aberrations, one can at least know what they are talking about. Whenever there is a need to deal with wavefront theory, optical software can help perform much of the analysis and make the task much easier.

11.1.2 Aberration polynomials

A lens focusing an off-axis object point is used to illustrate the case shown in Fig. 11.1. To simplify the analysis, the coordinate is chosen so that the object point is on the vertical y axis with height h. Consider a ray from the object

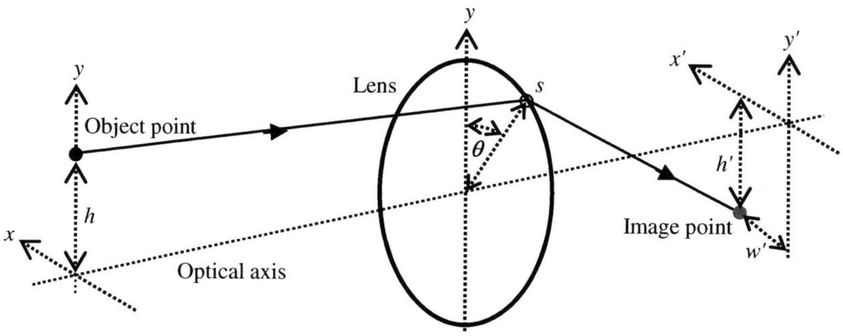

Figure 11.1 Illustration of wavefront aberration as a function of object height h and lens aperture size s.

point hitting the lens at a point with a radial distance s from the lens center and an angle θ to the y axis, as marked in the figure. The lens focuses this ray onto the image plane with horizontal and vertical distances w' and h' to the y' and x' axis in the image plane, respectively. The following two relations can be established:[1]

$$\begin{aligned}
h' = {} & A_1 \times s \times \cos(\theta) + A_2 \times h \\
& + B_1 \times s^3 \times \cos(\theta) + B_2 \times s^2 \times h \times [2 + \cos(2\theta)] + (3B_3 + B_4) \times s \times h^2 \\
& \times \cos(\theta) + B_5 \times h^3 \\
& + C_1 \times s^5 \times \cos(\theta) + [C_2 + C_3 \times \cos(2\theta)] \times s^4 \times h + [C_4 + C_6 \times \cos(\theta)^2] \\
& \times s^3 \times h^2 \times \cos(\theta) \\
& + [C_7 + C_8 \times \cos(2\theta)] \times s^2 \times h^3 + C_{10} \times s \times h^4 \times \cos(\theta) + C_{12} \times h^5 \\
& + \{\text{seventh-order terms}\} + \{\text{ninth-order terms}\} + \ldots,
\end{aligned} \tag{11.1}$$

$$\begin{aligned}
w' = {} & A_1 \times s \times \sin(\theta) \\
& + B_1 \times s^3 \times \sin(\theta) + B_2 \times s^2 \times h \times \sin(2\theta) + (B_3 + B_4) \times s \times h^2 \times \sin(\theta) \\
& + C_1 \times s^5 \times \sin(\theta) + C_3 \times s^4 \times h \times \sin(2\theta) + [C_5 + C_6 \times \cos(\theta)^2] \\
& \times s^3 \times h^2 \times \sin(\theta) \\
& + C_9 \times s^2 \times h^3 \times \sin(2\theta) + C_{11} \times s \times h^4 \times \sin(\theta) \\
& + \{\text{seventh-order terms}\} + \{\text{ninth-order terms}\} + \ldots,
\end{aligned} \tag{11.2}$$

where A_i are the coefficients of the terms containing the first order of either s or h, i is an integer, B_i are the coefficients of the terms containing the third order of either s or h or $s \times h$, and C_i are the coefficients of the terms containing the fifth order of either s or h or $s \times h$. It can be shown that because of the lens is symmetric about its axis, there are no terms containing even order of either s or h or $s \times h$. Equations (11.1) and (11.2) contain an infinite number of terms.

In Eqs. (11.1) and (11.2), all of the A_i terms are first order and describe the paraxial situation, and all of the B_i terms are third order and are called "primary aberrations" or "Seidel aberrations," including spherical aberration, coma, astigmatism, distortion, and field curvature. All of the C_i terms are fifth order.

The derivation process of Eqs. (11.1) and (11.2) are lengthy and complex. Much of the literature, including this book, cite only the results without proving it. It is more important to understand the nonlinear relations between various aberrations and s and h than to derive the relations, but interested readers should see Born and Wolf[2] for the mathematical details.

11.1.3 Zernike polynomials

Fourier series theory states that any 1D functions within a certain range can be mathematically expressed by the sum of an infinite series of 1D orthogonal functions, usually sine and cosine functions. Similarly, any 2D functions within a certain area can be mathematically expressed by the sum of an infinite series of 2D orthogonal functions. The Zernike polynomials are a set of infinite series of orthogonal polynomials with two variables and are commonly used to describe a 2D wavefront.[3–5] The two variables are azimuthal angle ϕ and normalized radial distance $\rho \leq 1$.

There are even and odd Zernike polynomials similar to sine and cosine functions in a Fourier series. The even polynomials are defined as

$$Z_n^m(\rho,\phi) = R_n^m(\rho)\cos(m\phi). \tag{11.3}$$

The odd polynomials are defined as

$$Z_n^{-m}(\rho,\phi) = R_n^m(\rho)\sin(m\phi), \tag{11.4}$$

where m and n are positive integers with $n \geq m$, and $R_n^m(\rho)$ is the radial polynomial.

The orthogonal condition of Zernike polynomials is

$$\iint Z_n^m(\rho,\phi) Z_{n'}^{m'}(\rho,\phi)\rho d\rho d\phi = \frac{\varepsilon_m \pi}{2n+2}\delta_{n,n'}\delta_{m,m'}, \tag{11.5}$$

where ε_m is a coefficient with $\varepsilon_m = 2$ for $m = 0$ and $\varepsilon_m = 1$ for $m \neq 0$. A 2D function $W(\rho,\phi)$, such as a wavefront, can then be expressed by the sum of Zernike polynomials as

$$W(\rho,\phi) = \sum_{m=0}^{\infty}\sum_{n=m}^{\infty}[A_n^m Z_n^m(\rho,\phi) + B_n^m Z_n^{-m}(\rho,\phi)]. \tag{11.6}$$

The two coefficients A_n^m and B_n^mcan be found by utilizing the orthogonal property of Zernike polynomials to be

$$A_n^m = \frac{n+1}{\varepsilon_{m,n}\pi}\iint W(\rho,\phi)Z_n^m(\rho,\phi)\rho d\rho d\phi, \tag{11.7}$$

$$B_n^m = \frac{n+1}{\varepsilon_{m,n}\pi}\iint W(\rho,\phi)Z_n^{-m}(\rho,\phi)\rho d\rho d\phi, \tag{11.8}$$

where $\varepsilon_{m,n} = 1/\pi$ for $m = 0$ and $n \neq 0$, and $\varepsilon_{m,n} = 1$ for any other values of m and n.

In reality, it is not necessary to use infinite terms of Zernike polynomials to describe a wavefront, several terms are sufficient to approximately describe the wavefront. Every Zernike polynomial has a certain optical meaning. The first eleven nonzero Zernike polynomials and their optical meanings are listed in Table 11.1.

The first term in Table 11.1 has a constant value of 1. If the corresponding parameters $A_0^0 \neq 0$ and all other parameters A_n^m and B_n^m equal zero, then the wavefront is planar. The second term in the table is $2\rho\text{con}(\phi) = 2x$. If the corresponding parameters $A_1^1 \neq 0$ and $A_0^0 \neq 0$, the wavefront is tilted in the x direction. The optical meaning of higher-order Zernike polynomials are difficult to conceive. A wavefront that contains several Zernike polynomials is complex. False-color pictures of the first 21 Zernike polynomials[6] are plotted in Fig. 11.2, which can help explain the optical meanings of Zernike polynomials.

Optical design software should use many more terms of Zernike polynomials to analyze a wavefront and have negligible errors. For example, Zemax uses 231 Zernike polynomials.

Table 11.1 The first eleven nonzero Zernike polynomials and their optical meanings.

Number	Radial Degree n	Azimuthal Degree m	$Z_n^m(\rho,\phi)$	Optical Meaning
1	0	0	1	Piston
2	1	1	$2\rho\cos\phi$	x-tilt
3	1	−1	$2\rho\sin\phi$	y-tilt
4	2	0	$3^{0.5}(2\rho^2 - 1)$	Defocus
5	2	2	$6^{0.5}\rho^2\cos 2\phi$	Oblique astigmatism
6	2	−2	$6^{0.5}\rho^2\sin 2\phi$	Vertical astigmatism
7	3	1	$8^{0.5}(3\rho^3 - 2\rho)\cos\phi$	Horizontal coma
8	3	−1	$8^{0.5}(3\rho^3 - 2\rho)\sin\phi$	Vertical coma
9	3	3	$8^{0.5}\rho^3\cos 3\phi$	Oblique trefoil
10	3	−3	$8^{0.5}\rho^3\sin 3\phi$	Vertical trefoil
11	4	0	$5^{0.5}(6\rho^4 - 6\rho^2 + 1)\sin\phi$	Primary spherical

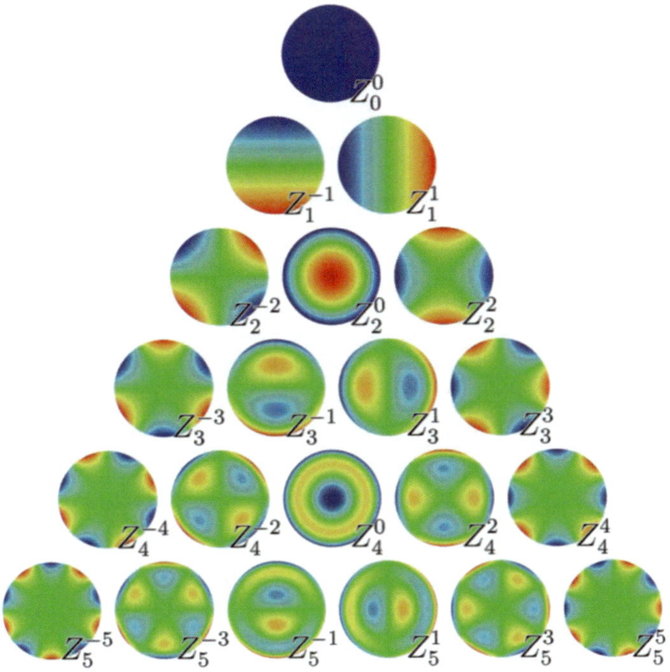

Figure 11.2 False-color pictures of the first 21 Zernike polynomials, arranged vertically by radial degree and horizontally by azimuthal degree.

11.1.4 Wavefront polynomials

It has been shown that a wavefront $W(\rho,\theta)$ can be approximately expressed by the sum of many Zernike polynomials with certain coefficients. Equation (11.9) shows a wavefront $W(\rho,\theta)$ approximately expressed by the first nine Zernike terms

$$\begin{aligned} W(\rho,\phi) = {} & Z_0 + Z_1 \times \rho \times \cos(\phi) + Z_2 \times \rho \times \sin(\phi) + Z_3 \times (2\rho^2 - 1) \\ & + Z_4 \times \rho^2 \times \cos(2\phi) + Z_5 \times \rho^2 \times \sin(2\phi) + Z_6 \times (3\rho^2 - 2) \\ & \times \rho \times \cos(\phi) \\ & + Z_7 \times (3\rho^2 - 2) \times \rho \times \sin(\phi) + Z_8 \times (6\rho^4 - 6\rho^2 + 1), \end{aligned} \tag{11.9}$$

where the conventional symbol Z_i, not A_n^m and B_n^m, are used to denote the coefficients.

Equation (11.9) can be rearranged with some algebra and triangular geometry to take the form

$$
\begin{aligned}
W(\rho,\phi) &= Z_0 - Z_3 + Z_8 && \text{Piston} \\
&+[(Z_1 - 2Z_6)^2 + (Z_2 - 2Z_7)^2]^{0.5}\rho\cos\left[\phi - \tan^{-1}\left(\frac{Z_2 - 2Z_7}{Z_1 - 2Z_6}\right)\right] && \text{Tilt} \\
&+[2Z_3 - 6Z_8 \pm (Z_4^2 + Z_5^2)^{0.5}]\rho^2 && \text{Focus} \\
&\pm 2(Z_4^2 + Z_5^2)^{0.5}\rho^2\cos^2\left[\phi - \frac{1}{2}\tan^{-1}\left(\frac{Z_5}{Z_4}\right)\right] && \text{Astigmatism} \\
&+3(Z_6^2 + Z_7^2)^{0.5}\rho^2\cos\left[\phi - \tan^{-1}\left(\frac{Z_7}{Z_6}\right)\right] && \text{Coma} \\
&+6Z_8\rho^4, && \text{Spherical}
\end{aligned} \tag{11.10}
$$

which presents the wavefront polynomials.[7] The optical meaning of each term is noted on the right side. Wavefront polynomials are equivalent to Zernike polynomials.

11.1.5 Interferograms

In optical testing, a planar or spherical test optical wave is incident on and reflected by the optical surface under test. The reflected wave contains the information about the profile of the optical surface under test and is combined with the incident or another planar or spherical wave to form a 2D interference diagram. If the profile of the optical surface under test is not complex, the details of the surface profile can be found from the interferogram by visual observation. Otherwise, specially developed software can be used to decipher the interferogram with a series of Zernike polynomials.

Figure 11.3 illustrates two planar wavefronts interfering. Interference fringes appear when there is an angle between the two wavefronts. The relative displacement between the two wavefronts does not affect the interference fringe patterns or the interferogram.

If the two wavefronts are not planar, such as spherical, the interference will be more complex. Both the relative tilt and displacement between the two wavefronts affect the interferogram. The difference between the two wavefront curvature radii also affects the interferogram.

Figure 11.4 shows Zemax-simulated interferograms for two spherical wavefronts with a radius of curvature of 10 mm and a diameter of 10 mm. These are fast-focusing *F*/1 wavefronts.

Figure 11.5 shows Zemax-simulated interferograms of two spherical wavefronts with a radius of curvature of 50 mm and a surface diameter of 10 mm. These are *F*/5 wavefronts. As the interferogram becomes more complex, it is more difficult to decipher. Special training is required to effectively analyze interferograms.

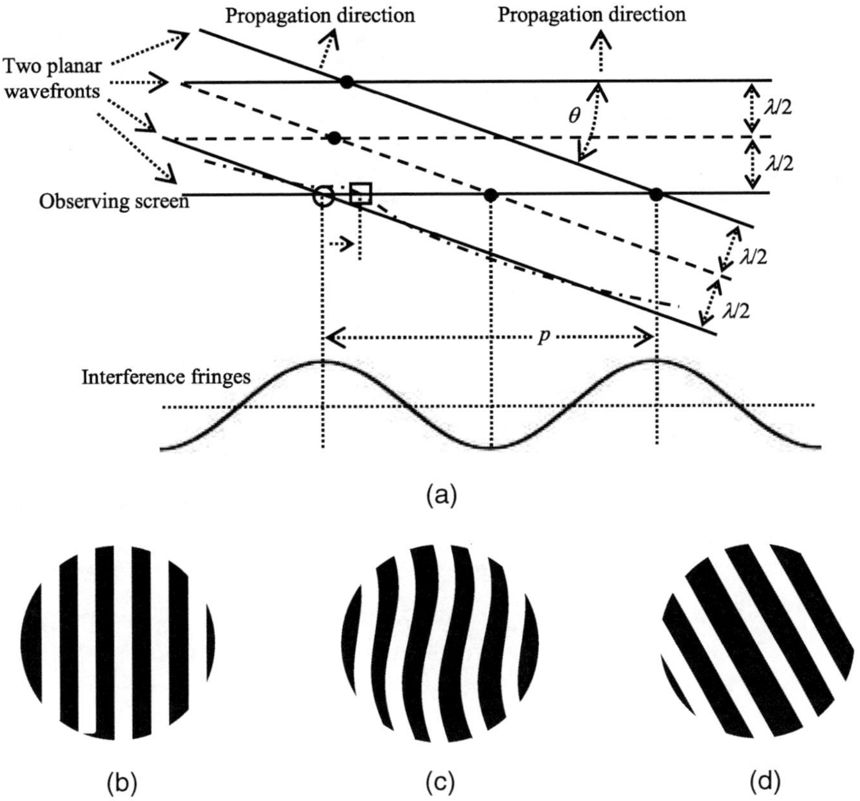

Figure 11.3 (a) A 2D illustration of two planar waves with an angle θ in between interfering. The solid lines represent the amplitude peak, and the dashed lines represent the amplitude valley. Consider an observed plane as marked; constructive interference occurs at any points where two peaks or two valleys meet, and destructive interference occurs at any points where one peak and one valley meet. The interference fringe period is $p = \lambda/\sin(\theta)$. For an aberrated wavefront represented by the dash-dotted curve, the interference maximum point moves from the open-circle position to the open-square position. (b) When the two wavefronts don't have aberrations, the interference fringes are straight. (c) When the two wavefronts have aberrations, the interference fringes are curved and have an irregular pattern. (d) When the two wavefronts are not planar, the interference fringes are tilted.

Most interferometers can generate such interferograms, such as Fizeau, shearing, Twyman–Green, and Mach–Zehnder interferometers, all of which are briefly discussed in this chapter.

Note that the percentages of spherical and astigmatism error in Figs. 11.4(e) and 11.5(a) are the same, while the fringe numbers are 1 and 7, respectively, because the fringe numbers (sag variation) are not proportional to the percentage variation of the wavefront radius of curvature $\Delta R/R$ but proportional to $\Delta R/R^2$ (see Section 14.3.1).

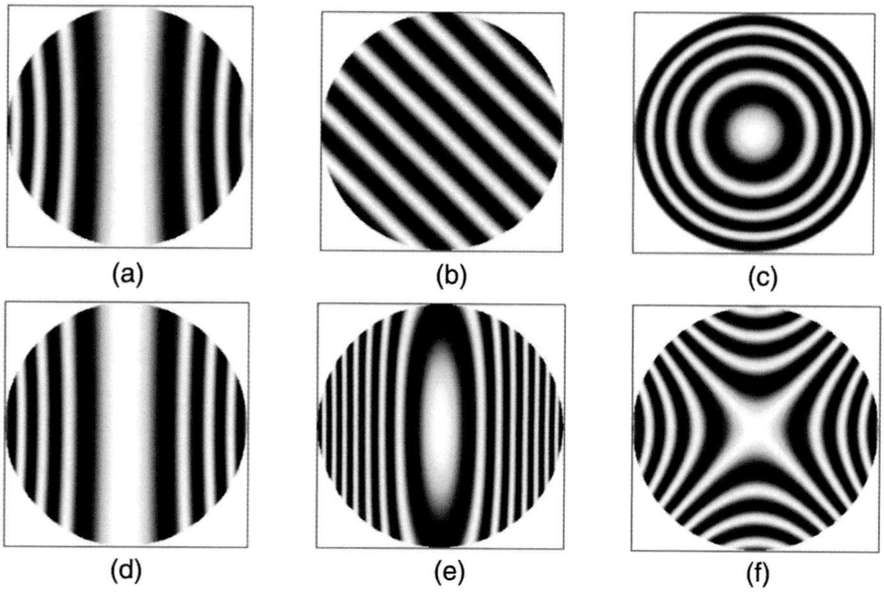

Figure 11.4 Interferogram of a two wavefronts with a curvature radius of 10 mm and diameter of 10 mm (*F*/1). (a) Two wavefronts have a 0.3-mm relative shift in the horizontal direction. (b) Two wavefronts have a 5° relative tilt in both horizontal and vertical directions. (c) One wavefront radius of curvature has 0.1% (0.01 mm) error (spherical). There are three fringes. (d) One wavefront radius of curvature has 0.1% error in the horizontal direction (astigmatism). (e) One wavefront radius of curvature has 0.1% and 0.2% error in the vertical and horizontal directions, respectively (spherical + astigmatism). (f) One wavefront radius of curvature has 0.1% error in the horizontal direction, and the two wavefronts have a 0.3-mm relative shift in the vertical direction.

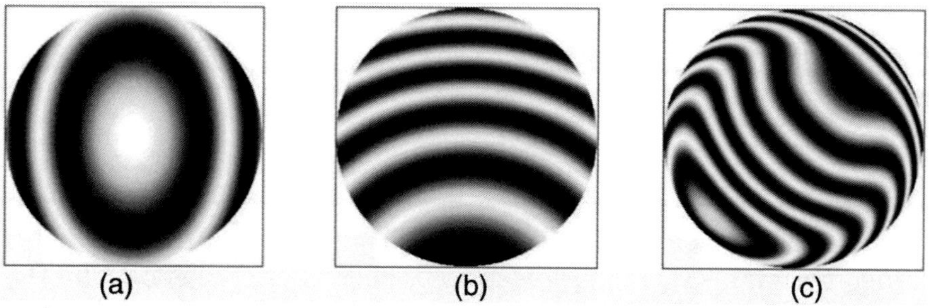

Figure 11.5 Interferogram of a two wavefronts with curvature radius of 50 mm and diameter of 10 mm (*F*/5). (a) One wavefront radius of curvature has 0.1% and 0.2% errors in the vertical and horizontal directions, respectively (spherical + astigmatism). (b) One wavefront radius of curvature has 0.1% error and the two wavefronts have relative tilt of 5° in the vertical direction. (c) One wavefront has aspheric surface and the two wavefronts have relative tilt of 5° in both horizontal and vertical directions.

11.2 Wavefront Analyzers

Wavefront analyzers analyze the wavefront of light output from an optical system, which can be as simple as a single lens or mirror. With the wavefront information, it is possible to evaluate the performance and quality of the optical system.

Most wavefront analyzers utilize an interference technique that lets the wavefront output from the optical system interfere with a known reference wavefront to generate an interferogram. Information about the wavefront being analyzed can be found by deciphering the interferogram. Some wavefront analyzers utilize other techniques; see Refs. 8–12 for more detailed descriptions.

11.2.1 Point diffraction interferometer

Point diffraction interferometers (PDIs) are simple devices often used to measure the wavefront quality of laser beams. The schematics of a PDI is shown in Fig. 11.6; it is somewhat similar to a spatial filter, shown in Fig. 9.23. The wave under measurement is focused by a spherical-aberration-free lens onto a mask with ~0.1% transmittance and a pinhole at the center. The pinhole size approximately equals the Airy disk size of the focusing lens. The spherical component of the incident wave can be focused to a diffraction-limited spot about the pinhole size, pass through the pinhole without being obstructed, and serve as the reference wave. The non-spherical components of the incident wave, which is the aberration part in the wave, cannot be focused to a sufficiently small spot, but it can transmit directly through the mask. The two waves meet on a screen and interfere with each other. The wavefront error can be found based on the interferogram. The transmission of the mask and the pinhole size are selected to balance the intensities of the two waves.

PDIs are a type of common-path interferometer, i.e., they generate their own reference wave. The wave under test and the reference wave travel the same or almost the same path. This working principle makes the PDIs simple, relatively easy to align, and resistant to environmental disturbance. As a comparison, amplitude-splitting interferometers, such as Twyman–Green

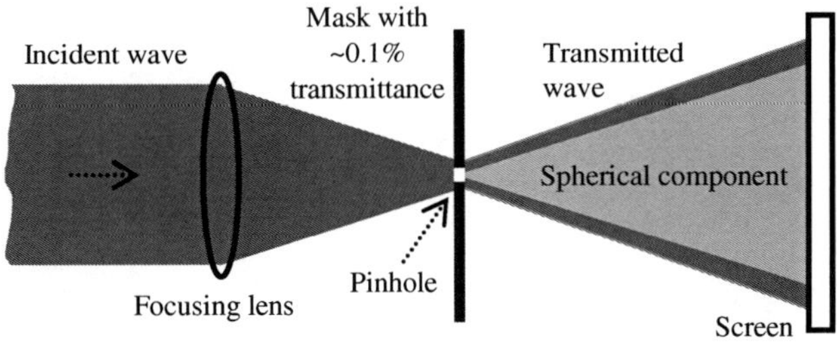

Figure 11.6 Schematics of a point diffraction interferometer for measuring wavefront error.

interferometers, separate an unaberrated wave and interfere it with the wave under test. The two waves travel two different paths.

11.2.2 Shack–Hartmann wavefront sensors

Shack–Hartmann wavefront sensors are the most widely used wavefront sensors to measure both coherent and incoherent optical waves. Compared with interferometer versions, which can only measure coherent wavefronts, Shack–Hartmann wavefront sensors have a simpler working principle and are easier to use, but have low accuracy.

A Shack–Hartmann wavefront sensor consists of a 2D micro-lens array and a 2D detector array, as illustrated in Fig. 11.7(a). The lens array focuses the optical wave or laser beam under test onto the detector array. The 3×3 lens array shown here is a simplified scheme. The actual lens array has hundreds of lenses. The wavefront under test can be reconstructed by special software based on the positions of the focused spots, as illustrated in Fig. 11.7(b).

More lenses in the lens array can provide more-accurate test results. The pixel numbers in the 2D detector array are always much larger than the lens numbers and are not a concern. The detector spectral response range determines the spectral range of the sensors. Commercial Shack–Hartmann wavefront sensors usually have a focal length from several to more than ten millimeters and an individual lens size of a few hundreds of microns, respectively, which produces an F-number >10. For a given individual lens size, a longer focal length provides a higher test resolution but smaller test range, as illustrated in Fig. 11.7(c). Shack–Hartmann wavefront sensors are also used to measure aspheric surfaces.[8]

11.2.3 Fizeau interferometers

A Fizeau interferometer is another type of common-path interferometer; it utilizes the reflections from the two surfaces of a Fizeau wedge to form an interferogram, as shown in Fig. 11.8(a). A Fizeau wedge can be either glass or air spaced. There is an angle in between the two reflected waves; this angle is the same as the angle of the wedge. If the incident wavefront is planar, then interference fringes are still formed, as illustrated in Fig. 11.3. The interferogram formed will be projected on a screen or a detector array for analysis. Fizeau interferometers are mostly used for test optical flats.

11.2.4 Shearing interferometers

A shearing interferometer is also a common-path interferometer, as shown in Fig. 11.8(b). The wave under test is incident on a glass- or air-space plate with two parallel surfaces. The two surfaces generate two reflected waves. These waves have a shearing displacement between them, but they are parallel,

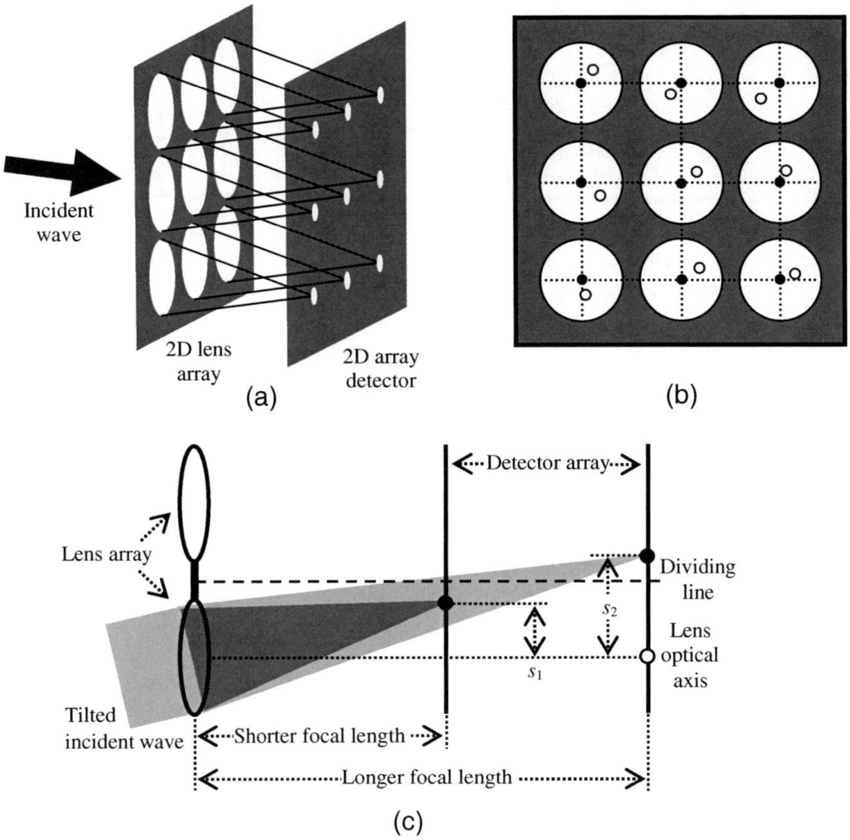

Figure 11.7 (a) A Shack–Hartmann wavefront sensor consists of a 2D micro-lens array and a 2D detector array. The lens array focuses the optical wave or laser beam under test onto the detector. (b) When the wavefront is planar, all of the focused spots of the lens array are at the optical axes of the lenses, as shown by the black dots. When the wavefront is not planar or has aberrations, the actual focused spots are off the optical axes, as shown by the open dots. The wavefront under measurement can be reconstructed based on the positions of the actual focused spots. (c) For the same tilted incident wave, the focused spot of a longer-focal-length-lens has a larger displacement s_2 than the focused spot displacement s_1 of a shorter-focal-length lens, but it can cross the dividing line into the neighboring lens territory and cause measurement error.

partially overlap, and interfere with each other. The interferogram formed will be projected on a screen or a detector array for analysis. Because the two waves are parallel, they will not form any interference fringes if the wave under test is planar. Therefore, shearing interferometers are mainly used to check the collimation of laser beams. A lack of fringes means that the laser beam is well collimated.

In both Fizeau and shearing interferometers, the wave under test is split to two and the two sub waves interfere with each other. If the wave under test

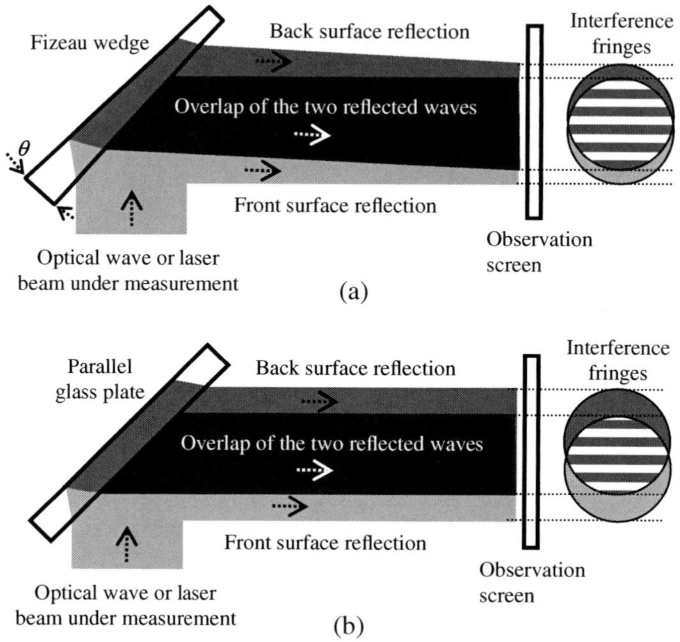

Figure 11.8 (a) Schematics of a Fizeau interferometer. (b) Schematics of a shearing interferometer.

has complex wavefront structure, it will be difficult to decipher the interferogram, since little is known about the wavefront. Therefore, Fizeau and shearing interferometers are often used to test relatively simple wavefronts, such as planar or spherical wavefronts.

11.2.5 Twyman–Green interferometers

Twyman–Green interferometers are suitable for testing optical surfaces with complex aberrations, since the reference wavefront is well known. Most commercial, high–quality interferometers are Twyman-Green interferometers.

Figure 11.9(a) shows the schematics of a Twyman–Green interferometer. A light wave from a monochromatic light source is collimated and split in two by a beamsplitter. One of the two waves serves as the reference wave, and the other is first reflected by the surface under test before recombining with the reference wave to form an interferogram. Different standard lenses must be used to pair with different shapes of optical surfaces under test, as explained in Figs. 11.9(b) and (c).

Twyman–Green interferometers are amplitude-splitting interferometers that have two optical arms. The reference wave and the testing wave

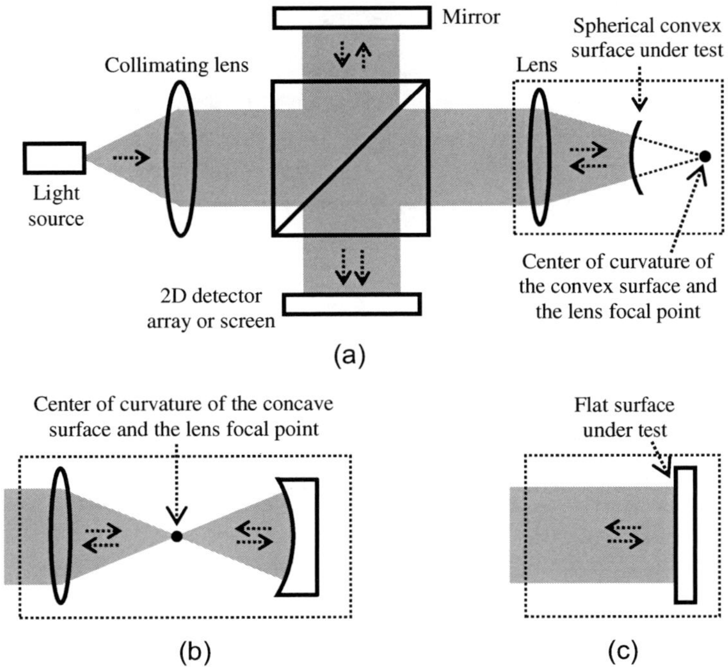

Figure 11.9 (a) Schematics of a Twyman–Green interferometer. A collimating lens collimates the light wave from a monochromatic light source. A cube beamsplitter splits the wave to two. One wave is reflected by a mirror, passes through the beamplitter, and reaches the 2D detector. This is the reference wave with planar wavefront. Another wave is the test wave and incident on the optical surface under test, either a lens or a mirror surface. The optical surface under test shown here is a convex surface only as an example. The test wave is focused by a lens onto the convex surface under test. The convex surface has a center of curvature that coincides with the lens focal point. The wave reflected by the convex surface is traced back, collimated by the lens, and reflected by the cube beamsplitter before it reaches the 2D detector and interferes with the reference wave. The 2D detector array outputs the inteferogram it receives to a computer for analysis. The setup on the right side inside the dotted frame can be changed to test optical surfaces with other shapes. (b) The setup for testing a concave surface of either a mirror or a lens to replace the setup shown in the dotted line frame in (a). (c) The setup for testing a flat surface of either a mirror or a glass plate to replace the setup shown in the dotted frame in (a).

propagate through these two optical arms, respectively. Such a structure requires precise alignment of all of the optical components used and is sensitive to environmental disturbance.

Twyman–Green interferometers are a variation of the earlier invented Michelson interferometers. The difference between these two interferometers is that the former use a collimator to collimate monochromatic light from a point source and the collimated beam is large enough to test optical surfaces of certain size, whereas the latter do not have such a structure and are intended for other applications.

11.2.6 Mach–Zehnder interferometers

Mach–Zehnder interferometers are another type of amplitude-splitting interferometer, as shown in Fig. 11.10. The light wave from a point light source is first collimated and then split in two; one wave is the reference wave, and the other is the test wave. The test wave first propagates through the samples under test and then is recombined with the reference wave to form an interferogram for analysis. Since the testing wave transmits through the sample, Mach–Zehnder interferometers are best for testing the optical properties of liquids, fluid dynamics, transparent objects, chemical processes such as combustion, etc.

11.3 Spectral Analyzers

11.3.1 Monochromators

Monochromators are the most commonly used instruments for optical spectrum analysis. Various types of monochromators have been developed; Fig. 11.11 shows the schematics of the most common diffraction-grating monochromator.

A laser beam or optical wave under test enters the monochromator through an entrance slit, is expanded by two concave mirrors, diffracted by a grating, and focused by another concave mirror on the exit slit before reaching an optical power meter. The grating is mounted on a rotator. For any given orientation of the grating, the wavelength that can pass through the exit slit is known by calibration. By rotating the grating and recording the power

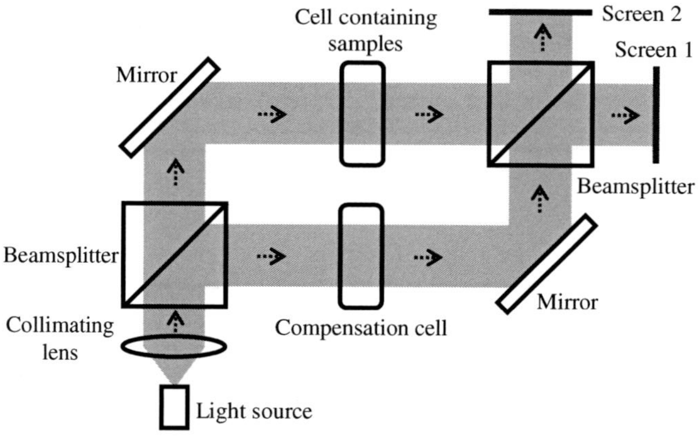

Figure 11.10 Schematics of a Mach–Zehnder interferometer. A collimating lens collimates a light wave from a point light source. A cube beamsplitter splits the wave in two. One wave propagates through a cell containing the sample under measurement. Another wave propagates through a compensation cell that is used to offset the effects caused by the sample cell (not the sample). Another cube beamsplitter combines the two waves to interfere. The interferogram can be observed at both screen 1 and 2.

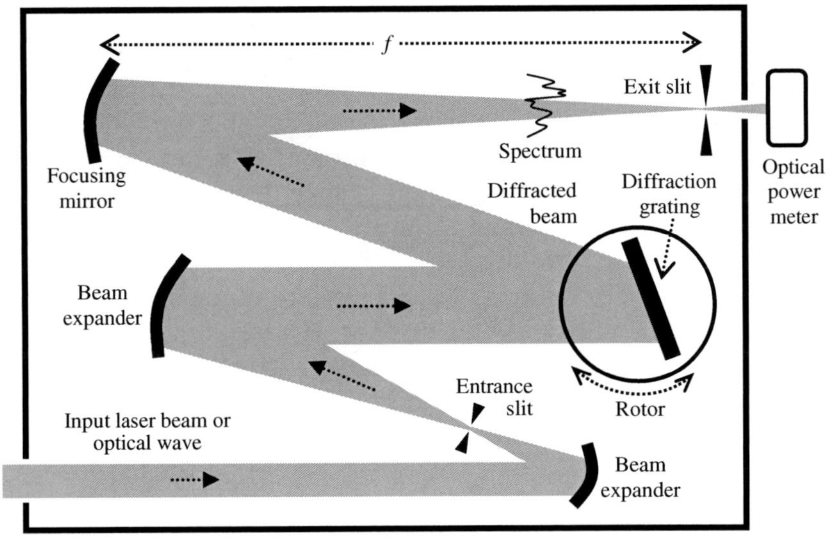

Figure 11.11 Schematics of a diffraction-grating-based monochromator. The laser beam or the optical wave under measurement is expanded by two concave mirrors, incident on a diffraction grating, diffracted by the grating, and focused by the focusing mirror. One certain wavelength component in the diffracted beam has the right orientation, can pass through the exit slit, and reach the optical power meter. The wavelength that can reach the power meter is known based on the grating angle and the calibration data. By rotating the diffraction grating, one can change the wavelength that can reach the power meter. The spectrum of the beam under measurement is thereby found by relating the power meter output, the grating angle, and the corresponding wavelength.

measured by the optical power meter as a function of the corresponding λ, the spectrum can be measured.

The reason for using an entrance slit is to make the light source a linear shape that can be well collimated in the direction perpendicular to the line. The reason for using an exit slit is to improve the measurement resolution. Since at the exit slit location, the laser beam/optical wave is spread in the direction perpendicular to the slit. A narrower exit slit can intercept a narrower spectral bandwidth. The reason for using mirrors to manipulate the beam/wave is to get rid of any color aberrations caused by lenses. All three mirrors must be off-axis parabolas to minimize the spherical aberration.

The widths of the entrance and exit slits Δ_{s1} and Δ_{s2} are adjustable. The last two mirrors image the entrance slit on the exit slit. The width of the image of the entrance slit Δ'_{s1} is proportional to Δ_{s1}. The measurement resolution R of this monochromator is given by

$$R = \frac{\Delta_s}{f} \frac{d\lambda}{d\theta_1}, \tag{11.11}$$

where Δ_s equals the larger of Δ'_{s1} and Δ_{s2}, f is the focal length of the focusing mirror as marked in Fig. 11.11, and $d\lambda/d\theta_1$ is the inverse of the angular dispersion resolution of the grating. Equation (11.11) shows that a reduction in the widths of the two slits can increase the measurement resolution. However, the slits must be wide enough to allow enough optical power to pass through for measurement. The grating rotation angular resolution also affects the measurement resolution of this monochromator. A monochromator often comes with a few gratings with different groove densities (dispersion power). The measurement resolution and range of the monochromator can be changed by adjusting the grating.

If a linear photodetector array is used in place of the optical power meter, the exit slit is no longer needed. The linear array can simultaneously detect the whole spectrum of laser beam or optical wave under test. Such an instrument is called a spectrometer.

The resolution of commonly used monochromators is about 0.1 nm, a little too low to measure the linewidth of lasers but adequate to measure the spectrum of lamps. There are some higher-resolution monochrmators available on the market. For example, a Horiba (www.horiba.com) iHR550 monochromator has a resolution of 0.025 nm, and a Spectral Products (www.spectralproducts.com) DK480 0.5-m monochromator has a resolution of 0.03 nm. These monochromators can be used to measure the linewdith of laser diodes. Reference 13 provides more detailed descriptions of monochromators.

11.3.2 Scanning Fabry–Pérot interferometers

Scanning Fabry–Pérot interferometers (SFPIs) are probably the most widely used instruments for analyzing the spectrum and measuring the linewidth of lasers and spectral lines. Their measurement ranges are too small, and their measurement resolutions are excessive for measuring the whole spectrum of lamps.

An SFPI consists of two slightly wedged glass plates with flat surfaces, as shown in Fig. 11.12. The two inner surfaces of the plates are set parallel to each other, usually polished to a flatness of better than $\lambda/8$, and high-reflection coated. The two outer surfaces of the plates are also optically polished, AR coated, and have an angle between them such that the reflections of the two outer surfaces are deflected and cannot interfere with the reflections of the two inner surfaces. The distance D between the two inner surfaces can be adjusted by a piezoelectric device. The medium between the two plates is usually air with a unit refractive index.

The laser beam or optical wave under measurement is expanded by two high quality lenses (groups) and incident on the SFPI. Multiple reflections take place at the two inner surfaces of the SFPI and produce a series of transmitted beams whose amplitudes fall off progressively. These multiple reflections plotted here are not parallel to each other for illustration purpose, they are actually parallel and travel forth and back along the same paths. The beams transmitted through the SFPI are focused by another lens onto a laser

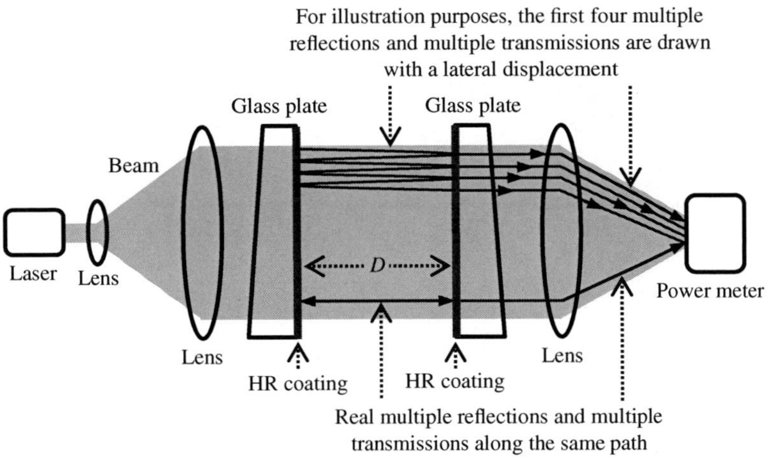

Figure 11.12 Schematics of a scanning Fabry–Pérot interferometer (SFPI). The SFPI consists of two wedged mirrors and is used to measure the spectrum and linewidth of lasers with high precision. The beam of a laser under measurement is expanded and incident on the SFPI. Multireflections occur between the two high-reflection-coated surfaces of the mirrors and lead to multitransmissions. All of the transmitted beams are focused onto a laser power meter where they interfere. The interference intensity changes as the spacing *D* between the two mirrors is varied, and the spectrum can be measured by plotting the power meter output vs. *D*.

power meter, where they interfere. This is a multiwave interference situation. This topic was discussed in detail in Section 2.4.2.

Figure 11.12 shows that the phase difference $\Delta\phi$ between two successive transmitted beams is

$$\Delta\phi = \frac{4\pi D}{\lambda}. \tag{11.12}$$

Interference occurs among the amplitudes of the transmitted beams. The intensity of the combination of all the transmitted beams is a function of $\Delta\phi$ and is given by[14]

$$\begin{aligned} I(\Delta\phi) &= I_0 \left| t^2 \sum_{k=0}^{\infty} r^{2k} \exp(-ik\Delta\phi) \right|^2 \\ &= I_0 \left| \frac{t^2}{1 - r^2 \exp(-i\Delta\phi)} \right|^2 \\ &= I_0 \frac{(1 - r^2)^2}{1 + r^4 - 2r^2 \cos(\Delta\phi)} \\ &= I_0 \frac{1}{1 + F \sin\left(\frac{\Delta\phi}{2}\right)^2}, \end{aligned} \tag{11.13}$$

where I_0 is the intensity of the beam before being incident on the SFPI, $r < 1$ and $t = (1 - r^2)^{0.5}$ are the amplitude reflection and transmission coefficient of the two inner surfaces of the SFPI, respectively, and $F = 4r^2/(1 - r^2)^2$ is a parameter called "fineness." Note that $R + T = 1$, where $R = r^2$ is the reflectance (power reflection), and $T = t^2$ is the transmittance (power transmission).

The FWHM of $I(\Delta\phi)$ can be found by letting $I(\Delta\phi) = 0.5I_0$ in Eq. (11.13) and solving for $\Delta\phi$:

$$\begin{aligned}\Delta\phi_{\text{FWHM}} &= 4\sin^{-1}\left(\frac{1}{F^{0.5}}\right)\\ &= 4\sin^{-1}\left(\frac{1-r}{2r}\right)\\ &\approx 2\frac{(1-r)}{r},\end{aligned} \tag{11.14}$$

where the approximation is taken because r is usually chosen to be close to 1. Equation (11.14) shows that the fringe width of $I(\Delta\phi)$ reduces as r is increased because a larger r results in more reflections between the two inner surfaces of the SFPI, more transmission beams, and more beams involved in interference. The FWHM of $I(\Delta\phi)$ in terms of wavelength can be found by combining Eqs. (11.12)–(11.14). The result is

$$\begin{aligned}\Delta\lambda_{\text{FWHM}} &= \frac{4\pi D}{\Delta\phi_{\text{FWHM}}}\\ &= \frac{\pi D}{\sin^{-1}\left(\frac{1}{F^{0.5}}\right)}.\end{aligned} \tag{11.15}$$

Constructive interference occurs when $\Delta\phi = 2m\pi$, which when plugged into Eq. (11.12), leads to

$$\lambda_m = \frac{2D}{m}, \tag{11.16}$$

where λ_m is the wavelengths at which the peak of $I(\Delta\phi)$ appears, and m is an integer. For any given D, there are many λ_m since m can take many values. The spacing between two adjacent peak wavelengths can be obtained by differentiating λ in Eq. (11.16) with respect to m, eliminating m, and letting $\Delta m = 1$; the result is

$$\Delta\lambda = \frac{\lambda^2}{2D}, \tag{11.17}$$

where $\Delta\lambda$ is known as the free spectral range (FSR) of a SFPI.

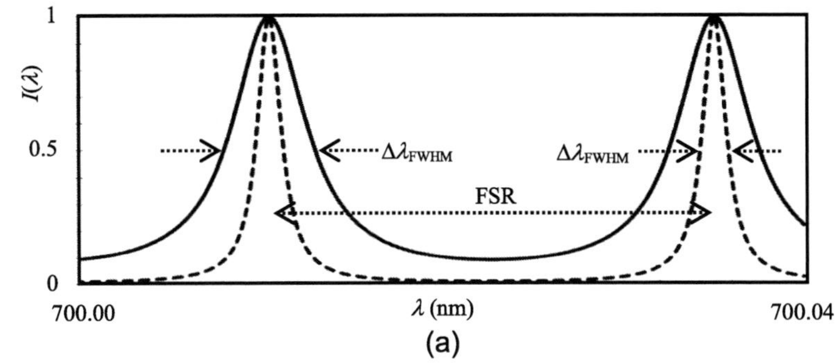

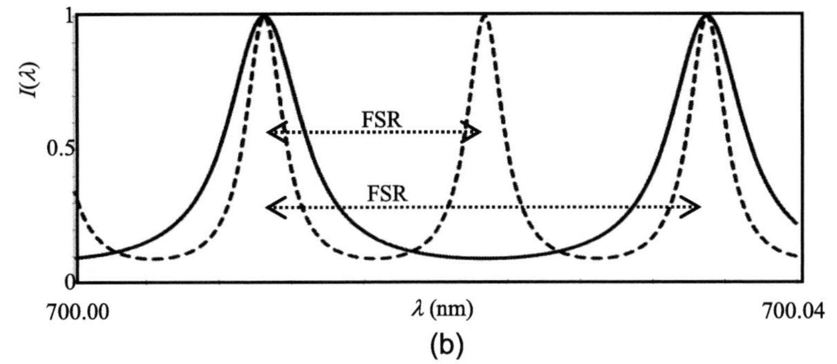

Figure 11.13 Normalized transmission intensity of a SFPI as a function of wavelength. (a) $D = 10$ mm, solid curve $F = 10$, and dashed curve $F = 100$. The FSR is determined by the D value, and a larger F value leads to a narrower fringe. (b) $F = 10$, solid curve $D = 10$ mm, and dashed curve $D = 20$ mm. The FSR is inversely proportional to the D value.

With Eq. (11.12) considered, the $I(\Delta\phi)$ given by Eq. (11.13) can be written as $I(\lambda)$ and is plotted in Fig. 11.13 with λ being the variable for several D and F values, and a normalized $I_0 = 1$. A SFPI acts like a bandpass comb. For a given D value, an increase in the values of F can reduce the bandwidth $\Delta\phi_{FWHM}$, as shown in Fig. 11.13(a). Most SFPIs have a $r > 0.95$ for a small $\Delta\phi_{FWHM}$. The measurement resolution of a SFPI is limited by $\Delta\phi_{FWHM}$.

For a given F value, an increase in the value of D will reduce the $\Delta\phi_{FWHM}$ at the cost of reducing the free spectral range $\Delta\lambda$, as shown in Fig. 11.13(b). When D is scanned, the transmitted wavelength is scanned according to Eq. (11.13), and the laser power meter measures the outputs of the SFPI as a function of wavelength and thereby measures the laser spectrum.

To properly measure the linewidth of a laser beam, some advance knowledge about the linewidth is required. Figure 11.14 shows the transmission bands of a SFPI and three laser lines with different linewidths. The line on the left has a width not much larger than the bandwidth of the SFPI; scanning the SFPI band over this laser line will lead to a low-accuracy measurement result. The central line, which has a width larger than the FSR of the SFPI, will transmit through two bands of the SFPI simultaneously. The laser power meter cannot tell which band generated the power it receives, and

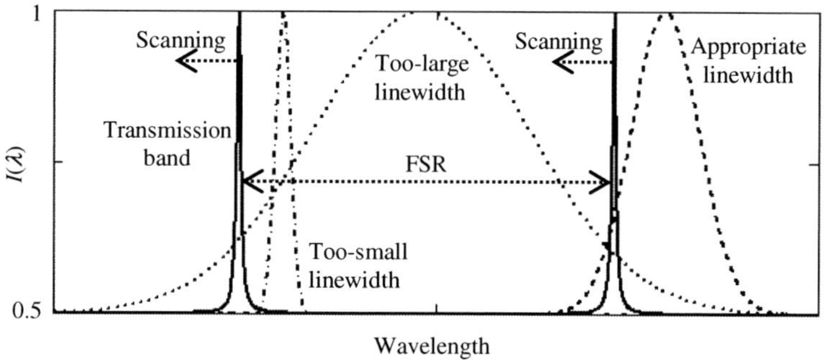

Figure 11.14 Solid curves are the transmission bands of a SFPI, and dashed or dotted curves are the three laser lines. The SFPI is only appropriate to measure the linewidth of the line on the right.

the measurement result will be erroneous. The line on the right has a width much larger than the bandwidth of the SFPI but smaller than the FSR, so the measurement result will have a high accuracy.

Commonly used commercial SFPIs, for example, the series sold by Thorlabs (www.thorlabs.com), have a FSR of ~0.01 nm and a resolution of about FSR/100 = 0.0001 nm. These values are good for measuring the linewidth of most lasers, but the FSR is too small to measure the linewidth of most laser diodes.

11.4 Other Optical Instruments

11.4.1 Measuring microscopes

Microscopes are used for the precise, non-contact measurement and inspection of optical components. For example, a measuring microscope can be used to check the quality of an optical surface for scratches and digs, chips and cracks, etc. A reticle is often mounted in the optical path of the microscope to facilitate the measurement of the physical size of the object under viewing or the defects. Many companies produce measuring microscopes, such as Nikon and Olympus.

11.4.2 Autocollimators

An autocollimator is an optical instrument for the non-contact measurement of small angles. They are typically used to align components and measure deflections in optical or mechanical systems.

Figure 11.15 illustrates how an autocollimator works. The measurement range is inversely proportional to the focal length f of the objective, whereas the measurement accuracy is proportional to f. If distance x is measured by visual observation with the help of a reticle, the measurement accuracy can be as high as 0.1 mR. If distance x is measured by a 2D sensor, the measurement

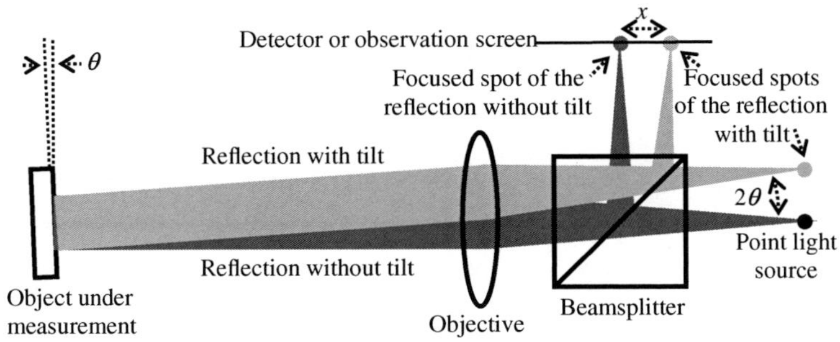

Figure 11.15 Schematic of an autocollimator. Light from a point source travels through a cube beamsplitter before being collimated by an objective and projected onto an object under measurement. A mirror is often attached to the object to provide specular reflection. The reflection is focused by the objective and split by the beamsplitter. A detector or a screen is used to observe the focused spot. If the object or mirror is perpendicular to the collimated light, the position of the focused spot of the reflected light is known *a priori* by calibration. If the object or mirror is tilted by an angle of θ from perpendicular to the light, the position of the focused spot of the reflected beam will be off by a distance x. θ can be calculated by $\theta = x/(2f)$, where f is the focal length of the objective.

accuracy can be around 1 mR. Reference 15 provides concise descriptions of several optical measurements performed using autocollimators.

11.4.3 Radiometers and photometers

A radiometer measures electromagnetic radiation across the total spectrum; the measured radiation is an absolute power. A photometer measures electromagnetic radiation across the total spectrum and scales the measured radiometric power by the spectral response (photopic curve) of the human eye. From a physics perspective, they both measure the same thing but display the results using different units.

The unit system for radiometry is relatively complex and somewhat confusing, and some parameters appear to be obsolete. People working in this area should pay particular attention to the definition and conversion of parameters.[16] Table 1.1 in Section 1.15.7 lists some SI units for radiometry that are widely used metric units.

The working principle of radiometers and photometers is the same as that of laser power meters. They detect the input optical or radio power, and an electronic system processes the output of the detector and displays the measurement results in proper units. Several type of radiometers and photodetectors are used to handle different optical spectra, incident power levels, minimum detectable power levels, etc.

The detector size is sometimes needed in order to calculate the solid angle the detector imposes on the emission source. A large-aperture lens can be used to collect more optical power and focus the collected optical power onto the detector, in which case the detector size does not matter; the collection lens size determines the solid angle to the emission source. Section 1.15 describes the physics involved with radiometry.

References

1. W. Smith, "The Aberration Polynomials and the Seidel Aberrations," Section 3.2 in *Modern Optical Engineering*, 3rd ed., McGraw-Hill, New York, pp. 62–64 (2000).
2. M. Born and E. Wolf, "Geometrical Theory of Aberrations," Chapter 5 in *Principles of Optics*, 7th ed., Cambridge University Press, Cambridge, UK, pp. 228–260 (2002).
3. Wikipedia, "Zernike Polynomials," wikipedia.org/wiki/Zernike_polynomials
4. M. Born and E. Wolf, "The Circle Polynomials of Zernike," Appendix VII in *Principles of Optics*, 7th ed., Cambridge University Press, Cambridge, MA, pp. 905–910 (2002).
5. M. Born and E. Wolf, "Expansion of the Aberration Function," Section 9.2 in *Principles of Optics*, 7th ed., Cambridge University Press, Cambridge, UK, pp. 523–527 (2002).
6. User Zom-B, "The First 21 Zernike Polynomials," wikipedia.org/wiki/Zernike_polynomials#/media/File:Zernike_polynomials2.png
7. J. C. Wyant and K. Creath, "Basic Wavefront Aberration Theory for Optical Metrology," Chapter 1 in *Applied Optics and Optical Engineering, Vol. Xl*, wp.optics.arizona.edu/jcwyant/wp-content/uploads/sites/13/2016/08/Zernikes.pdf
8. *Basics of Interferometers*, cmi.epfl.ch/metrology/files/Wyko/Interfero metery_Basics.pdf
9. *Interferometer*, www.interferometer.info
10. *Wavefront Sensors*, www.trioptics.com/products/wavefront-sensors
11. M. Hausner, *Optics Inspections and Tests: A Guide for Optics Inspectors and Designers*, SPIE Press, Bellingham, WA (2017) [doi: 10.1117/3.2237066.]
12. D. Malacara, *Optical Shop Testing*, 3rd ed., Wiley-Interscience, Hoboken, NJ (2007).
13. Wikipedia, "Monochromator," wikipedia.org/wiki/Monochromator
14. J. C. Wyant, *Multiple Beam Interference*, wyant.optics.arizona.edu/MultipleBeamInterference/MultipleBeamInterference.pdf
15. *Examples for Applications of Collimators, Telescopes, Visual, and Electronic Autocollimators*, www.vermontphotonics.com/VT_Photo_Applications.pdf
16. *Radiometry and Photometry FAQ*, www.physics.muni.cz/~jancely/PPL/Texty/IntegracniKoule/Radiometry%20and%20photometry%20FAQ.pdf

Chapter 12
Computer-Aided Optical Modeling

12.1 General Comments

Various types of optical design software are widely used currently. These software can model and analyze optical systems much faster and with more accuracy and detailed results than manual modeling and analysis. Therefore, optical software should be used whenever possible, and familiarity with using it is an important part of the optical-engineering skill set.

This chapter describes the basic processes of modeling various optical components with the optical design software Zemax. All other types of optical software are more or less similar, so familiarity with one program will facilitate learning others.

Generally speaking, there are three types of optical modeling:

1. Sequential raytracing, used primarily to model imaging optics.
2. Non-sequential raytracing, used primarily to model illumination optics.
3. Physical optics, used primarily to model laser beams.

This chapter is preparation for the design of optical components and systems, which is the topic of Chapter 13. The difference between "modeling" and "design" is that modeling only involves typing parameters of optical components/systems into software to see how the components/systems perform. Design means varying the parameters of optical components/systems to push their performance to the preset goals.

12.2 Commercially Available Optical Design Software

Several types of commercially available optical design software are widely used: CODE V, OpticStudio (Zemax), OSLO, ASAP, FRED, GLAD, etc. Opinions on these software vary depending on personal experiences and preferences, hence the variety. This section assesses these programs based on the author's experience, which may be incomplete or biased.

12.2.1 CODE V

CODE V is assumed to be the best general-purpose optical design software for the design of imaging optics, but it can also be used to design illumination optics and physical optics. The main advantage of CODE V is that it is convenient to write macros to handle special tasks. It is disputable whether CODE V has better optimization power than Zemax.

CODE V is available only with an annual subscription, and the price is the highest among the various types of general purpose optical software. Synopsys Corp. (www.synopsys.com) is the company that develops and sells CODE V.

12.2.2 OpticStudio (Zemax)

OpticStudio is the new name of Zemax, although most people still refer to it by its original name. This software is developed by a company named Zemax, which is now part of Radiant Imaging Inc. but still maintains its website (www.zemax.com).

Zemax is probably the most widely used general-purpose optical software, primarily to design imaging optics, but it can also be used to design illumination optics and physical optics. The cost effectiveness and relatively ease of use are the main advantages of Zemax, although the price has risen over the past few years.

Zemax is adequate for most optical design purposes (the author has used it for the past twenty years), and it provided the illustrations, analysis, and design included in this book. However, some people consider Zemax as "an entry-level package suited to less experienced designers."[1]

The program is sold as a one-time purchase plus an annual subscription. Forgoing the recurring subscription restricts access to new updates, but the purchased version can be used after the initial period expires.

Reference 1 provides a brief review of Zemax.

12.2.3 Other types of optical software

OSLO is a program developed and marketed by Lambda Research Corp. (www.lambdares.com) that can perform both sequential and non-sequential raytracing to design imaging optics and illumination optics. It has relatively fewer functions, costs much less than Zemax, and can be considered to be a simpler version.

ASAP is a general-purpose optical software developed and marketed by Breault Research Corp. (www.breault.com) that is effective at handling non-sequential raytracing for the design of illumination optics, though it can also be applied to imaging optics design and physical optical analysis for handling laser beams.

FRED is developed and marketed by Photon Engineering Corp. (photonengr.com). Its main strength is its strong ability to design opto-mechanical parts and its ease to interface with mechanical design software. FRED can also design imaging optics and illumination optics.

GLAD is developed and marketed by Applied Optics Research Corp. (www.aor.com) to handle physical optics. It is very effective at analyzing laser beam propagation through optical systems. Although it cannot be used for imaging optical design, GLAD is a necessity when faced with complex laser beam analysis because all of the other optical software mentioned previously are not very capable of this function.

12.3 Sequential Raytracing Modeling

12.3.1 General comments

Sequential raytracing is mainly used to model and design imaging optics and is the most widely used technique. For brevity, this book will briefly introduce some of the most basic functions of Zemax for sequential raytracing; consult the Zemax user manual for a complete and rigorous description of all functions. In this chapter and the following chapter, *italicized* words are terms used by Zemax with specially defined meanings.

Several books about optical design have been published,[2–9] but most of them explain the optical principles without providing a step-by-step guide about how to use optical software. Conversely, the software user manuals rigorously describe how to use the program but do not offer a straightforward optical-design guide. (Another book by the author, *Lens Design: A Practical Guide*,[10] is recommended for readers looking for a detailed guide to most aspects of lens design.)

12.3.2 Set optical parameters

To model an optical component, the first task is to set the spectral and field ranges in which the component will be used. After opening Zemax and then clicking *System Explorer|Wavelengths*, a *Wavelength Data* box will appear. Any desired wavelengths (up to 12) can be entered. The *Wavelength Data* box shown in Fig. 12.1(a) has three RGB wavelengths. The *Weight* boxes numerically set the importance of the wavelengths; a weight value of 1 is selected for all three, which means that the three wavelengths have equal importance. Note that the value of the weight only has a relative meaning not only for wavelengths, but also for all parameters of an optical system. Selecting a weight of 1 for the three wavelengths is equivalent to selecting a weight of 2 for the three wavelengths.

After clicking *System Explorer|Fields*, a *Field Data Editor* box will appear. There are a few field type options: *Angle*, *Object Height*, *Image Height*, etc. Select *Angle*. If the optical component being modeled is symmetric about the optical axis, which is the case for most optical components, only the field angle in one direction must be set, usually in the positive *y* direction because

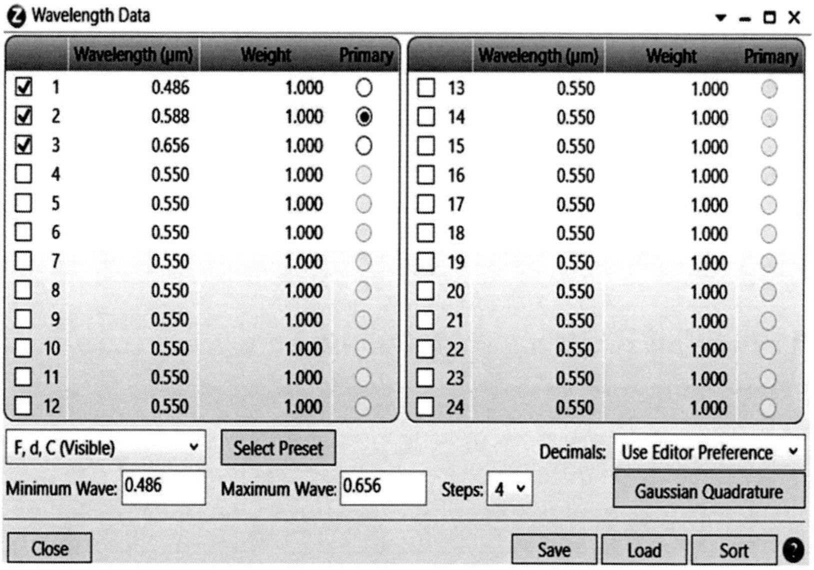

(a)

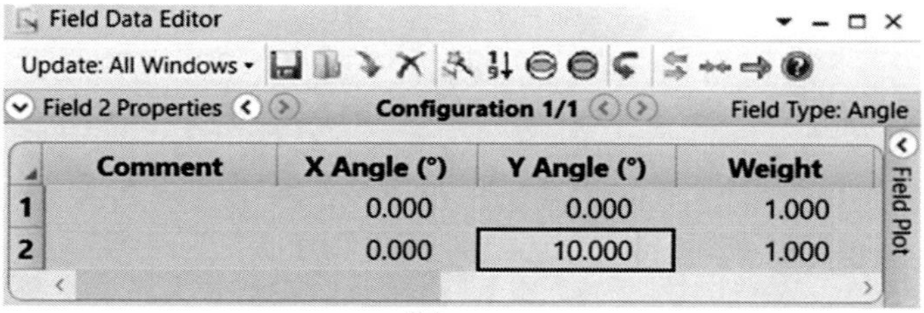

(b)

Figure 12.1 (a) Zemax *Wavelength Data* box for typing in wavelengths. (b) Zemax *Field Data Editor* box for typing in fields.

that direction is defined as being in the plane of the computer screen or book page and can be easily seen. The *Field Data Editor* box shown in Fig. 12.1(b) has 0° and 10° typed in the *Y Angle* column. A weight of 1 is selected for the two fields.

12.3.3 Modeling a single lens

The lens selected for modeling is an Edmund Optics off-the-shelf (in-store) lens, Part Number 32-490, that has a diameter of 25 mm with a 24-mm clear aperture, a focal length of 25 mm, equi-convex surfaces with radii of curvature 31.94 mm, a central thickness of 8.0 mm, and is made of Schott glass N-SF5. This lens has an *F*-number of 24/25 $\approx$ 1.

The modeling process proceeds as follows:

1. After clicking *Setup/Sequential/Lens Data*, a *Lens Data* box will appear for typing in the lens data. New rows can be inserted by clicking a box

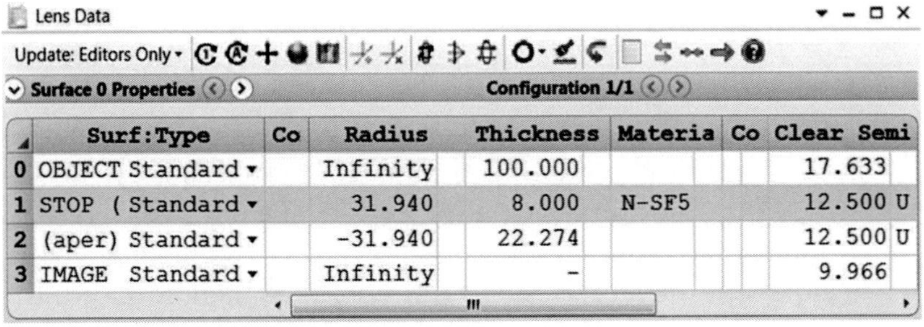

Lens Data

Update: Editors Only

Surface 0 Properties | Configuration 1/1

	Surf:Type	Co	Radius	Thickness	Materia	Co	Clear Semi	
0	OBJECT Standard		Infinity	100.000			17.633	
1	STOP (Standard		31.940	8.000	N-SF5		12.500	U
2	(aper) Standard		-31.940	22.274			12.500	U
3	IMAGE Standard		Infinity	-			9.966	

Figure 12.2 Zemax *Lens Data* box with the parameters of an Edmund PN 32-490 lens. Shown here is only the left half of the *Lens Data* box. The right half of the box is for other parameters, depending on the surface type.

in the *Lens Data* box and then pressing the *Insert* key. Figure 12.2 shows the *Lens Data* box with all of the data about the Edmund lens already typed into the box. The details about the *Lens Data* box are thus:

1. Row 0 defines the object, which is assumed to be a plane at a distance of 100 mm away from the front surface of the lens. *Infinity* is thus typed in the *Radius* (of curvature) box, and 100 is typed in the *Thickness* box. The default length unit is millimeter.
2. Row 1 defines the front surface of the lens, etc. The typed numbers of *Radius*, *Thickness*, and glass type are self-explanatory. The 12.5 in the *Clear-Semi* field defines the half-diameter of the lens. The clear aperture size will be defined by the aperture stop size:
   A. Zemax offers many types of surface, such as *Standard* (mainly for a spherical surface), *Asphere*, *Toroidal* (for a cylindrical surface), etc. Here, only the most basic *Standard* surface is used. The positive radius number means the vertex of this surface is at the left side of the other part of this surface, as shown in Fig. 12.3.

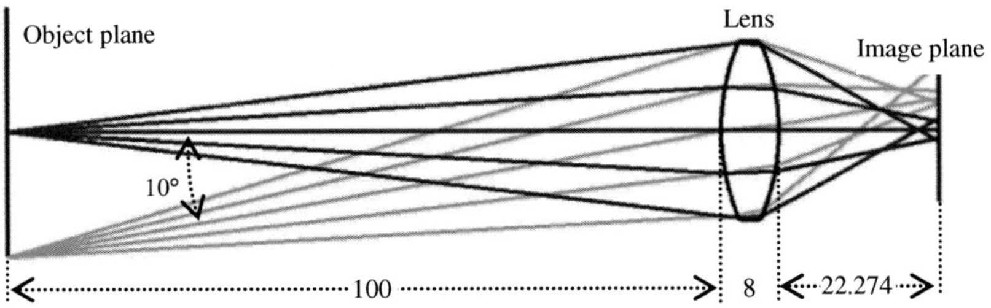

Figure 12.3 Layout or raytracing diagram of the Edmund equi-convex lens whose parameters are shown in Fig. 12.2. The rays cannot be well focused because of the severe spherical aberrations.

B. Zemax has many hundreds of glasses made by over ten vendors saved in its *Material Catalog* and recognizes the standard names of all these glasses.

C. The *Stop* sign shown in the first box means that the front surface of the lens is chosen to be the aperture stop, which is customary. If the *Stop* sign appears at another row, it can be moved to row 1 by double clicking the first box in row 1, clicking *Type* in the appeared *Surface 1 Properties* box, and then checking the *Make Surface Stop* box.

D. In this case, the aperture stop is the entrance pupil, and the 24-mm clear aperture of the lens is the size of the entrance pupil. This can be set by clicking *System Explorer/Aperture*, selecting *Entrance Pupil Diameter* in the *Aperture Type* box, and typing 24 in the *Aperture Value* box.

3. Row 2 defines the back surface of the lens:

A. The negative radius number means that the vertex of this surface is at the right side of the other part of the surface, as shown in Fig. 12.3, since this lens is an equi-convex lens.

B. The lens must be focused by adjusting the value in the *Thickness* box, which is the image distance, defined as the distance between this lens surface and the image plane. By clicking *Optimize/Quick Focus*, selecting *Spot Size Radial* in the *Quick Focus* box, and then clicking *OK*, Zemax will adjust the value in the *Thickness* box to 22.274, which will move the image plane to the focal plane of the lens.

2. Plot the raytracing diagram. After clicking *Analyze/Cross-Section*, a *Layout* box will pop up, as shown in Fig. 12.3. Five rays are plotted here. The ray number can be changed by clicking the *Setting* button in the *Layout* box and selecting the ray number in the *Number of Rays* box. The large spherical aberration is visually noticeable in the raytracing diagram.

3. Check the performance. For such a single lens, the focused spot size is the main measure of performance. Clicking *Analyze/Rays & Spots/Standard Spot Diagram*, a *Spot Diagram* will pop up, as shown in Fig. 12.4. The default unit is micron. The RMS radii of the focused spots are 1147.46 μm and 1741.84 μm for a field angle of 0° and 10°, respectively, which are huge compared with the Airy disk radius of 0.7466 μm. The three RGB colors each have a slightly different focused spot size, and the 10° field spot shows a coma-type aberration.

   If the *Airy Radius* sign does not appear, click the *Setting* button in the *Spot Diagram* box and then check the *Show Airy Disk* box.

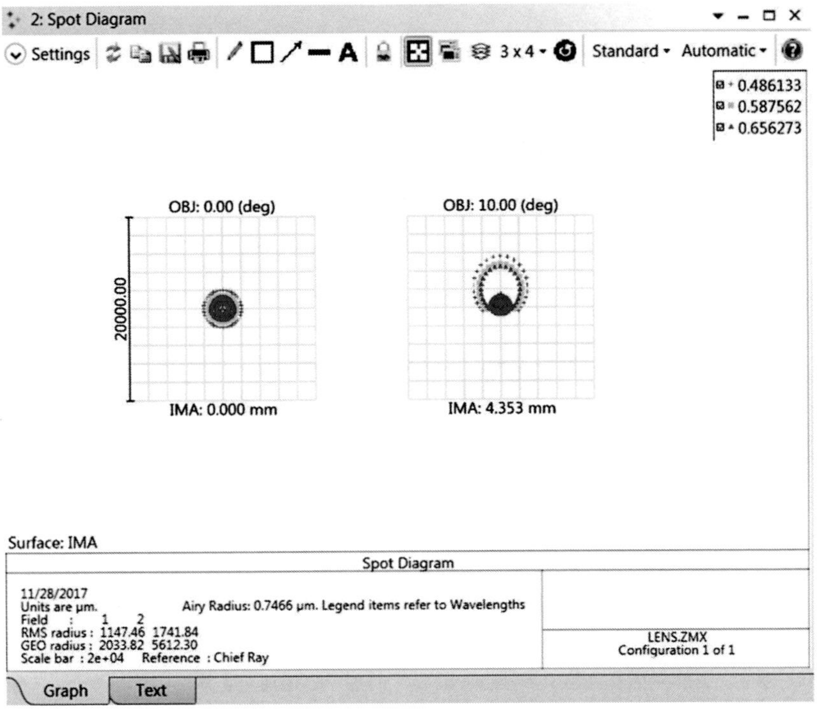

Figure 12.4 Spot diagram of the lens shown in Fig. 12.3.

If there is only one wavelength, the focused spot size remains about the same. The modeling results state that an equi-convex single lens with an *F*-number around 1 cannot focus well and the spherical aberration is the main aberration. Be sure to save this Zemax file since it will be improved in the next chapter as an example of optical design.

12.3.4 Modeling a mirror

A mirror with a 25-mm diameter, 24-mm clear aperture, and 25-mm focal lens is modeled in this section. The modeling process is described thus:

1. The wavelengths and field angles used to model the Edmond equi-convex lens are used here to model this mirror.
2. The aperture stop and the aperture value and position (row number) are the same as those for modeling the Edmond equi-convex lens.
3. The *Lens Data* box is opened and these numbers are entered, as shown in Fig. 12.5; the details are explained below:
   1. Row 0 is exactly the same as those for modeling the equi-convex lens shown in previous section. The back surface of the mirror does not matter and is not specified in the *Lens Data* box.

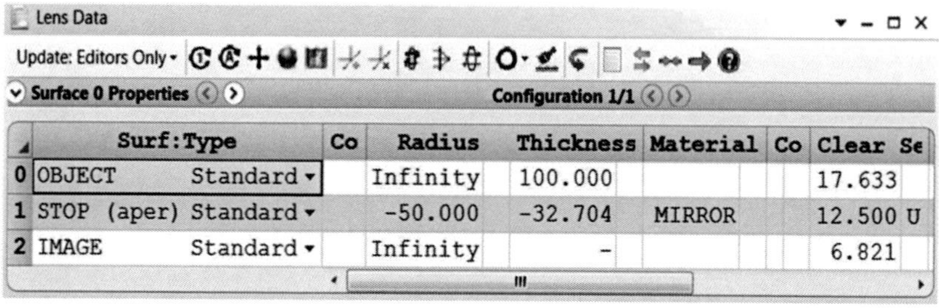
Lens Data

Update: Editors Only

Surface 0 Properties

Configuration 1/1

	Surf:Type		Co	Radius	Thickness	Material	Co	Clear Se
0	OBJECT	Standard		Infinity	100.000			17.633
1	STOP (aper)	Standard		-50.000	-32.704	MIRROR		12.500 U
2	IMAGE	Standard		Infinity	-			6.821

Figure 12.5 Zemax *Lens Data* box with the parameters of a mirror.

2. Row 1. For a mirror with a focal length of 25 mm, the surface radius of curvature is 50 mm. The negative value of the radius means that the surface vertex is at the right side of the other part of the surface, and the mirror is a concave mirror, as shown in Fig. 12.6. The word "Mirror" is typed in the *Material* box because Zemax recognizes it.
4. When modeling the Edmond equi-convex lens, the values in all of the *Thickness* boxes are positive, which means that the rays travel all the way from left to right. However, a mirror reflects rays, which then travel from right to left. The value in this *Thickness* box must thus be negative. Using the same way of focusing and then clicking *Optimize/ Quick Focus*, Zemax will adjust the image distance value to –32.704.
5. The raytracing diagram or layout of this mirror is plotted with five rays in Fig. 12.6. The spherical aberration is noticeably smaller than the spherical aberration for the single lens.
6. The spot diagram is shown in Fig. 12.7. The RMS radii of the focused spots are 110.127 μm and 366.475 μm for field angles 0° and 10°, respectively, which are much larger than the Airy disk radius of 1.022 μm but much smaller than the RMS radii of the focused spots of the Edmond equi-convex lens modeled earlier.

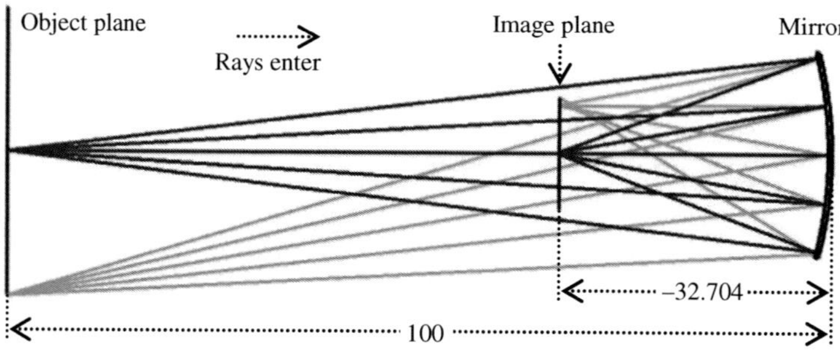

Figure 12.6 Raytracing diagram or layout of the mirror whose parameters are shown in Fig. 12.5. The image plane blocks most of the rays traveling from the object plane to the mirror, but this does not affect the modeling exercise. The situation is reciprocal: if the rays are emitted at the image plane, they will be focused at the object plane.

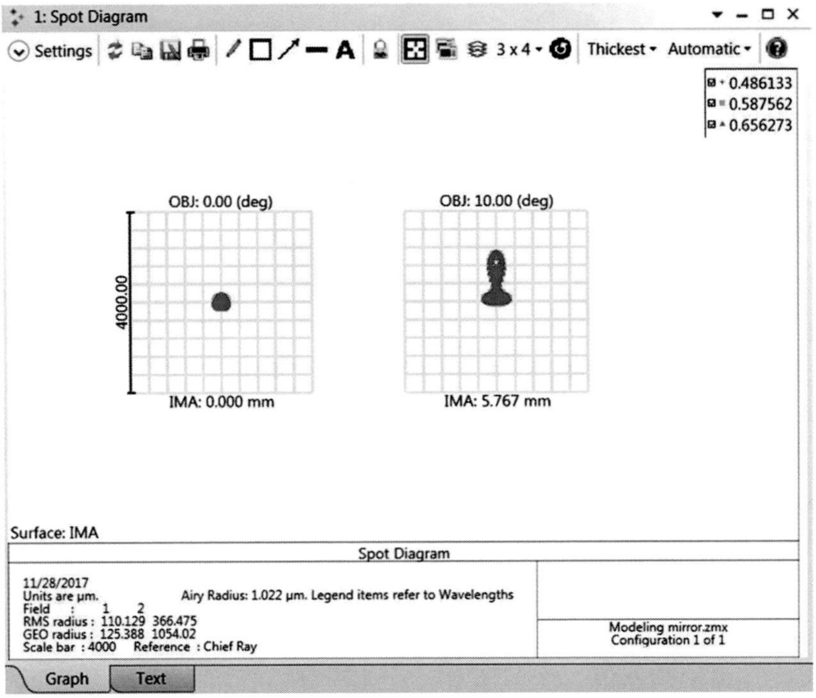

Figure 12.7 Focused spot diagram of the mirror shown in Fig. 12.6.

12.3.5 Modeling coatings

Zemax has a *Coating Catalog* that contains typical transmission curves of tens of standard coatings offered by several coating vendors, such as Thorlabs, Optimax, and Edmund Optics, as well as tens of ideal coatings such as I.xx, where I.0 means 0 transmission for any wavelengths, and I.99 means 99% transmission for any wavelengths.

To apply a coating on one optical surface, e.g., the third surface, double click the first box in the third row in the *Lens Data* box so that a *Surface 3 Properties* box will appear. Clicking the *Coating* button will open a *Coating* box, and then click the *Coating* box and select the desired coating from the dropdown list, as shown in Fig. 12.8, to select the I.50 coating.

12.3.6 Modeling a multi-element lens

When the Zemax file of a lens is available, it is easy to click the file to open it. Everything should already be set, such as wavelengths and field angles. If the box is not already open, click the *Setup/Lens Data* button to open the *Lens Data* box to edit. Click *Analyze/Rays & Spots* or *Analyze/MTF* to plot a focused spot diagram or MTF (CTF) curves.

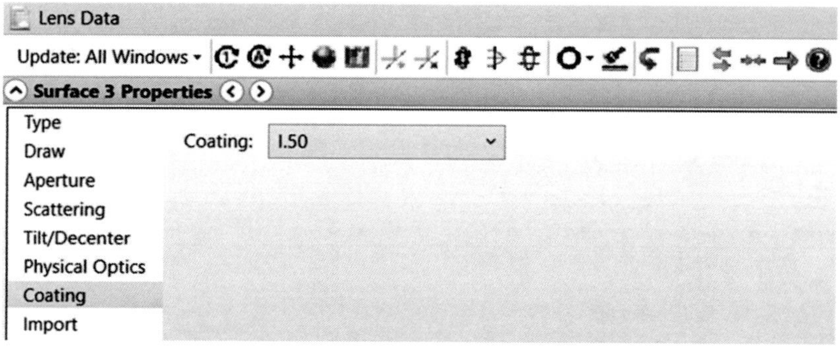

Figure 12.8 Apply a coating "I.50" in the *Surface 3 Properties* box in the *Lens Data* box.

If a Zemax file is not available, at least a text file that contains all of the optical details should be available. Here is an example:

Surface	Radius	Thickness	Glass
0	Infinity	Infinity	
1	54.15325	8.746658	SK2
2	152.5219	0.5	
3	35.95062	14	SK16
4	Infinity	3.776966	F5
5	22.26992	14.25306	
STO	Infinity	12.42813	
7	–25.68503	3.776966	F5
8	Infinity	10.83393	SK16
9	–36.98022	0.5	
10	196.4173	6.858175	SK16
11	–67.14755	57.31454	
IMA	Infinity		

This is basically a *Lens Data* box in a text format. *STO* is shown in surface 6, which indicates the aperture stop is at this surface. All of the radius, thickness, and glass data shown in the text can be typed into the *Lens Data* box. Some thicknesses are glass thicknesses. Some thicknesses are air-space thicknesses. Surfaces 3 and 4, as well as surfaces 7 and 8, have two successive glasses, which means that both are doublets. The wavelengths, field type and values, and aperture stop type and value must be provided elsewhere so that they can be set in the Zemax file.

The *Lens Data* box and the raytracing diagram of the lens, whose parameters are listed in the text file provided earlier, are shown in Figs. 12.9(a) and (b), respectively. This is the double Gauss lens that was used several times before in this book. The three field angles are 0°, 10°, and 14°, and the wavelengths are RGB.

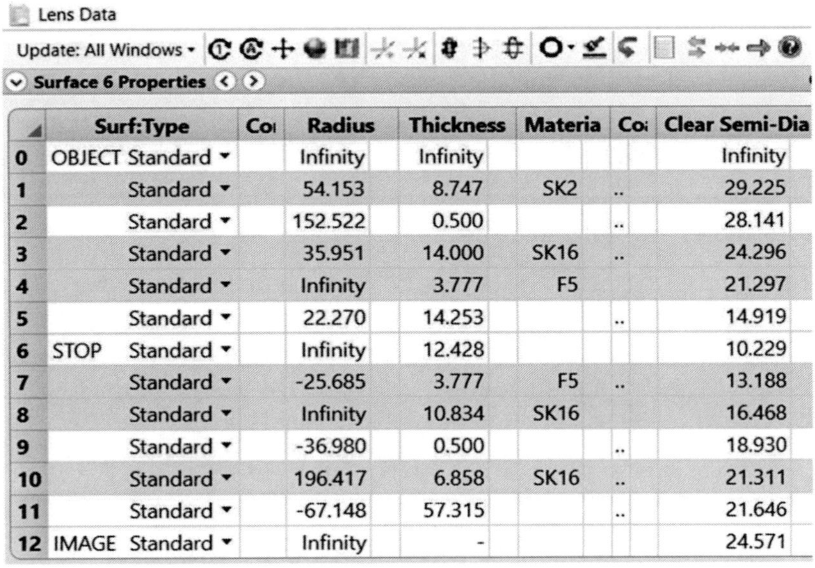

	Surf:Type		Co	Radius	Thickness	Materia	Co	Clear Semi-Dia
0	OBJECT	Standard		Infinity	Infinity			Infinity
1		Standard		54.153	8.747	SK2	..	29.225
2		Standard		152.522	0.500		..	28.141
3		Standard		35.951	14.000	SK16	..	24.296
4		Standard		Infinity	3.777	F5		21.297
5		Standard		22.270	14.253		..	14.919
6	STOP	Standard		Infinity	12.428			10.229
7		Standard		-25.685	3.777	F5	..	13.188
8		Standard		Infinity	10.834	SK16		16.468
9		Standard		-36.980	0.500		..	18.930
10		Standard		196.417	6.858	SK16	..	21.311
11		Standard		-67.148	57.315		..	21.646
12	IMAGE	Standard		Infinity	-			24.571

(a)

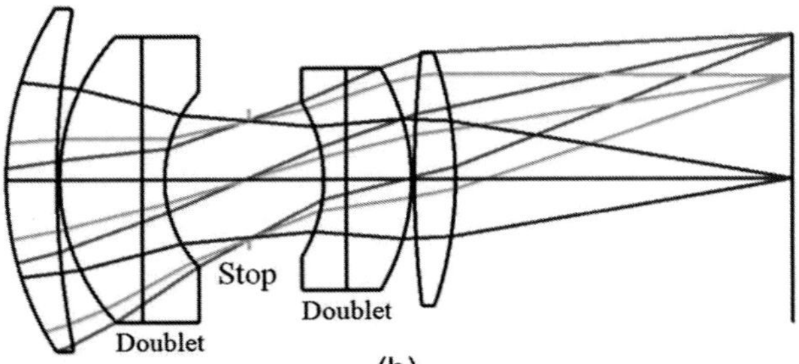

(b)

Figure 12.9 (a) *Lens Data* box for the double Gauss lens used in this book. (b) Raytracing diagram of the double Gauss lens.

12.4 Non-sequential Raytracing Modeling

12.4.1 General comments

Non-sequential raytracing is mainly used to model or design illumination optics. The difference between sequential raytracing and non-sequential raytracing is explained in Section 1.14. Since the majority of optical designs involve imaging optics, optical professionals tend to be less familiar with non-sequential modeling/design, and most types of optical software, such as CODE V and Zemax, are less capable of performing non-sequential raytracing than sequential raytracing.

In non-sequential modeling or design, one must first select a light source that is equivalent to a set wavelength and field in sequential raytracing, then select a detector that is equivalent to set the image plane in sequential raytracing, and place certain optical components between the light source and detector. The rays launched by the light source pass through the optical components and reach the detector. The detector output is the only information provided by the software to evaluate the performance of the optical components.

12.4.2 Modeling a light source

This section describes how to set a point light source with *Source Point*. After opening Zemax and clicking *Setup/Non-Sequential*, a *Non-Sequential Component Editor* box appears. This box is too wide for the book page, so it is broken into three sections and shown in Fig. 12.10(a), where the light source is already selected. The selection process works as follows:

1. Click the arrow sign marked by the circle, as shown in Fig. 12.10(a), to open an *Object 1 Properties* box, as shown in Fig. 12.10(b). Then click the *Type* button and select a *Point Source* from many available objects in the *Type:* box.
2. The point source is intended to be placed at the origin and emits rays towards the right direction. Thus, type 0 in all six boxes of the *X, Y, Z Position* and *Tilt About X, Y, and Z.*
3. For a light source, the *Material* box is disabled.
4. The # *Layout* box is used to select the number of rays for layout; usually tens of rays are sufficient. The # *Analysis* box is used to select the number of rays for analysis. More analysis rays lead to more accurate results but take a longer time to run. Usually, tens of thousands rays to a couple million rays are adequate. Here, one million is typed in the box.
5. A normalized power of 1 W is typed in the *Power* (*Watts*) box.
6. The wavelength can be selected by clicking *System Explorer/ Wavelengths*, which is the same as selecting a wavelength in sequential mode. One wavelength of 0.588 μm is selected.
7. There are different selections for the *Wavenumber* box. 0 means that all of the wavelengths selected are used. *i* means the ith wavelength is used. In this case, there is only one wavelength. The two available selections 0 and 1 are equivalent.
8. The *Color* # box is used to select the color for raytracing and is cosmetic.
9. The *Cone Angles* box is specific for the *Source Point*. For this example, 28.00 (the half-angle in degrees) is entered, although for other types of light source, several boxes shown in Fig. 12.10(a) will be different.

The performance of this light source will be evaluated together with the lens and the detector, which will be set later.

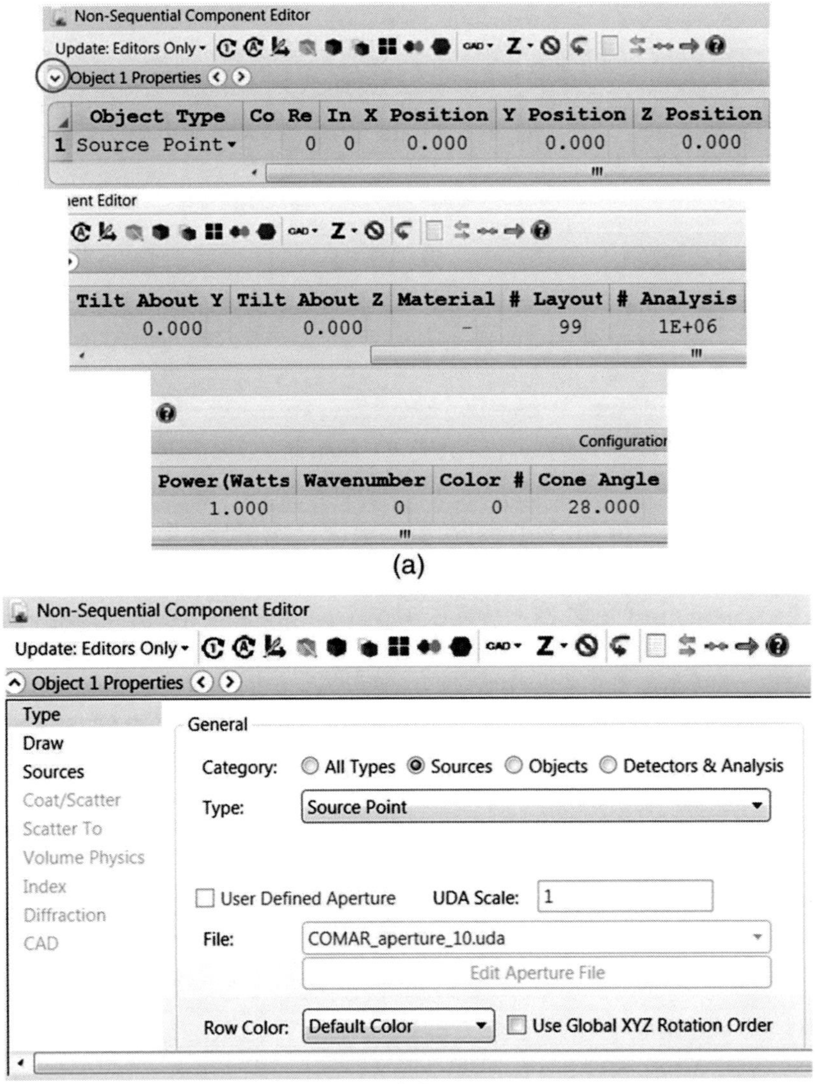

(b)

Figure 12.10 (a) *Non-Sequential Component Editor* box for a *Source Point* light source. (b) The process of selecting a light source.

12.4.3 Modeling a detector

The *Source Point* was set up in the first row of the *Non-Sequential Component Editor* box. The last row is used to set up a detector, e.g., a rectangular detector, is the *Detector Rectangle*. All of the rows between the first and last rows are for other optical components. The process to set up a *Detector Rectangle* is similar to that for a *Source Point*. The only difference is the selection of a *Detector Rectangle*, not a *Source Point*, in the *Type:* box.

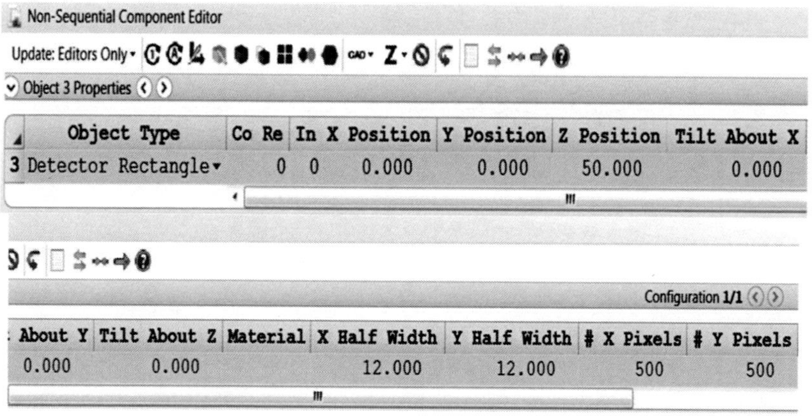

Figure 12.11 *Non-Sequential Component Editor* box for a *Detector Rectangle*.

The *Non-Sequential Component Editor* box, which is too long for the book page, is broken into two sections and shown in Fig. 12.11:

1. Typing 50 in the *Z Position* box places the detector at 50 mm from the origin, where the *Source Point* is placed.
2. The detector half-width is selected to be 12 by 12 in the *X* and *Y* directions, respectively. This is a square detector since the case considered here is symmetric about the optical axis.
3. The pixel numbers in both *X* and *Y* directions are selected to be 500. More pixels can provide more detection details, but a longer time is required to trace the rays.

The performance of this detector will be evaluated along with the light source and the lens, which will be set later.

12.4.4 Modeling a lens

The lens to be modeled here is the same as the lens modeled in Section 12.3.3. A *Standard Lens* is selected at the second row of the *Non-Sequential Component Editor* box, which is shown in Fig. 12.12 along with the *Source Point* and *Detector Rectangle*, which were set up earlier.

Note that the displayed contents of the top row (in bold) can vary depending on the type of the object. If any box in the first row is clicked, the contents shown are all about *Source Point*, and the same are for the second and third rows. The contents shown in Fig. 12.12 are for *Standard Lens*; some of the contents are common for the *Source Point*, the *Detector Rectangle*, and other components.

The lens is placed at $Z = 18$ (mm). The chosen *Material* is N-SF5. The two surface curvature radii, *Radius 1* and *Radius 2*, are set the same way as in sequential mode. The *Thickness* is the central thickness of the lens, which is

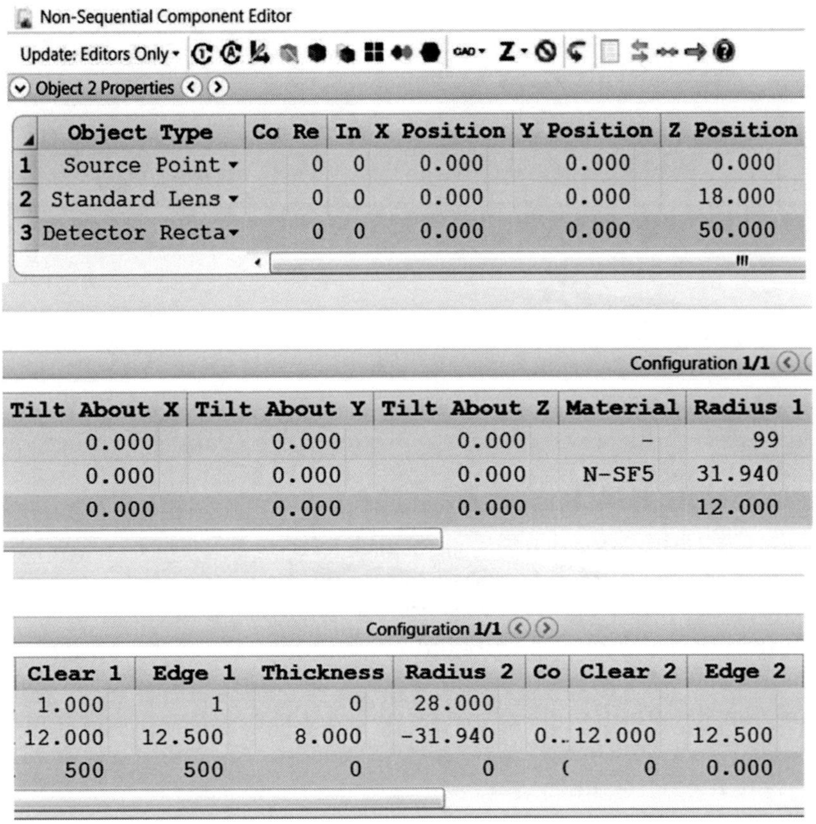

	Object Type	Co	Re	In	X Position	Y Position	Z Position
1	Source Point		0	0	0.000	0.000	0.000
2	Standard Lens		0	0	0.000	0.000	18.000
3	Detector Recta		0	0	0.000	0.000	50.000

Tilt About X	Tilt About Y	Tilt About Z	Material	Radius 1
0.000	0.000	0.000	-	99
0.000	0.000	0.000	N-SF5	31.940
0.000	0.000	0.000		12.000

Clear 1	Edge 1	Thickness	Radius 2	Co	Clear 2	Edge 2
1.000	1	0	28.000			
12.000	12.500	8.000	-31.940	0..	12.000	12.500
500	500	0	0	(	0	0.000

Figure 12.12 A *Non-Sequential Component Editor* box contains a *Standard Lens*, a *Source Point*, and a *Detector Rectangle*. The box is too long for the book page and is thus broken into three sections. The contents of the top row shown here are for the *Standard Lens*. Some contents are common for all three components.

8 (mm). *Clear 1* and *Clear 2* are the half-clear-aperture of the front and back surfaces of the lens, respectively. *Edge 1* and *Edge 2* are the half-aperture of the front and back surfaces of the lens, respectively. It always holds that *Edge i* $\geq$ *Clear i*.

12.4.5 Detector viewer and raytracing

Now to show the raytracing diagram and illumination intensity profile on the detector. After clicking *Setup|NSC 3D Layout*, a layout window will appear, as shown in Fig. 12.13(a). Clicking *Setup|NSC Shaded Model*, a layout window will appear, as shown in Fig. 12.13(b). Both layouts are equivalent and provide direct views of how the optical system (source + lens + detector) looks. The lens collimates rays from a point source. Note that in non-sequential mode, Zemax will trace rays forward only when there is a detector

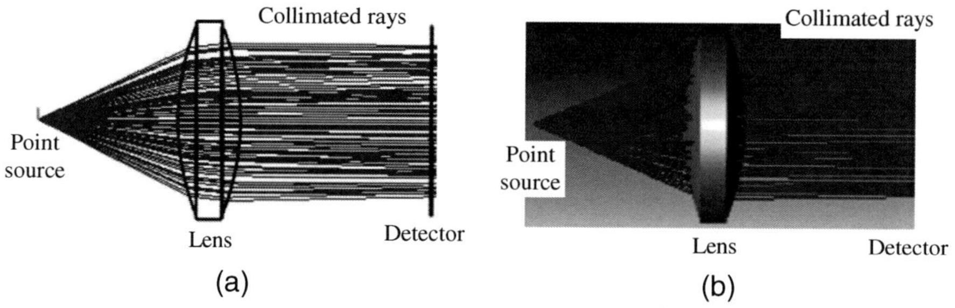

Figure 12.13 (a) *NSC 3D Layout* of the optical system set up in the *Non-Sequential Component Editor* box shown in Fig. 12.12. (b) The *NSC Shaded Model* of the same optical system.

ahead to intercept the rays. If there is no detector, Zemax will trace rays only for a short distance and then stop.

Clicking *Analyze/Detector Viewer* will prompt a *Detector Viewer* window to appear. Click the *Settings* button in the *Detector Viewer* window and select *False Color* in the *Show As:* box. After clicking *Ray Trace* (in the main window toolbar, not in the *Detector Viewer* window) and clicking the *Clear & Trace* button in another pop-up window, Zemax will start to trace rays. The raytracing process can take from a few seconds up to a few minutes, depending on the number of rays to be traced, the detector pixel numbers, and the complexity of the optical system involved. For this simple optical system, it takes only a few seconds.

After raytracing is completed, the *Detector Viewer* window will display a 2D false-color graph, as shown in Fig. 12.14(a), which is the illumination intensity pattern that the detector sees. There is a bright ring at the edge of the circular illumination pattern caused by spherical aberration. The horizontal or vertical cross-section profile of the illumination pattern can be displayed by changing the *False Color* selection to *Cross Section Row* or *Cross Section Column*; the result is shown in Fig. 12.14(b). There are other display selections, such as *Grey Scale*, *Contour*, *Phase*, etc.

12.4.6 Modeling a mirror

The mirror being modeled is the same as the mirror modeled in Section 12.3.4, continuing from the *Non-Sequential Component Editor* box shown in Fig. 12.12 but replacing the *Standard Lens* in the second row with *Standard Surface*. The *Non-Sequential Component Editor* box is shown in Fig. 12.15 and explained here:

1. The word "Mirror" is typed in the *Material* box, and Zemax recognizes this word.
2. –50 is typed in the *Radius* box to define the mirror concave surface curvature. The *Z Position* of the mirror is 25, and then the *Source Point*

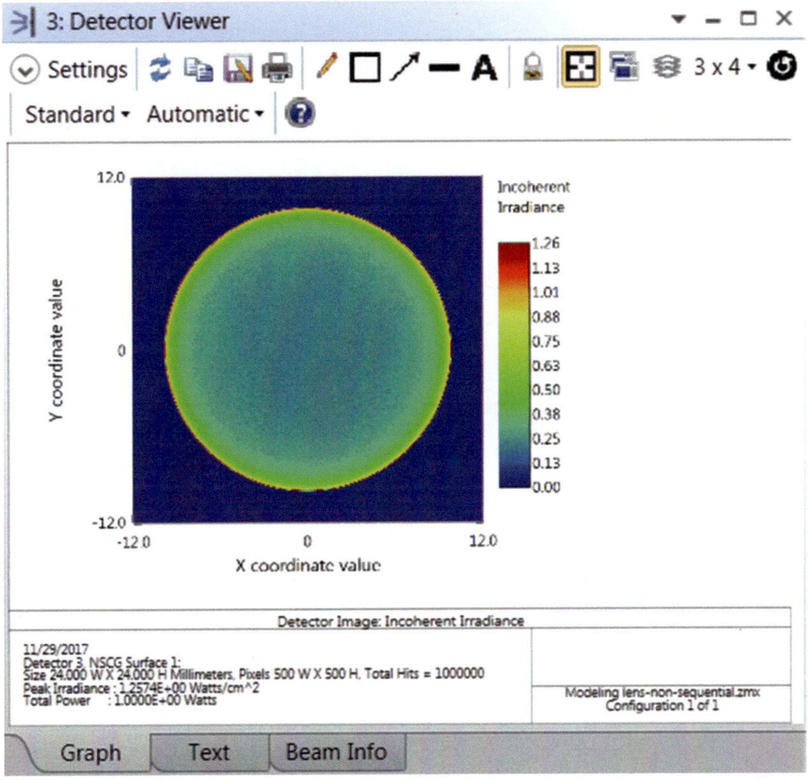

(a)

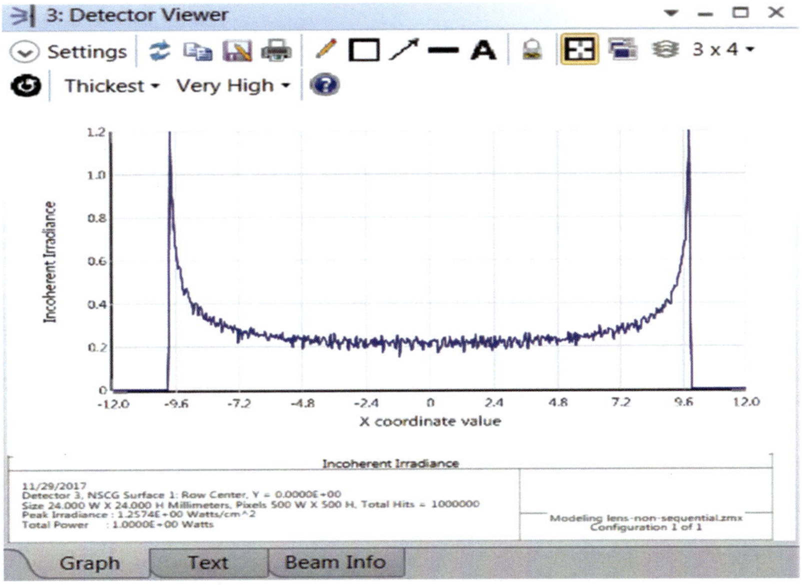

(b)

Figure 12.14 (a) A false-color view of the illumination intensity pattern generated by the *Source Point* and the lens on the detector, as shown in Fig. 12.13. (b) Intensity profile of the same illumination pattern.

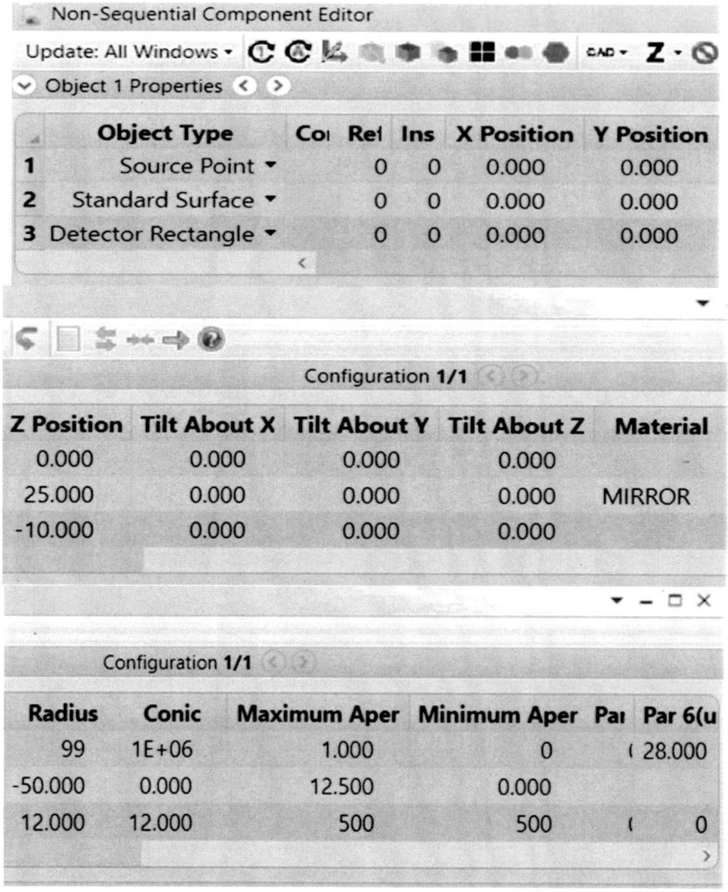

Figure 12.15 *Non-Sequential Component Editor* box for a *Standard Surface*, a *Source Point*, and a *Detector Rectangle*. The box is too long for the book page and is broken into three sections.

is at the focal point of the mirror, which uses the mirror to collimate the rays from the source.

3. The *Maximum Aper* is the half-diameter of the mirror, and the value entered is 12.5. The *Minimum Aper* is the half-diameter of the hole in the mirror center. Since this mirror does not have a hole, 0 is typed in this box.
4. The *Z Position* of the *Detector Rectangle* is changed from 50 to –10 since the rays reflected by the mirror travel to the left.

The *NSC 3D Layout* and the *NSC Shaded Model* of the optical system (source + mirror + detector) are plotted in Fig. 12.16. The false-color view of the illumination intensity pattern generated by the *Source Point* and the mirror on the detector is shown in Fig. 12.17(a). The intensity profile of the same illumination pattern is plotted in Fig. 12.17(b), with "*Smooth: 5*." Since

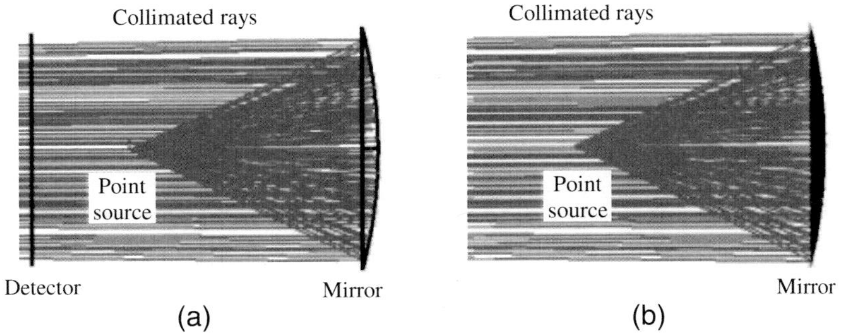

Figure 12.16 (a) *NSC 3D Layout* of the optical system set up in the *Non-Sequential Component Editor* box shown in Fig. 12.15. (b) The *NSC Shaded Model* of the same optical system.

the numbers of detector pixels and analysis rays are limited, the intensity profile obtained often contains a lot random up and downs, as if there is a big noise in the profile, as shown in Fig. 12.17(c). This big "noise background" is false and can be removed by using an artificial "*Smooth*" function provided by Zemax. There are 25 levels of smooth. Caution must be taken: if over-smoothed, the true details of the profile may be smoothed out.

12.4.7 Modeling a dispersion prism

The modeling steps described earlier will not be repeated here. The goal of this section is to let rays from a Gaussian source pass through a prism and be incident on a rectangular detector.

1. Change the light source to *Source Gaussian*; the *Non-Sequential Component Editor* box is shown in Fig. 12.18(a). The 0.5 *Beam Size* selected is the beam-waist radius size. The 0 in *Position* means that the beam waist position is the beam position. In the *System Explorer/ Wavelengths*, select three colors of 0.45 μm, 0.55 μm, and 0.65 μm.

   The *Non-Sequential Component Editor* box for the prism is shown in Fig. 12.18(b). A prism is called a *Polygon Object* in Zemax, and many *Polygon Object* are available. First, select *Polygon Object* in the *Object Type* box, and then select the type of prisms in the pop-up *Data File:* box. The prism selected here is a 45-45° prism. The prism position and orientation are adjusted to intercept the Gaussian beam at an angle for a large dispersion effect. All of the *Polygon Object* values provided by Zemax do not specify the glass type and have certain sizes. The user must enter the glass type in the *Non-Sequential Component Editor* box and select a *Scale* number to adjust the size of the prism. The glass used here is Ohara S-NPH3, and the *Scale* factor selected is 3.
2. The *Non-Sequential Component Editor* box for *Detector Rectangle* is shown in Fig. 12.18(c). The position and orientation of the detector are

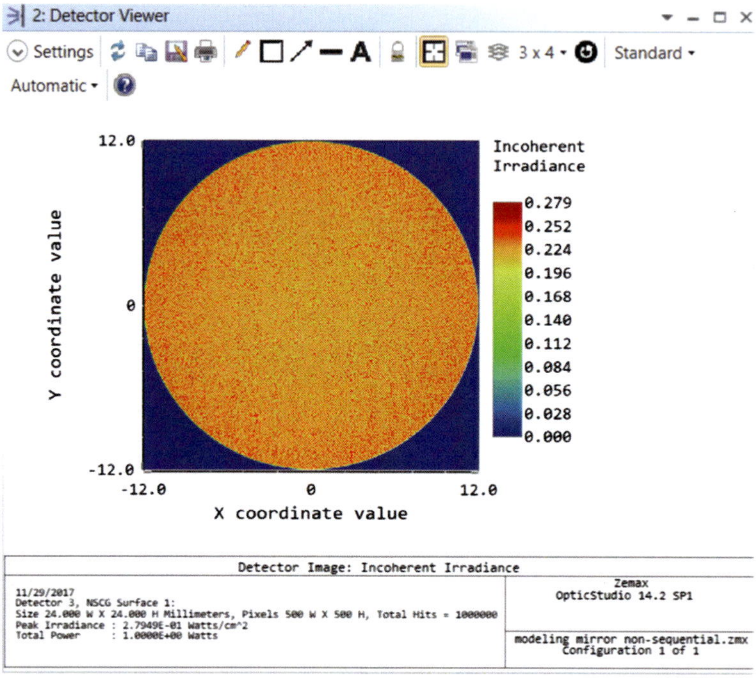

(a)

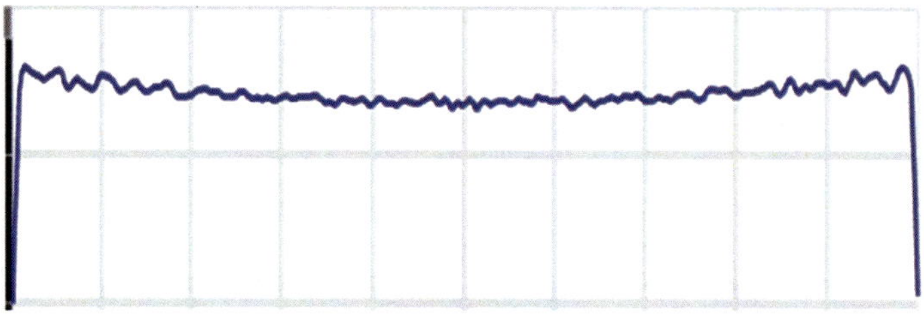

(b)

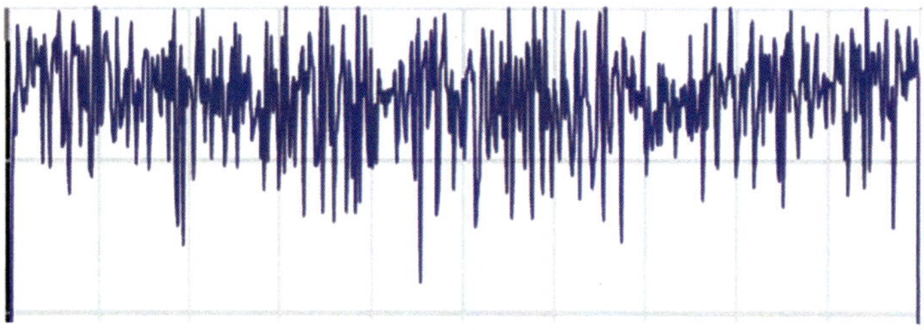

(c)

Figure 12.17 (a) A false-color view of the illumination intensity pattern generated by the *Source Point* and the lens on the detector, as shown in Fig. 12.15. (b) Intensity profile of the same illumination pattern with "*Smooth: 5.*" (c) Intensity profile of the same illumination pattern without "*Smooth.*"

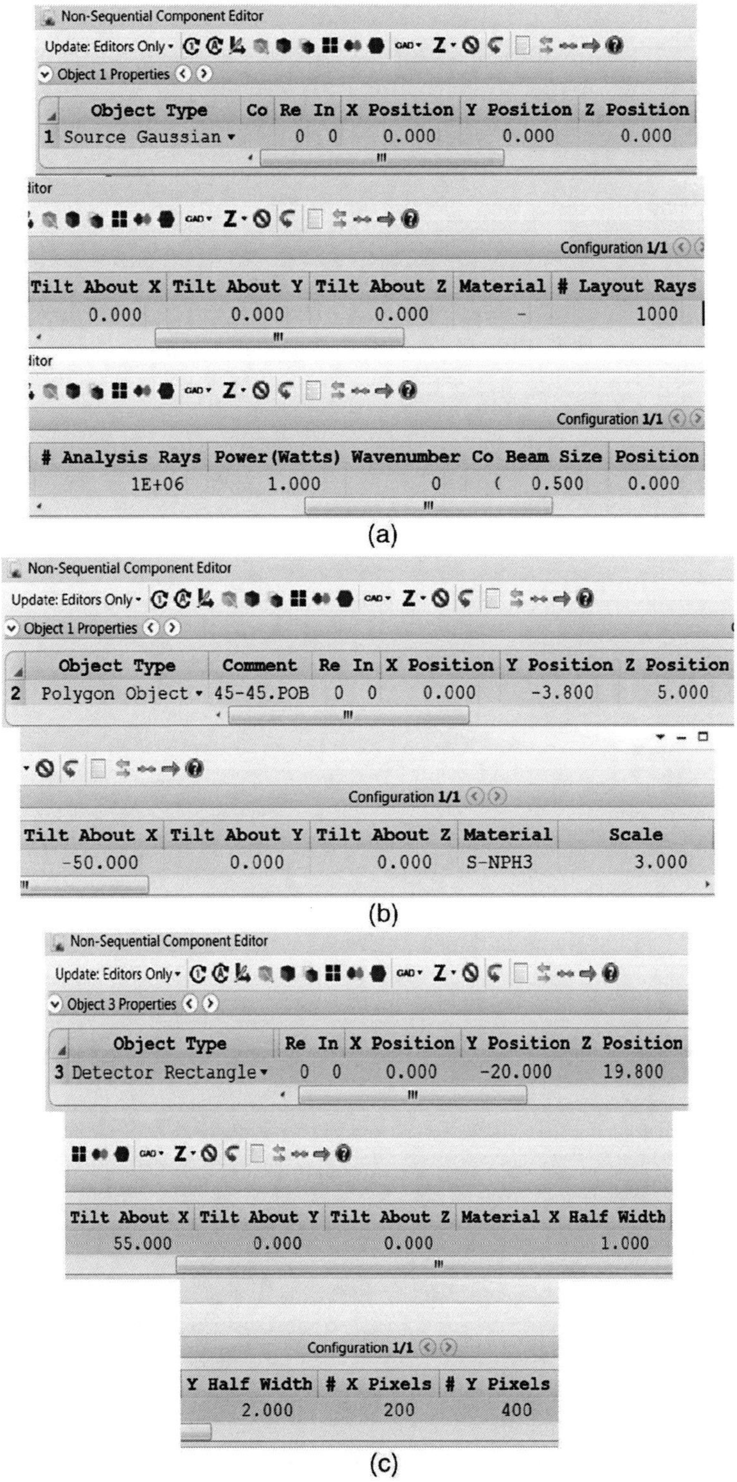

Figure 12.18 (a) *Non-Sequential Component Editor* box for the *Source Gaussian*. (b) *Non-Sequential Component Editor* box for the dispersion prism (*Polygon*). (c) *Non-Sequential Component Editor* box for the *Detector Rectangle*.

adjusted to intercept the beam after it passes through the prism. The correct position and orientation can be found after several tries. The detector has a rectangular and different numbers of pixel in the two directions since dispersion only occurs in one direction.

After all three components are set, the layout of the optical system (source + prism + detector) is shown in Fig. 12.19. The RGB colors are dispersed.

The false-color view of the RGB color beams incident on the detector is shown in Fig. 12.20(a). The vertical intensity cross-profile is shown in Fig. 12.20(b). Note that the RGB colors have the same optical power. After being dispersed, the B color beam has a narrower spatial size and larger peak.

12.4.8 Modeling a cube beamsplitter with a beamsplitting coating

This modeling is performed by revising the *Non-Sequential Component Editor* box shown in Fig. 12.18.

1. Change the RGB colors to one color of 0.55 μm.

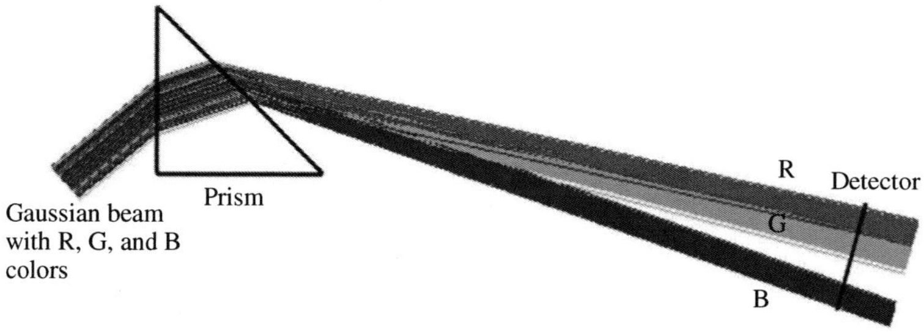

Figure 12.19 Layout of a dispersion prism dispersing three RGB colors.

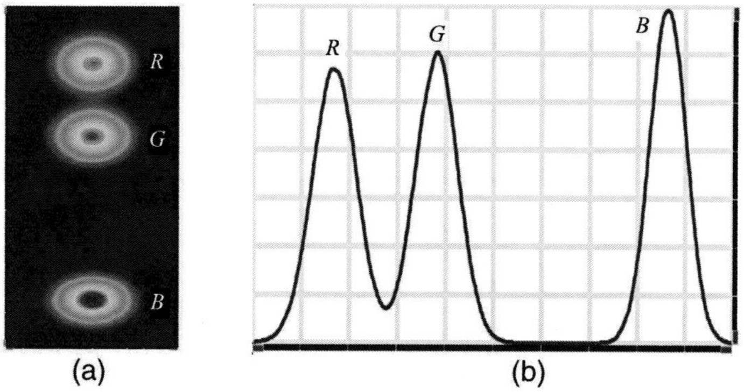

Figure 12.20 (a) A false-color view of the illumination intensity pattern with "*Smooth: 10*" generated by the three-color *Source Gaussian* and the dispersion prism lens on the detector, as shown by the raytracing diagram in Fig. 12.19. (b) Intensity profile of the illumination pattern with "*Smooth: 10*," the same as (a).

2. Add the second 45-45° prism and adjust the orientations and positions of the two prisms so that they form a cube, as shown in Fig. 12.21. Use N-BK7 glass for both prisms.
3. Add one more *Detector Rectangle* and adjust the orientations and positions of the two detectors, as shown in Fig. 12.21.

Figure 12.22 shows part of the *Non-Sequential Component Editor* box for this optical system after the three steps. The raytracing result shows that the

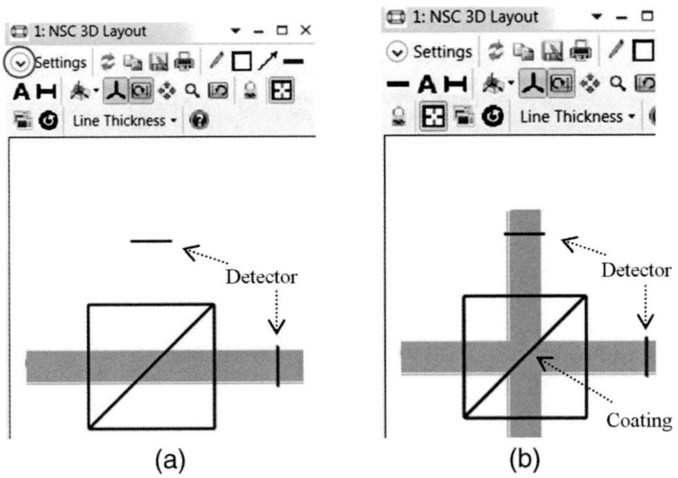

Figure 12.21 (a) Raytracing diagram of two prisms without beamsplitting coating. The beam passes through the interface of the two prisms. (b) Raytracing diagram of two prisms with a I0.50 (50% transmittance) coating at the hypotenuse of one prism. The beam is split in two.

	Object Type	Co	Re	In	X Position	Y Position	Z Position	Tilt About X	Tilt About Y	Tilt About Z	Material	Scale
1	Source Gaussian ▾		0	0	0.000	0.000	0.000	0.000	0.000	0.000	-	1000
2	Polygon Object ▾	4...	0	0	0.000	0.000	3.000	0.000	0.000	0.000	N-BK7	3.000
3	Polygon Object ▾	4...	0	0	0.000	0.000	9.000	180.000	0.000	0.000	N-BK7	3.000
4	Detector Rectangle ▾		0	0	0.000	0.000	12.000	0.000	0.000	0.000		1.000
5	Detector Rectangle ▾		0	0	0.000	6.000	6.000	90.000	0.000	0.000		1.000

Figure 12.22 Part of the *Non-Sequential Component Editor* box for one *Source Gaussian*, two 45-45° prisms, and two *Detector Rectangles*. The top-row contents displayed are for the prisms.

rays will pass through the interface of the two prisms as shown in Fig. 12.21(a) because there is no beamsplitting coating. A I.50 (50% transmittance) coating must be applied at the hypotenuse of either prism to make a 50:50 split. The process is as follows:

1. Double click the first box of the row of either prism in the *Non-Sequential Component Editor*, shown in Fig.12.22.
2. Click the *Coat/Scatter* button in the pop-up *Object 2* (or *3*) *Properties* box.
3. Make selections as shown in Fig. 12.23.

Beamsplitting raytracing is performed somewhat differently than the raytracings used before; a few special steps are required:

1. Click the arrow sign marked by the red circle in Fig. 12.21(a) to open an additional window, as shown in Fig. 12.24.
2. Check the boxes of *Use Polarization* and *Split NSC Rays,* as shown in Fig. 12.24. Then clicking the button marked by the red circle to perform raytracing. The raytracing diagram with beamsplitting is shown in Fig. 12.21(b).

Now there are two detectors specified at row 4 and row 5 of the *Non-Sequential Component Editor* box, respectively, as shown in Fig. 12.22. Either one or two *Detector Viewers* can be open to display the contents of either one or both the detectors. Which detector to display can be specified, as shown in Fig. 12.25(a). When performing raytracing, the two boxes of *Use Polarization* and *Split NSC Rays* in *Ray Trace Control box* must be checked to trace beamsplitting analysis rays, as shown in Fig. 12.25(b).

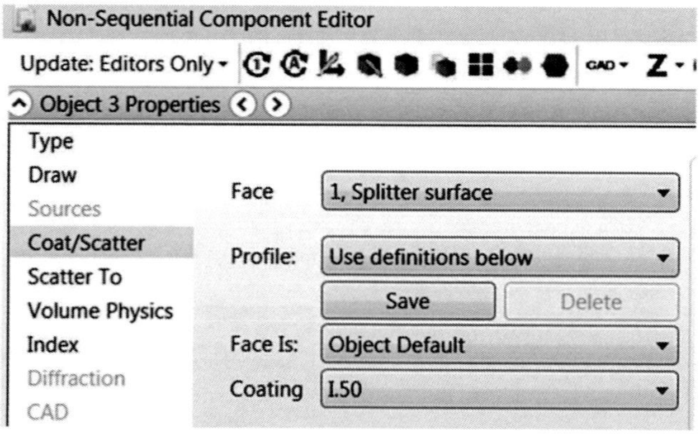

Figure 12.23 Applying a I.50 coating to the *Splitter Surface* (hypotenuse surface) of one prism.

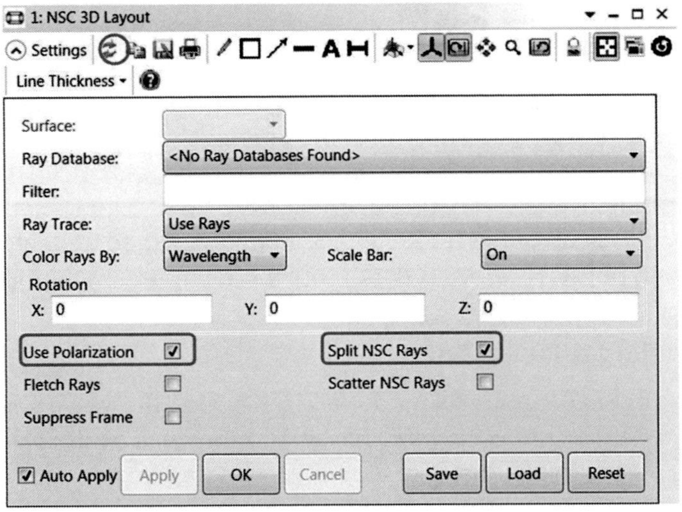

Figure 12.24 The two boxes of *Use Polarization* and *Split NSC Rays* must be checked to trace split layout rays.

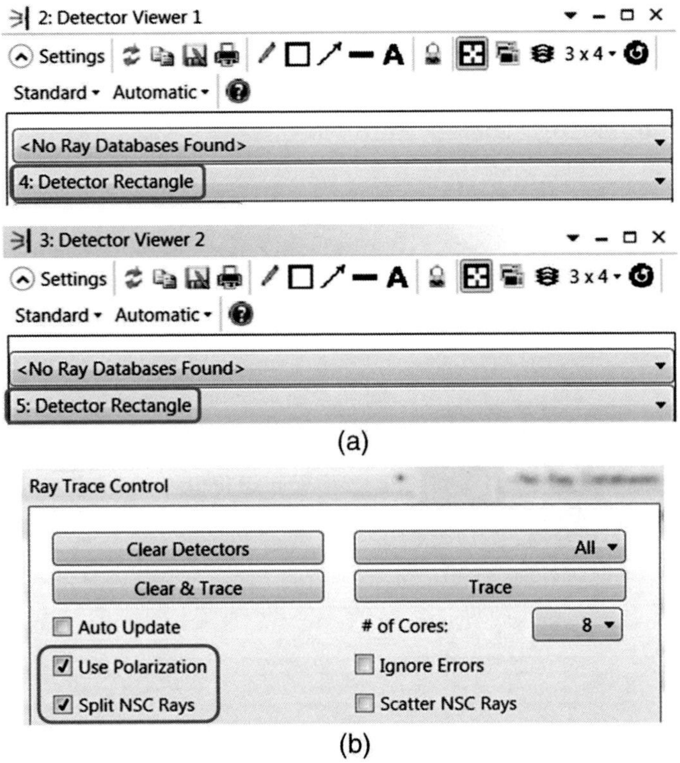

Figure 12.25 (a) Select a detector to display. (b) The two boxes of *Use Polarization* and *Split NSC Rays* must be checked in order to trace split analysis rays.

12.5 Physical Optics Modeling

The sequential and non-sequential raytracing modeling discussed earlier involve geometrical optics, where all of the rays traced are straight lines. Geometrical optics can model laser beams only in some limited circumstances. If the concept of rays is used to describe a laser beam, then in the near field the "rays" are curved, not straight lines. Treating curved "rays" as straight lines will result in large errors. Physical optics modeling is mainly used to model laser beams. Zemax offers a physical optics function for modeling laser beam propagations.

12.5.1 Modeling a laser diode beam

When using physical optics to model a laser beam, the wavelength and field angle are still set in the *System/Explorer/Wavelengths* and *System/Explorer/Fields* boxes, respectively, just like the sequential and non-sequential modeling described earlier. One wavelength of 0.65 μm and one field of 0° are selected here. The field angle is the incident angle of the beam.

Zemax associates the *Physical Optics Propagation* box with the sequential *Lens Data* box. The process of modeling a laser beam is as follows:

1. Set a *Lens Data* box as shown in Fig. 12.26(a), where 1 is typed in the *Thickness* box in row 0 to avoid placing the aperture stop at the object. Any number >0 can be entered; it makes no difference for physical optics modeling. The aperture type and value are irrelevant. The beam

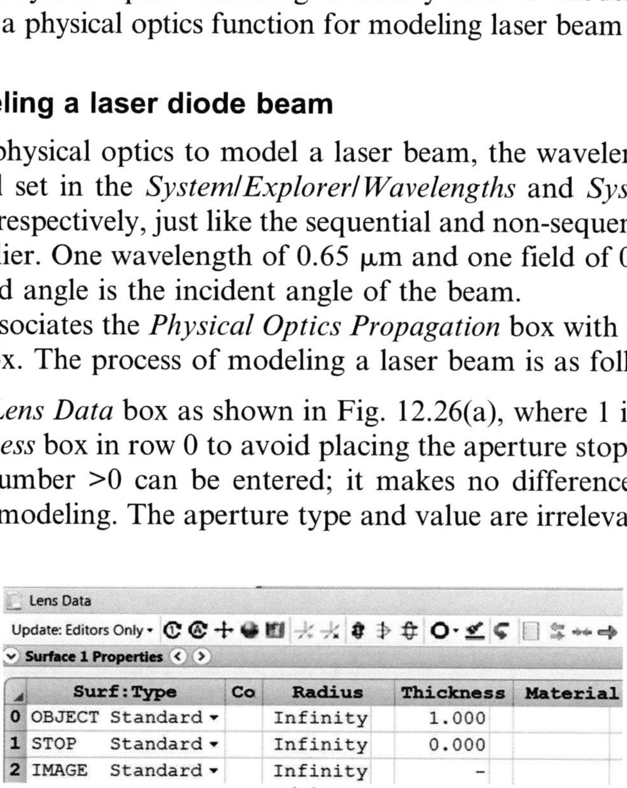
Lens Data
Update: Editors Only
Surface 1 Properties

	Surf:Type		Co	Radius	Thickness	Material
0	OBJECT	Standard		Infinity	1.000	
1	STOP	Standard		Infinity	0.000	
2	IMAGE	Standard		Infinity	-	

(a)

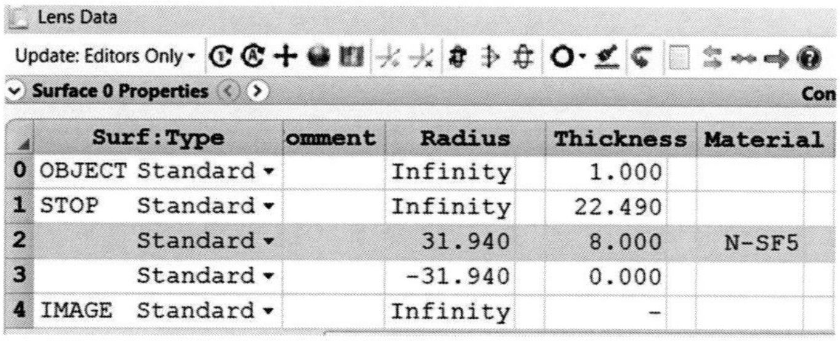
Lens Data
Update: Editors Only
Surface 0 Properties
Con

	Surf:Type		omment	Radius	Thickness	Material
0	OBJECT	Standard		Infinity	1.000	
1	STOP	Standard		Infinity	22.490	
2		Standard		31.940	8.000	N-SF5
3		Standard		-31.940	0.000	
4	IMAGE	Standard		Infinity	-	

(b)

Figure 12.26 (a) *Lens Data* box for modeling laser diode beam propagation. (b) *Lens Data* box for modeling a laser diode beam propagating through a lens.

waist is usually chosen to be placed at row 1. The value in the *Thickness* box in row 1 specifies the image plane location, where the beam characteristics will be checked. A number of 0 is entered, which means the beam waist characteristics will be checked. This number will be changed later.

2. Click *Analyze/Physical Optics* to open a *Physical Optics Propagation* window, and then click the *Setting* button at the top-left corner of the box to set the details as follows:
   1. The setting in the *General* box is shown in Fig. 12.27(a). The *Start Surface* is where the laser is placed, and the *End Surface* is where the laser beam is evaluated. There is only one wavelength and one field of 0°, which are already selected.
   2. The *Beam Definition* box is shown in Fig. 12.17(b). The *X-Sampling* and *Y-Sampling* numbers are selected based on the shape of the beam to be evaluated and the resolution required. More sampling numbers require more time to run. The numbers in the *X-Width* and *Y-Width* boxes are determined by the sampling numbers and the beam size, which varies at different locations. Since this issue is somewhat complex, click the *Automatic* button to let Zemax determine these numbers. The numbers typed in the *Waist X* and *Waist Y* boxes are 0.0015 (mm) and 0.0005 (mm), respectively. This is a laser diode beam with an elliptical shape and tiny waist size. Since there is no aperture to truncate the beam, type any numbers much larger than the bean waist size in the *Aperture X* and *Aperture Y* boxes. Also, the laser beam is not decentered.
   3. *False Color* and *Irradiance* are selected in the *Display* box, as shown in Fig. 12.17(c). There are other selections, such as *Cross X*, *Cross Y*, *Contour*, *Grey Scale*, etc.

After completing all of these settings, clicking the *Setting* button at the top-left corner of the box, a false-color beam intensity pattern will be displayed as shown in Fig. 12.28(a), which is the intensity pattern of the beam waist. The beam shape is elliptical with a major axis in the horizontal direction. The $1/e^2$ diameter is 3 μm horizontal (x) and 1 μm vertical (y). All of these values are expected.

Now type a number 3 in the *Thickness* box in surface 1, as shown in Fig. 12.26(a). The beam intensity pattern 3 mm away from the beam waist is shown in Fig. 12.28(b). The beam shape is elliptical but with a major axis in the vertical direction. The $1/e^2$ diameter is about 1 mm horizontal (x) and about 3 mm vertical (y). Since this laser diode beam naturally has larger divergence in the fast (vertical) axis direction, after propagating 3 mm the beam shape changes to vertical elliptical.

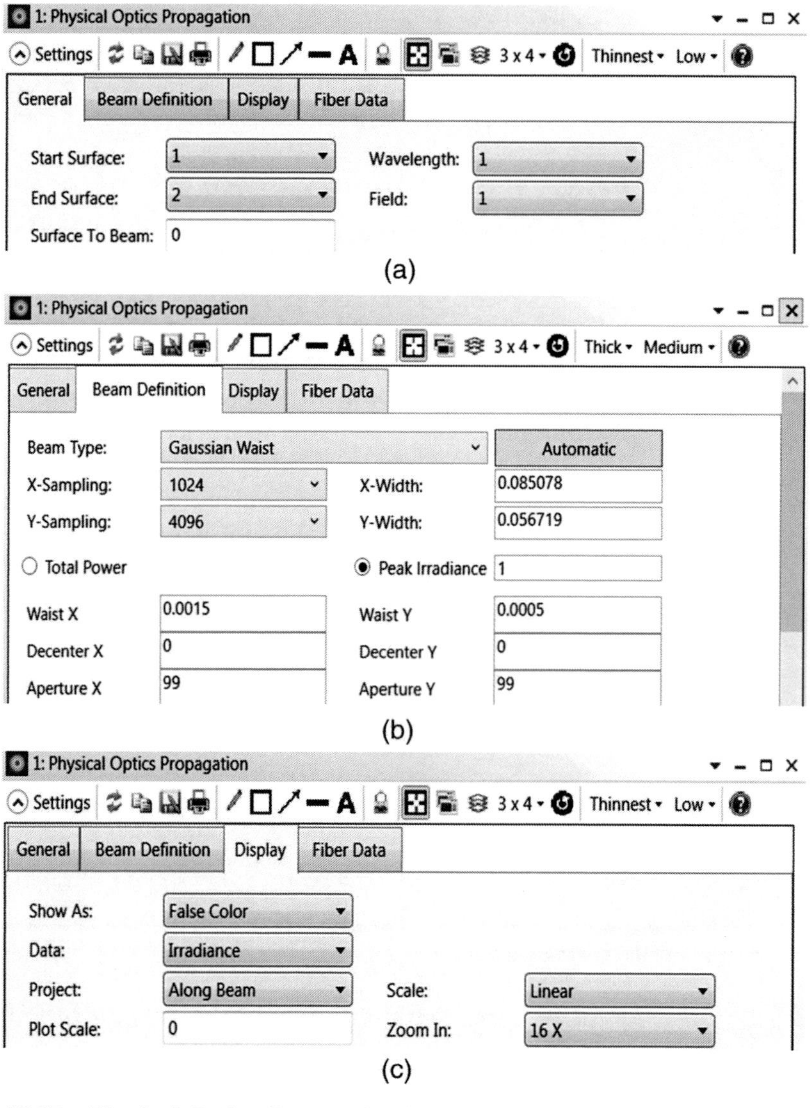

Figure 12.27 *Physical Optics Propagation* box setting: (a) *General* settings, (b) *Beam Definition* settings, and (c) *Display* settings.

12.5.2 Modeling a laser beam propagating through an equi-convex spherical lens

In this section, the laser diode beam modeled earlier will propagate through the Edmund equi-convex lens, which was modeled in Section 12.3.3. The *Lens Data* box is set as shown in Fig. 12.26(b). This lens has a back focal length of 22.49 mm. The number 22.49 in the *Thickness* box in row 1 means the laser diode is placed at the focal point of the lens, and the beam emitted from the laser diode is collimated by the lens. The value in the *Thickness* box

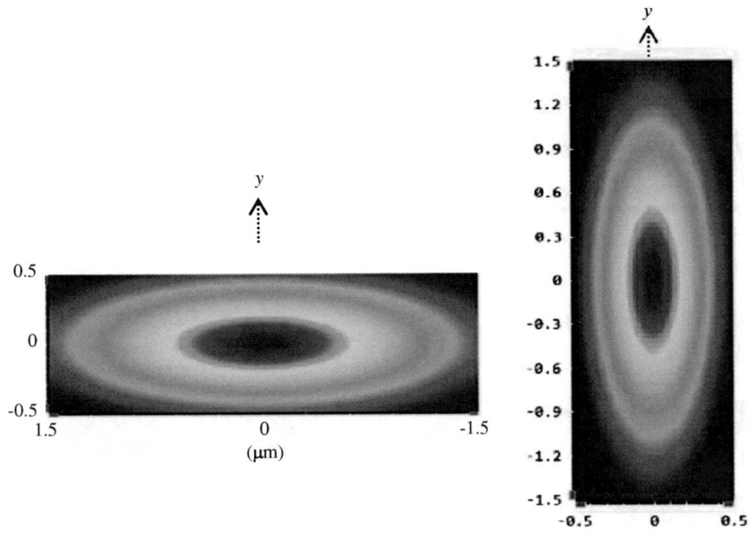

(a) (b)

Figure 12.28 (a) The intensity pattern at the beam waist. (b) The intensity pattern 3 mm away from the waist. The unit is mm.

in row 3 is 0, which means that the characteristics of the beam immediately after the back surface of the lens will be checked. This number will be changed later. Since the laser beam is unchanged, nothing in the *Physical Optics Propagation* box must be changed, although the *Automatic* button must be clicked once so that Zemax can adjust the display width.

The beam intensity pattern just after the beam propagating through the lens is plotted in Fig. 12.29(a); it remains nicely Gaussian. Then, the value in the *Thickness* box in row 3 in Fig. 12.26(b) is changed from 0 to 100, which means that the beam intensity pattern at a distance of 100 mm from the back surface of the lens is checked. The beam intensity pattern is plotted in Fig. 12.29(b); it is messy and has a lot of interference patterns because the lens used has large spherical aberration and is not suitable for collimating a laser diode beam. The beam intensity pattern at a distance of 300 mm from the back surface of the lens is plotted in Fig. 12.29(C), and it is even messier—the central lobe disappears.

12.5.3 Modeling a laser beam propagating through an aspheric laser diode lens

As a comparison, an aspheric lens developed to collimate a laser diode beam made by LightPath Technologies, part number 354453, is used to collimate the beam of this diode. This lens, used in Fig. 9.7, has a clear aperture of 4.8 mm. The *Lens Data* box shown in Fig. 12.26(b) must be changed for this aspheric lens but is omitted here. The beam intensity pattern just after the lens

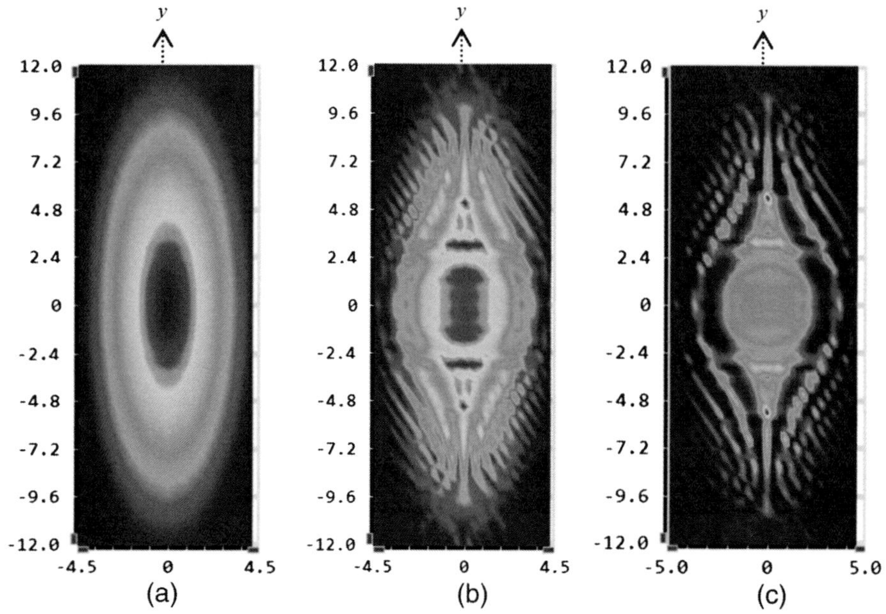

Figure 12.29 Intensity pattern of a laser diode beam after it propagates through an equiconvex glass spherical lens; the unit here is mm: (a) just after the lens, (b) 100 mm after the lens, and (c) 300 mm after the lens.

is plotted in Fig. 12.30(a). When the beam exits the lens, the beam diameter in the fast (vertical) direction exceeds 4.8 mm in size and causes diffractions, as can be seen in Fig. 12.30(a). The beam intensity pattern 10 m after the lens is plotted in Fig. 12.30(b), and it is still nicely Gaussian.

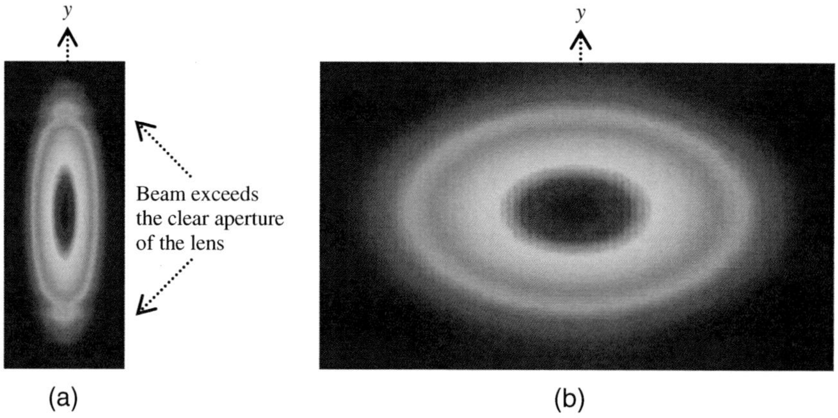

Figure 12.30 Intensity pattern of a laser diode beam after it propagates through an aspheric lens specially developed to collimate laser diode beams: (a) just after the lens and (b) 10 m after the lens.

References

1. Wikipedia, "Zemax," wikipedia.org/wiki/Zemax
2. R. R. Shannon, *The Art and Science of Lens Design*, Cambridge University Press, Cambridge, UK (1997).
3. G. H. Smith, *Practical Computer Aided Lens Design*, Willmann-Bell, Inc., Richmond, VA (1998).
4. J. M. Geary, *Introduction to Lens Design with Practical Zemax Examples*, Willmann-Bell, Inc., Richmond, VA (2002).
5. W. Smith, *Modern Lens Design*, McGraw-Hill, New York (2004).
6. M. Laikin, *Lens Design*, 4th ed., CRC Press/Taylor and Francis, Boca Raton, FL (2000).
7. R. Fischer, *Optical System Design*, 2nd ed., McGraw-Hill Professional, New York (2008).
8. R. Kingslake and R. B. Johnson, *Lens Design Fundamentals*, 2nd ed., Academic Press, New York (2009).
9. C. Velzel, *A Course in Lens Design*, Springer, Dordrecht, Germany (2014).
10. H. Sun, *Lens Design: A Practical Guide*, CRC Press, Boca Raton, FL (2016).

Chapter 13
Computer-Aided Optical Design

13.1 Optical Engineering and Design

13.1.1 Optical engineering versus electrical engineering

A common misconception about optical engineering and optical design underestimates their complexity because they only deal with "a few pieces of glass." A few pieces of glass may look simple, but they can also be difficult. For example, by using proper machine tools, such as a lathe, one can make a nice metal part. The machine tools are complex and feature thick operational manuals, but they provide a means to get things done. The same results could not be achieved with only simple tools, such as hammers and pliers. Even though the latter are simple and don't require a manual before operating them, they do not provide an effective means to get things done.

In electrical engineering, the conversion of AC to DC or vice versa, changing the frequency of an AC signal to other frequencies (not extreme changes), and amplifying a signal are all simple tasks. Compared with electrical engineering, optical engineering lacks effective means. All tasks similar to these are difficult for optical engineering. For example, changing the frequency of an optical signal to other frequencies is mostly impossible with current available technologies, although a few optical frequencies can be doubled or tripled with the help of some special materials. Amplifying an optical signal is also a complex task. In short, "simple" optical engineering often means "difficult."

13.1.2 The unique nature of optical design

Optical modeling involves typing the given parameters for a lens or an optical system into optical software to see its performance. Optical design goes a step further to vary all of the parameters to change the performance of the lens in order to meet certain goals. Six parameters are needed to describe a single optical element, such as a lens:

- Two parameters to describe the two surface radii of curvature,
- One parameter to describe the central thickness,

- One parameter to describe the size,
- One parameter to describe its position relative to other elements, and
- One high-order polynomial of wavelength to describe the index of the glass used.

If a lens contains five elements, a total of thirty parameters are needed to describe this lens.

Furthermore, the lens must work up to the specifications under several "boundary" conditions, such as

- A field range (angle or object height);
- A wavelength range;
- An object distance or distance range;
- An image size;
- A certain back working distance;
- Certain limitations on the size, weight, cost, etc.; and
- A temperature range, etc.

Nothing is perfect, i.e., every parameter, including the glass index of a real fabricated optical element, contains certain errors, and every element is mounted with certain axial, transverse, and angular errors. All of these errors can be considered as small perturbations. The lens must perform up to specifications with the effects of all of these perturbations.

Mathematically, the design of a lens, e.g., a five-element lens, must solve an equation with thirty variables, including high-order variables that each contain a small perturbation, and under several (e.g., 10) boundary conditions. This is a daunting task.

Before the invention of computers, only several grandmasters in the world were capable of designing multi-element lenses. They used a pencil and ruler to calculate and draw rays on paper through all of the optical elements and tried to focus the rays. It often took several hours or days to make one ray probably correct. Their great knowledge, experience, and instincts guided them through the optical maze.

However, a modern computer and optical software can calculate and draw tens of rays in one second and can significantly improve older optical designs within a few seconds. They have dramatically lowered the threshold of entering the optical design field. Many people can now design multi-element lenses.

An equation with tens of high-order variables does not have an analytical solution, and it often has more than one numerical and approximate solution. All optical design software have a certain level of artificial intelligence to help them search for the best solutions. For a complex multi-element lens, say, with more than ten elements, even a modern computer and software cannot guarantee a reasonably good solution. The optical software can be stuck during the searching process and fail to navigate the maze. Optical designers must use all of their experience and knowledge to direct the software, which is where the caliber of the optical designer has an effect.

This chapter discusses the designs of a few simple imaging lenses. Illumination optics design is less used and thus not addressed in this book. In-depth discussion of optical design is also beyond the scope of this book; see Refs. 2–10.

13.1.3 Design process: merit function

The performance goals for the design of a lens or optical system are set in the "merit function" defined as

$$M = \left| \sum_i W_i (g_i - G_i) \right|, \tag{13.1}$$

where W_i is the weight of the ith operand, G_i is the target for this operand, g_i is the real value of this operand obtained by the optical software (e.g., Zemax), and i is an integer covering all of the operands. Both W_i and G_i are assigned by the optical designer. For example, the target effective focal length (operand EFFL) and the weight of a lens are set to be 25 and 10, respectively, as shown in the first row of Fig. 13.2(a). If the real focal length value of the lens under design is 26 mm, then the merit function value for this operand is $M = |10 \times (26 - 25)| = 10$. If the focal length of the lens changes to 25.1 when the design is completed, then the M value drops from 10 to 1, and the lens performs closer to the desired status or better.

If there is more than one operand, their weight values only have a relative meaning. For example, if there are two operands, then assigning weights of 1 and 10, respectively, is equivalent to assigning weights of 10 and 100, respectively. Thus, the M value also has a relative meaning. The merit function provides a tool to numerically evaluate the specifications and performances of a lens or optical system against the desired goals.

13.1.4 Design process: optimization

The design of a lens (optical system) involves minimizing its M value, which is called "optimization." Zemax varies all variables during optimization to find the minimum M value. Mathematically, the process seeks the numerical solution(s) of a multivariable function under multiboundary conditions (see Section 13.1.2).

For a simple single lens under optimization, the exclusive best solution can be found in a few seconds by the software. For a complex lens, there is no guarantee that the best solution can be found. There may exist multiple equivalently approximate solutions. Whether or not a reasonable solution has been found is determined by the designer based on personal knowledge and experience. In most cases, the final value of every parameter can be more or less off the target values.

13.2 Design a Focusing Single Lens

13.2.1 Set a *Lens Data* box

To design a lens, a *Lens Data* box must be opened to enter certain optical elements and the system parameters, such as fields, wavelengths, and aperture type and values.

The design goal here is to reduce the very large focused spot of the Edmund equi-convex lens that was modeled in Section 12.3.3. The Zemax file, including the *Lens Data* box, the fields, and the wavelengths, is already set. After opening the Zemax file, clicking one of the surface radii, and then pressing Control–Z, the selected radius is set to a variable, as indicated by the symbol *V* at the right of this variable. Do the same to the other radius and the thickness in row 2; the *Lens Data* box will look like that shown in Fig. 13.1(a). During optimization, these three variables will be varied by Zemax for the smallest-possible focused spot.

Lens Data
Update: Editors Only
Surface 1 Properties — Conf

	Surf:Type		Co	Radius	Thickness	Material	Conic
0	OBJECT	Standard		Infinity	100.000		0.000
1	STOP (aper)	Standard		31.940 V	8.000	N-SF5	0.000
2	(aper)	Standard		-31.940 V	22.459 V		0.000
3	IMAGE	Standard		Infinity	-		0.000

(a)

Lens Data
Update: Editors Only
Surface 1 Properties — Con

	Surf:Type		Co	Radius	Thickness	Material	Conic
0	OBJECT	Standard		Infinity	100.000		0.000
1	STOP (aper)	Standard		21.830 V	8.000	N-SF5	0.000
2	(aper)	Standard		-64.163 V	23.064 V		0.000
3	IMAGE	Standard		Infinity	-		0.000

(b)

Lens Data
Update: Editors Only
Surface 3 Properties — Conf

	Surf:Type		Co	Radius	Thickness	Material	Conic
0	OBJECT	Standard		Infinity	100.000		0.000
1	STOP (aper)	Standard		20.726 V	8.000	N-SF5	-1.436 V
2	(aper)	Standard		-75.314 V	28.584 V		-0.723 V
3	IMAGE	Standard		Infinity	-		0.000

(c)

Figure 13.1 (a) *Lens Data* box of an Edmund equi-convex single lens. Three variables are set as the variables for optimization. (b) *Lens Data* box after optimization. All three variables are adjusted by Zemax for the smallest-possible focused spots. (c) *Lens Data* box after further optimization with the two conic parameters also set as variables.

13.2.2 Construct a merit function

To build a merit function, click *Optimize/Merit Function Editor*, and a *Merit Function Editor* box will appear. There are many *Operands* available for setting the detailed targets. Figure 13.2(a) shows the *Merit Function Editor* box that has already been set for this design. The following steps are next:

1. In the *Merit Function Editor* box, highlight the *Type* (first) box in row 1, and then click the *Wizards and Operands* button at the top-left corner. Click the *Operands* button, select the operand, and enter the numbers for

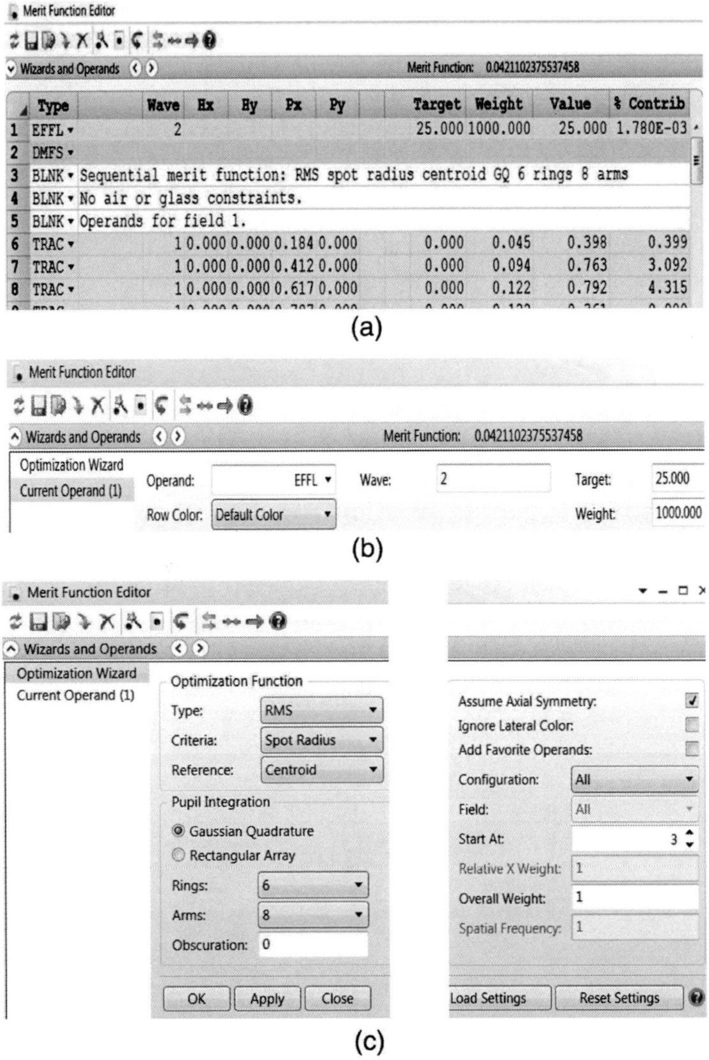

Figure 13.2 The process of setting a merit function: (a) the final *Merit Function Editor* box, (b) the *Current Operand* button sets the individual operand, and (c) the *Optimization Wizard* button sets the default merit function.

target, weight, and others. Figure 13.2(b) shows an example where operand *EFFL* (effective focal length) is selected: the target value is 25, the number 2 in the *Wave* box means that the focal length is calculated at the second wavelength (there are three RGB wavelengths), and a large number (1000) is selected for weight to guarantee that the focal length will be very close to 25 after optimization.

2. Click the *Optimization Wizard* button and make selections as shown in Fig. 13.2(c). Zemax has saved a few default merit functions for common goals. Since the optimization goal in this case is to obtain a small focused spot, *Spot Radius* is selected in the *Criteria* box to call one of these default merit functions. The numbers in the *Rings* and *Arms* boxes determine the spatial fineness of optimization. For the simple optimization here, these numbers do not matter very much, so just select any numbers in the middle of the ranges. The default merit function should *Start At: 3* (row 3) to avoid overriding the *EFFL* operand. The weight of the default merit function is chosen to be 1.

After these steps, the merit function should look like that shown in Fig. 13.2(a). Default merit functions often contain many lines, from tens up to thousands, and can be treated as a black box. The contents of the default merit functions are irrelevant and should not be altered. The function shown in Fig. 13.2(a) includes only the first few lines; there are many lines following them.

13.2.3 Optimize

Now to run the optimization. Click *Optimize*|*Optimize!*, and a *Local Optimization* box will appear, as shown in Fig. 13.3. Then click the *Start* button to start the process. The value of the merit function drops as shown by the value in the *Initial Merit Function:*. The optimization process takes a few seconds or so to complete, and then Zemax will stop.

After optimization, the *Lens Data* box will look like that shown in Fig. 13.1(b); the raytracing diagram is shown in Fig. 13.4(a), and the focused spot diagram is plotted in Fig. 13.5. The two RMS spot sizes are 880.094 μm

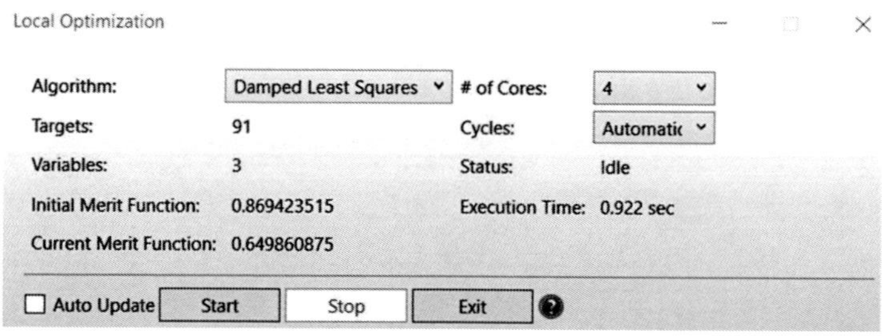

Figure 13.3 *Local Optimization* box. The optimization process reduces the value of the merit function.

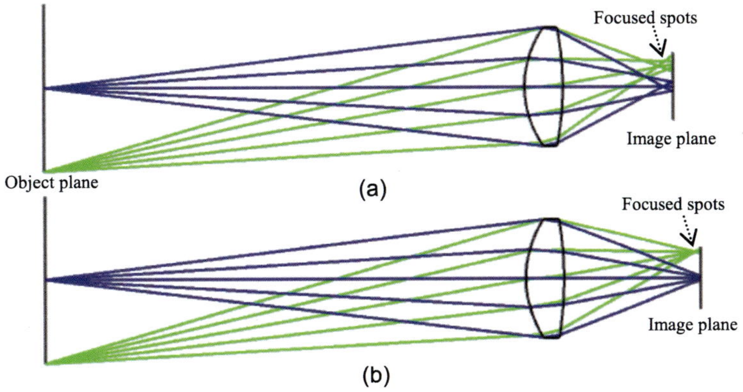

Figure 13.4 (a) Raytracing diagram of the optimized single lens. The radii of the two surface curvatures are different now. (b) Raytracing diagram of the optimized single lens that has two conic surfaces. The focusing is much sharper. Compared with the raytracing in (a), the back working distance is increased, but the focal length remains unchanged.

Figure 13.5 Focused spot diagram of the optimized spherical singlet. The two RMS radii are 880.094 μm and 912.977 μm, respectively. For comparison, the two RMS radii before optimization, shown in Fig. 12.4, are 1147.46 μm and 1741.84 μm, respectively.

and 912.977 μm, respectively—smaller than the spot size before optimization as shown in Fig. 12.4, but still too large. Such a single spherical lens of this glass type cannot do any better.

13.2.4 Use aspheric surfaces

Aspheric surface(s) can be used to further reduce the focused spot size of this singlet. In Zemax, the standard surface includes a spherical surface with a conic surface, although the latter is actually a simple aspheric surface. To change a standard surface to a formal aspheric surface, double click the first column of the lens surface, select *Type* in the pop-up window, and then select *Even Asphere* in the *Surface Type* box. Here, we only vary the conic parameter for optimization.

Make the two conic parameters in the *Lens Data* box in Fig. 13.1(b) variable, and then optimize. The *Lens Data* box after optimization is shown in Fig. 13.1(c). The raytracing diagram is shown in Fig. 13.4(b). The focusing is much sharper now. The two RMS spot sizes are 284.896 μm and 311.413 μm. These spot sizes are much smaller than those of the purely spherical surface shown in Fig. 13.5 but still much larger than the Airy disk size of <1 μm. To further reduce the focused spot size, an aspheric surface must be used with several aspheric parameters as variables.

13.3 Design a Focusing Doublet

13.3.1 Set a *Lens Data* box

Since the focused spots of the optimized single spherical lens are still too large, as shown in Fig. 13.5, this section discusses how to add one more spherical lens to the *Lens Data* box to form a doublet, as shown in Fig. 13.6(a). A doublet can mainly reduce the color aberration and also reduce spherical aberration.

Zemax *Hammer Optimization*, rather than *Local Optimization*, will be used for this purpose. *Hammer Optimization* can change the glass as well as vary the values of parameters, perform extensive searches for solutions if the lens is complex, and is a much more powerful optimization tool than *Local Optimization*. The vast majority of lens designs are performed using *Hammer Optimization*.

Local Optimization, used previously, only searches for a solution within a small scope, has a high chance of missing some better solutions, and is sufficient only for very simple optimization, such as the single-lens design discussed here. To design more complex lenses, *Hammer Optimization*—a much more exhaustive optimization—is needed.

Since *Hammer Optimization* can change the glass, the initial glass type of the added lens does not matter much, so N-BK7 is entered. All three surface radii and the thickness in row 3 are set as variables. The initial values of the

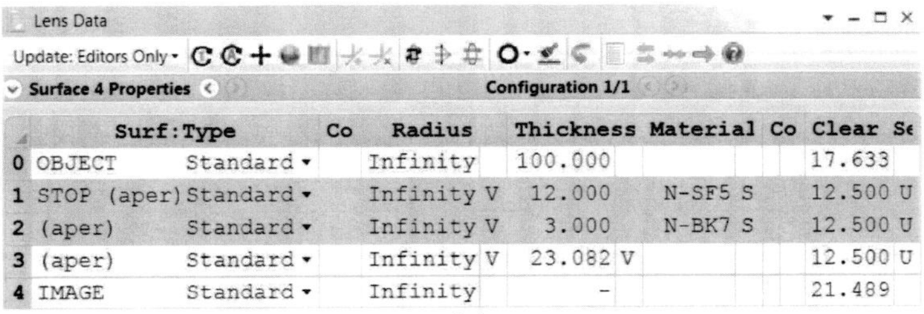

(a)

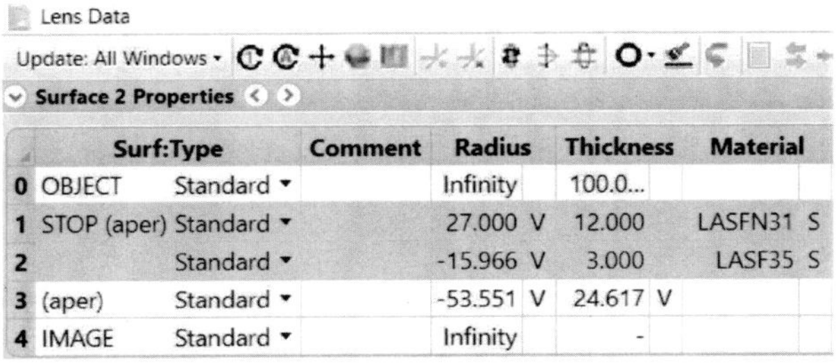

(b)

Figure 13.6 (a) *Lens Data* box of a doublet before optimization. (b) *Lens Data* box of a doublet after *Hammer Optimization*. The two glasses have been changed by Zemax.

radius of curvature do not matter much since Zemax will vary them. However, small values of radius make the lens surfaces very curved, and Zemax may lose the track and be unable to start optimization. Therefore, it is suggested to type 0 (0 radius is impossible, so Zemax will treat 0 as default and change it to infinity, i.e., a flat surface) in the radius box. The thicknesses of the two glasses are manually increased. The two glasses must be set as substitutable so that *Hammer Optimization* can change them. This can be done by clicking the small box at the right side of the glass box and selecting *Substitute* in the pop-up box. Finally, the *Lens Data* box will look like that shown in Fig. 13.6(a).

13.3.2 Construct a merit function and optimize

The merit function constructed for the single lens shown in Fig. 13.2 can be used here. Nothing else is necessary. Click *Optimize/Hammer Current/Start*, and the *Hammer Optimization* starts. *Hammer Optimizations* takes somewhat longer to run than local optimization, and the state of the optimization is updated in a gradually slower manner. For such a simple doublet, *Hammer Optimization* will find the best design within several minutes after tens of updates. Once *Hammer Optimization* has exhausted all of the possibilities, it will no longer update, and the optimization process can be halted.

The *Lens Data* box after optimization is shown in Fig. 13.6(b). The raytracing diagram after optimization is shown in Fig. 13.7; the focusing is visually improved. The *Spot Diagram* is shown in Fig. 13.8. The RMS spot

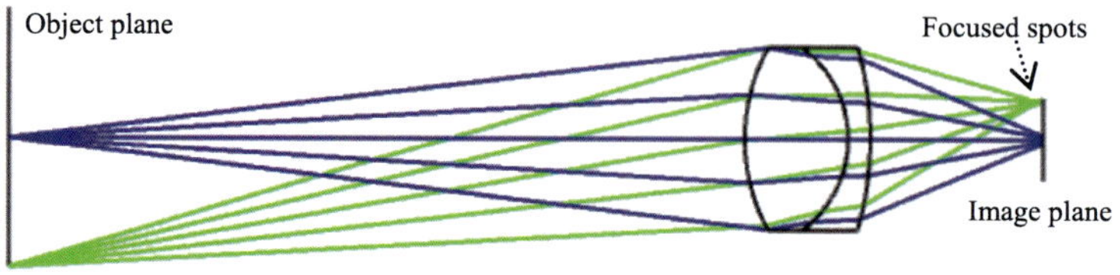

Figure 13.7 Raytracing diagram of the optimized doublet. The focusing is visually sharper than that shown in Fig. 12.3.

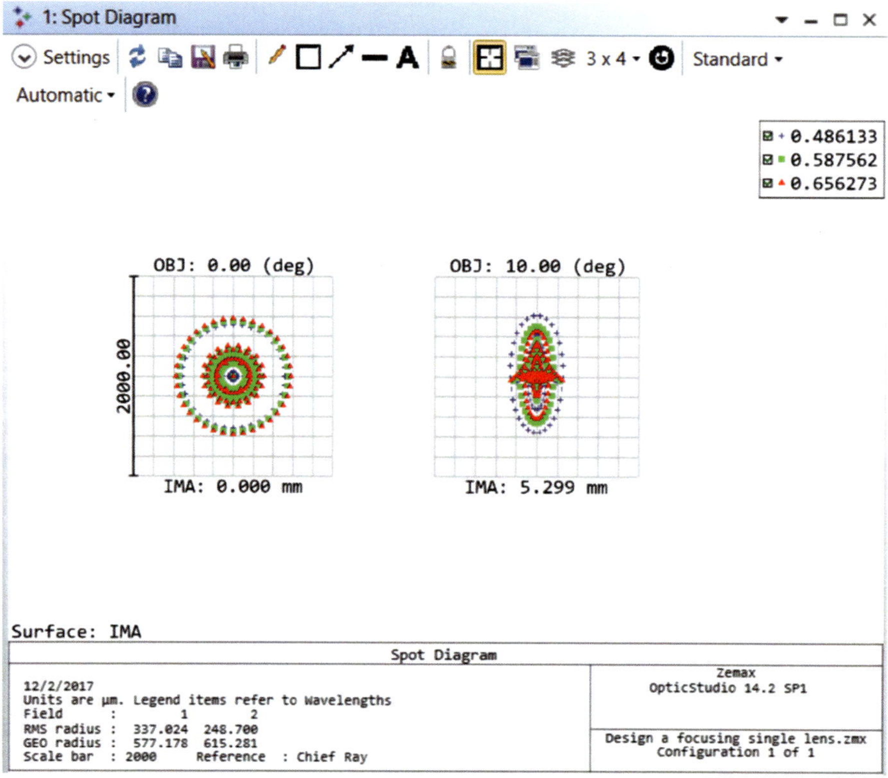

Figure 13.8 Focused spot diagram of the optimized doublet. The RMS spot radii are 337.024 µm and 248.700 µm, respectively, smaller than those shown in Fig. 13.5 for the singlet.

radii are 337.024 µm and 248.700 µm, respectively, which are much smaller than the RMS spot radius of the optimized singlet shown in Fig. 13.5 but much larger than the Airy disk size of <1 µm.

It is possible to obtain results different than those shown her, because there is more than one glass combination that has similar performance. Regardless, the final spot radii should be close to those in Fig. 13.8.

13.4 Design a Simple Eyepiece

13.4.1 Set system parameters

The complexity of designing an eyepiece is a significant step forward from designing a doublet. The first thing to do is to set the system parameters. Assume that this eyepiece will have a 20-mm field of view, 7X magnification, 10-mm working distance, 12-mm eye relief distance, 6-mm exit pupil diameter, $<$50-mm total length, $\leq$5% image distortion, as high as possible CTF values, four glass spherical elements (one doublet counts as two), and work in the visible range. The first two parameters are set in *System Explorer/Fields* and *System Explorer/ Wavelengths*; all other parameters are set in the *Lens Data* box.

The wavelengths selected are RGB, the same as those for modeling a single lens, so the process is not repeated here. The type of field of view selected here is the *Object Height*, the largest field (object half-height) selected is 10 mm, and a total of six fields are selected, as shown in Fig. 13.9(a). The

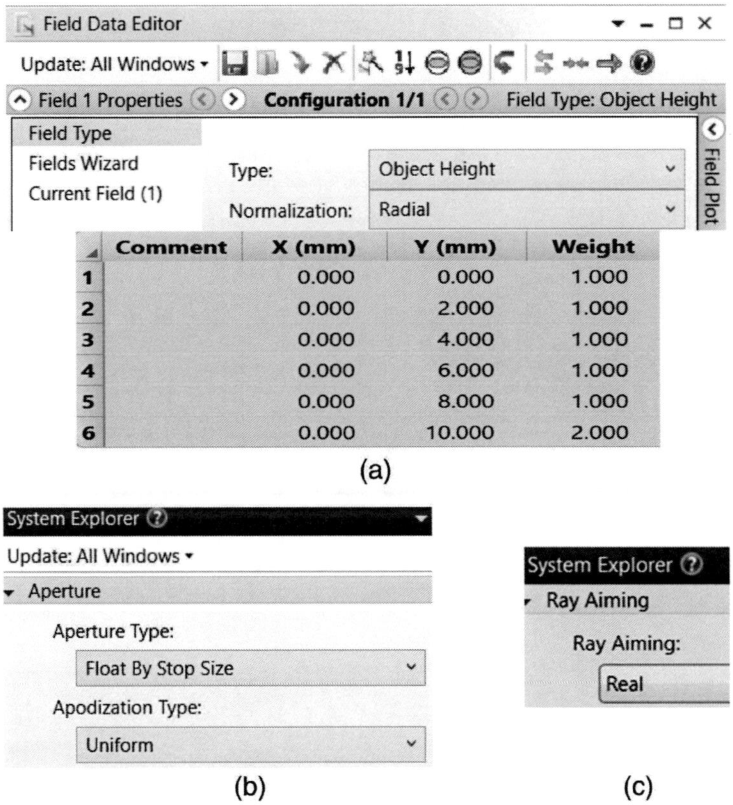

Figure 13.9 (a) *Field Data Editor* of the eyepiece. Select *Object Height*. Six field values are used up to 10 (mm). (b) *Float By Stop Size* is selected as the aperture stop type. (c) *Ray Aiming* must be activated.

weight of the last field is selected to be 2, whereas the weight of all other fields is 1. The performance of most lenses drops as the value of the field increases. In order to obtain as even as possible performance over the entire field, the weight of the large field is often larger than the weight of the small field.

13.4.2 Set a *Lens Data* box

The *Lens Data* box and the raytracing diagram of the eyepiece after the design is completed are shown in Figs. 13.10 and 13.11, respectively. The raytracing

	Surf:Type	Co	Radius	Thickness	Material	Co	Clear Semi
0	OBJEC' Standard ▾		Infinity	10.000			10.000
1	Standard ▾		124.882 V	6.000	H-LAF50B S		11.857
2	Standard ▾		-25.814 V	15.346 V			11.993
3	Standard ▾		-13.610 V	1.500	ZF50 S		5.798
4	Standard ▾		15.151 V	1.605 V			5.879
5	Standard ▾		26.633 V	6.000	H-ZLAF68C S		6.358
6	Standard ▾		-9.723 V	1.500	H-ZBAF4 S		6.562
7	Standard ▾		-22.683 V	12.000			6.464
8	STOP Standard ▾		Infinity	-250.000			3.000 U
9	IMAGE Standard ▾		Infinity	-			70.022

Figure 13.10 The *Lens Data* box of the eyepiece after the design is completed.

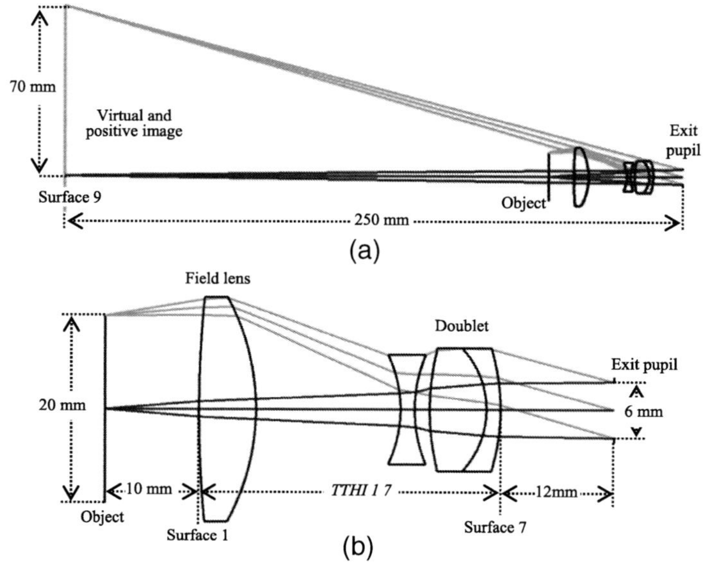

Figure 13.11 Raytracing diagram of the eyepiece after the design is completed. Only two fields are plotted for clarity. (a) The whole optical train. The image formed is positive and virtual. (b) The details of the eyepiece. The user's eye should be placed at the exit pupil to view the image.

only contains two fields, not all the fields, for clarity. The details of the set *Lens Data* box are explained are as follows:

1. The row-0 *Thickness* value is 10, which means the object distance is 10 mm, as marked in Fig. 13.11(b).
2. All of the air-gap and glass thicknesses, and the element surface radii of curvature, are set to be variables during the optimization process. After the optimization is completed, the glass thicknesses are manually rounded up to the nearest two digits and fixed (no longer variable). Then local optimization is performed once more for a few seconds to adjust the values of the air gaps and surface radii of curvature to match the new glass thicknesses.
3. The row-7 *Thickness* value is 12 and defines the eye relief distance, as marked in Fig. 13.11(b).
4. The aperture stop is set at row 8, and the selected type is *Floating By Stop Size*, as shown in Fig. 13.9(b). The stop size must be manually entered.
5. Manually type 3 in the *Clear Semi-Dia* box in row 8, which sets the aperture stop size. Zemax will align the rays with the exit pupil, not the front element, and it will not launch any rays that have a height larger than the exit pupil radius when the ray reaches the exit pupil.
6. If the aperture stop is at the front of the lens, it is easy for Zemax to align the rays with the aperture. In case that the aperture stop is at the rear part of the lens, such as the present case, the *Ray Aiming* must be activated so that Zemax will spend some time and effort to align rays to this "hide-behind" aperture stop. This can be done in the *System Explorer*, as shown in Fig. 13.9(c).
7. The *Thickness* value in row 8 is set to be −250, which extends the conceived rays from exit pupil leftward to form a virtual image, as shown in Fig. 13.11(a).
8. All glasses are set to be substitutable. The glasses in this *Lens Data* box are CDGM brand. There are over ten optical glass manufacturers. The glasses made by several major manufacturers, such as Schott, Ohara, Hoya, and CDGM, are mostly equivalent. For most lenses, one brand of glasses provides enough choices. The glass brand can be selected by clicking *Libraries|Materials Catalog* and making a selection in the pop-up *Materials Catalog* box.

13.4.3 Construct a merit function

The merit function constructed to optimize this eyepiece is shown in Fig. 13.12 and is explained item by item or row by row here:

1. The operand *TTHI 1 7* in row 1 will return the thickness value between surfaces 1 and 7, as marked in Fig. 13.11(b). But no target value is set.

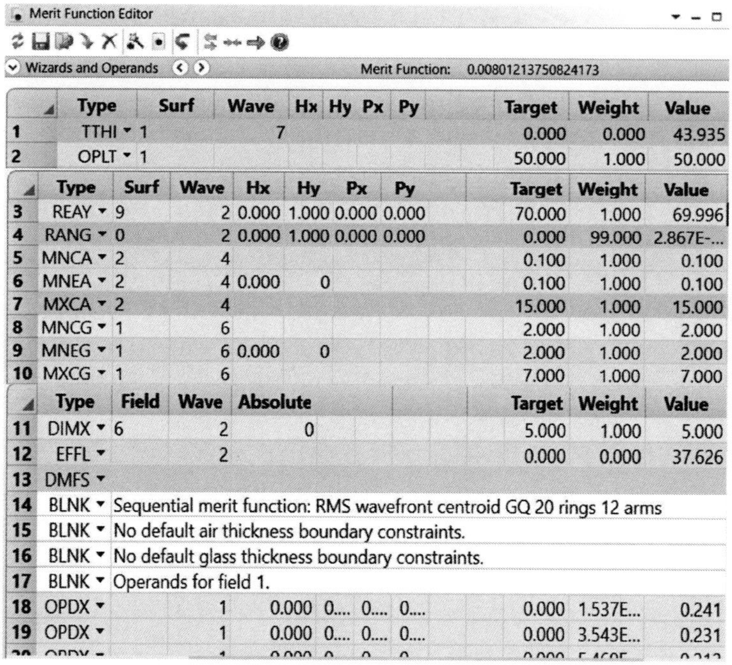

Merit Function Editor

Wizards and Operands — Merit Function: 0.00801213750824173

	Type	Surf	Wave	Hx	Hy	Px	Py	Target	Weight	Value
1	TTHI	1	7					0.000	0.000	43.935
2	OPLT	1						50.000	1.000	50.000
	Type	**Surf**	**Wave**	**Hx**	**Hy**	**Px**	**Py**	**Target**	**Weight**	**Value**
3	REAY	9	2	0.000	1.000	0.000	0.000	70.000	1.000	69.996
4	RANG	0	2	0.000	1.000	0.000	0.000	0.000	99.000	2.867E-...
5	MNCA	2	4					0.100	1.000	0.100
6	MNEA	2	4	0.000	0			0.100	1.000	0.100
7	MXCA	2	4					15.000	1.000	15.000
8	MNCG	1	6					2.000	1.000	2.000
9	MNEG	1	6	0.000	0			2.000	1.000	2.000
10	MXCG	1	6					7.000	1.000	7.000
	Type	**Field**	**Wave**	**Absolute**				**Target**	**Weight**	**Value**
11	DIMX	6	2	0				5.000	1.000	5.000
12	EFFL		2					0.000	0.000	37.626
13	DMFS									
14	BLNK	Sequential merit function: RMS wavefront centroid GQ 20 rings 12 arms								
15	BLNK	No default air thickness boundary constraints.								
16	BLNK	No default glass thickness boundary constraints.								
17	BLNK	Operands for field 1.								
18	OPDX		1	0.000	0....	0....	0....	0.000	1.537E...	0.241
19	OPDX		1	0.000	0....	0....	0....	0.000	3.543E...	0.231

Figure 13.12 The merit function for optimizing the eyepiece.

2. The operand *OPLT 1 50 1* in row 2 limits the return value of operand 1 (the thickness between surfaces 1 and 7) to be less than 50. This action sets an upper limit for the eyepiece length. Any length below 50 is fine. The actual length is 43.935, as shown in row 1. No lower limit is set since a small length is always desired.
3. Operand *REAY* returns the value of a specified ray height at a certain surface. The entire row 3 means that the ray height at surface 9 (image plane as marked in Fig. 13.11(a)) calculated at wavelength 2 (the G color among the RGB) of full field ($Hy = 1$) should be 70. This sets the image half-height. The magnification of the eyepiece is defined as the ratio of image size/object size and is $70/10 = 7$X in this case. The actual image half-height is 69.996, as shown at the end of row 3, which is accurate enough.
4. Operand *RANG* returns the angle between a specified ray and the optical axis. The entire row 4 means that the angle of the full-field ($Hy = 1$) chief ray ($Py = 0$) at surface 0, calculated at wavelength 2 (the G color among the RGB), should be 0. This pushes the chief rays from the object to be horizontally incident on the front surface of the eyepiece so as to increase the power collection of the eyepiece. Since most objects have their maximum radiance in the normal (horizontal in this case) direction, the final *RANG* value will be slightly different than 0.

5. In row 5, operand *MNCA* sets the minimum central air gap between surfaces 2 and 4 to be 0.1 (mm).
6. In row 6, operand *MNEA* sets the minimum edge air gap between surfaces 2 and 4 to be 0.1 (mm).
7. In row 7, operand *MXCA* sets the maximum central air gap between surfaces 2 and 4 to be 15 (mm).
8. In row 8, operand *MNCG* sets the minimum central glass thickness between surfaces 1 and 6 to be 2 (mm).
9. In row 9, operand *MNEG* sets the minimum edge glass thickness between surfaces 1 and 6 to be 2 (mm).
10. In row 10, operand *MXCG* sets the maximum central glass thickness between surfaces 1 and 6 to be 7 (mm).
11. Operand *DIMX* in row 11 sets the maximum absolute image distortion calculated at wavelength 2 to be 5%.
12. Operand *EFFL* in row 12 returns the effective focal length calculated at wavelength 2. Since there is no target value and no weight, this row only monitors the focal length. The actual focal length of this eyepiece is 37.626 (mm).
13. In image optics design, the minimization of the wavefront error, not the focused spot size, is the most effective way of improving the image resolution. Thus, the *RMS Wavefront* is selected as the criterion in the default merit function. A maximum *Ring* and *Arm* number of 20 and 12, respectively, are selected for fine optimization. The process of setting the default merit function was explained in Fig. 13.2(c).

13.4.4 Optimize

The optimization of a complex lens should proceed gradually and methodically, but there is no fixed pattern; the designer must be flexible. Before running any type of optimization, save the Zemax file since the optimization can mess up the file. Generally speaking, the optimization steps are as follows:

1. Run *Local Optimization* until the value of the merit function stops decreasing, and then check the layout of the lens. If the layout looks erroneous or does not make sense, such as a large negative air gap or glass thickness, discard the result and open the file saved earlier before changing the problematic variables and repeating *Local Optimization.* This process should prevent the file from going awry again. Then return the fixed problematic parameters to variables and run *Local Optimization* again. Note that small negative values in the air gap and/or glass thickness are acceptable at this design stage.
2. After completing *Local Optimization*, run *Hammer Optimization*. For a lens with four elements, one hour or so or runtime should be sufficient. *Hammer Optimization* will update the design result in a

gradually slower and random pace. There is no clear indication when to halt the optimization process. Generally speaking, if the *Hammer Optimization* has not updated the design for a "long" time, it is advisable to stop the optimization. To process a simple lens, such as the eyepiece here, the word "long" means tens of minutes. To process a complex lens, e.g., one with ten elements, "long" means several hours or one day. If the designer is not in a rush, it does not hurt to wait longer for the software to optimize.

3. Clean up the design. Change the small negative air gap to 0.1 mm or so. Check the optical element edge thickness, which should be at least 2 mm or so. Since Zemax only plots the optical elements to their clear aperture, the real elements fabricated will be larger than the clear aperture, and there should be enough edge thickness to mount the elements. If some elements appear to be too thin, the thickness can be manually increased somewhat. After adjusting some air gaps and glass thicknesses, change all of these parameters from variables to fixed and then run local optimization again. The design is done.
4. Save the Zemax file before making any major changes in case the changes harm the file. If the file is not saved and must be repaired, try to manually increase the shortest radius of curvature of one or two elements a bit. This can more or less relax the tension of the lens and give Zemax a better chance to launch raytracing again.

13.4.5 Design results

The image CTF and distortion curves are the two performances to be checked. Click *Analyze/MTF/FFT MTF*, and a *FFT MTF* box will appear. Click *Settings*, select 10 in the *Maximum Frequency* box, choose *Square Wave* in the *Type* box, and check the *Show Diffraction Limit* box; the *FFT MTF* box will look like that shown in Fig. 13.13(a).

The CTF curves shown are for both tangential and sagittal directions and all fields. The values of all curves are >0.5 at 7 line pairs/mm, which is the human-eye resolution limit for an image 250 mm away. The solid curve at the top of all the CTF curves is the diffraction limit CTF, which is the theoretically highest possible CTF.

Click *Analyze/Aberrations/Field Curvature and Distortion* so that a *Field Curv/Dis* box appears, as shown in Fig. 13.13(b). The maximum image distortion is about 0.6%, appearing at 10 mm, which is the edge of the object.

13.4.6 Design strategies

Lens design is a complicated skill and requires a lot of practical experience. It is unlikely that one can become a lens design expert by only reading textbooks. This section summarizes some design experiences to speed up the

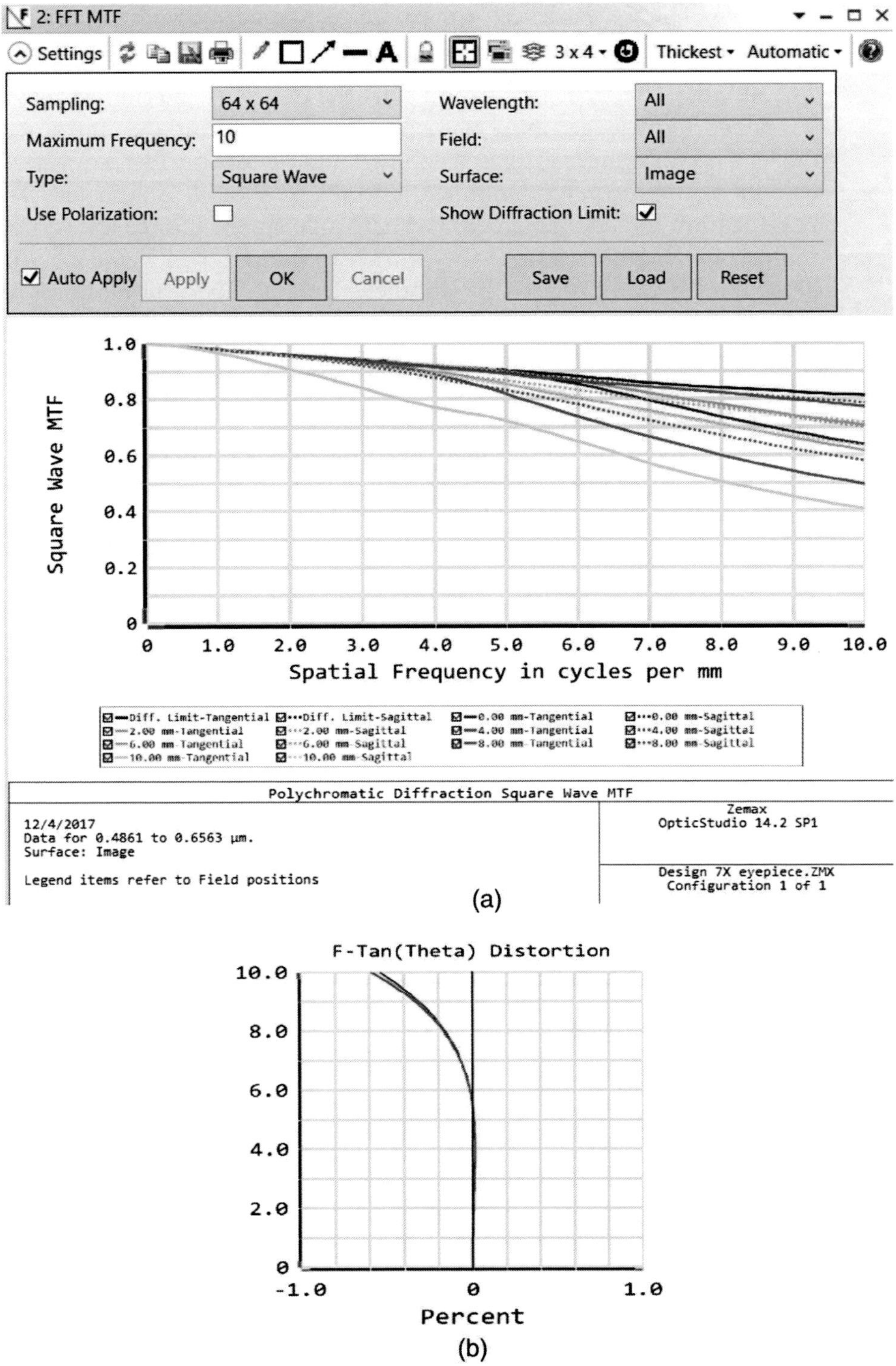

Figure 13.13 Performance of the eyepiece: (a) CTF curves and (b) distortion diagram.

learning process. These experiences are general guidance and will not apply to every instance. There are always some exceptions.

1. The number of elements to use to start the design is determined by the designer's knowledge and experience. If after longtime optimization, the lens performance is still not good enough, one more element should be added, and the whole optimization process should be repeated. The best location to add an element is where an element surface has a short radius of curvature because the burden placed on this element is too heavy and must be shared by another element. The best material for the added element is the same as the material of the burdened element.
2. If the lens easily exceeds the performance target after brief optimization, one can try to remove one element and run the whole optimization process again. The element being removed should be the one with a long surface radius of curvature since this element has little work to do.
3. Field elements are those at which the rays of different fields separate. In the raytracing diagram, the leftmost element shown in Fig. 13.11 is the field element. Field elements are usually singlets with a large refractive index.
4. For simple imaging lens that cover the RGB range, one doublet is necessary and enough. The best location to place the doublet is where the rays of different fields overlap, as shown by the raytracing diagram in Fig. 13.11. For a positive lens, such as the eyepiece being designed here, the positive element in the doublet should have a relatively larger Abbe number; the negative element in the doublet should have a relatively small Abbe number. Although the final glass selection will be determined by Zemax *Hammer Optimization*, the initial glass selection made by the designer can affect the optimization process.

13.5 Design a Simple Microscope Objective

13.5.1 Set system parameters

Assume that this microscope objective will have a 0.5-mm field of view, 0.5 numerical aperture, 40X magnification, 2-mm working distance, <40-mm lens train length, ≤2% image distortion, as high as possible CTF values, five glass spherical elements (one doublet counts as two), and work in the visible range. The field, aperture type, and wavelength are set in *System Explorer|Fields* and *System Explorer|Wavelengths*, whereas all other parameters are set in the *Lens Data* box.

The field selected here is the *Object Height*, same as the one shown in Fig. 13.9(a), only the largest field (object half-height) selected is 0.25 mm. A total of six fields, all with weight 1, are selected. All of the weights can be adjusted after optimization if the result is not satisfactory.

13.5.2 Set a *Lens Data* box

The *Lens Data* box and the raytracing diagram of the objective after the design is completed are shown in Figs. 13.14 and 13.15, respectively. The raytracing only show two fields for clarity.

The front thin layer of H-K9L glass is a protective window. The aperture stop is set at row 3, which is the front surface of the first element. The aperture stop type selected is *Object Space NA* with a value of 0.5, as shown in Fig. 13.15(d). Such a NA is small for a microscope object and is relatively easy to design. The thickness value of row 11 is 160, which places the image plane 160 mm behind the last surface of the last element, as shown in Fig. 13.15(a), and is common for microscope objectives.

13.5.3 Construct a merit function

The merit function constructed to optimize this objective is shown in Fig. 13.16. It is similar to the merit function shown in Fig. 13.12 for the eyepiece. Rows 1 and 2 control the lens length to be below 40 mm. Row 3 forces the image half-height to be –10 (mm). The magnification of this objective is thus 10/0.25 = 40. The negative sign means that the image orientation is opposite to that of the object and is a negative image.

13.5.4 Optimization and design results

The optimization process is the same as that for the eyepiece, but the *Hammer Optimization* must be run longer than that for the eyepiece—at least a couple of hours—because there are more elements in the objective.

Lens Data

Update: Editors Only

Surface 12 Properties — Configuration 1/1

	Surf:Type		Co	Radius	Thickness	Material	Co	Clear Se
0	OBJECT	Standard		Infinity	2.000			0.250
1	(aper)	Standard		Infinity	0.200	H-K9L		5.000 U
2	STOP (aper)	Standard		Infinity	0.082 V			5.000 U
3		Standard		-9.202 V	3.460 V	H-ZLAF68N S		1.231
4		Standard		-3.861 V	0.100			2.335
5		Standard		12.968 V	5.000 V	H-ZPK5 S		2.480
6		Standard		-3.783 V	1.000	D-ZF93 S		2.508
7		Standard		-11.030 V	0.100			2.784
8		Standard		4.962 V	4.539 V	H-ZLAF2A S		2.906
9		Standard		2.963 V	10.558 V			1.814
10		Standard		-62.979 V	5.000 V	H-ZF7LAGT S		3.112
11		Standard		-21.978 V	160.000			3.570
12	IMAGE	Standard		Infinity	-			10.136

Figure 13.14 *Lens Data* box of the microscope objective after the design is completed.

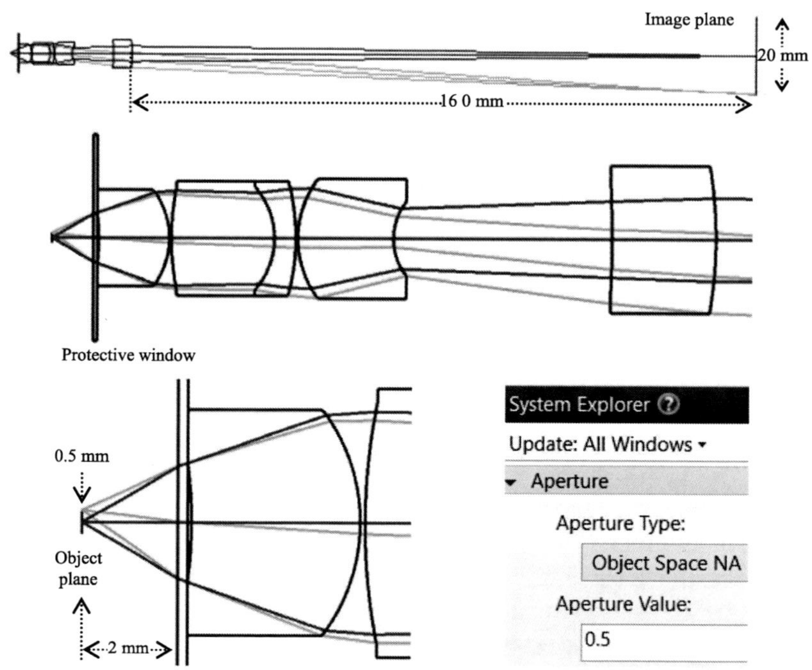

Figure 13.15 Raytracing diagram of the objective after the design is completed. Only two fields are plotted for clarity. (a) The whole optical train. (b) The detail of the objective. (c) The front part of the objective. (d) An *Object Space NA* of 0.5 is selected for the aperture stop.

	Type	Surf	Wave	Hx	Hy	Px	Py		Target	Weight	Value
3	REAY	12	2	0.000	1.000	0.000	0.000		-10.000	1.000	-10.000
4	EFFL		2						5.000	0.000	4.332
5	DIMX	6	2	0					2.000	1.000	2.000
6	MNCG	3	10						0.800	1.000	0.800
7	MNEG	3	10	0.000	0				0.800	1.000	0.800
8	MXCG	3	10						5.000	1.000	5.000
9	MNCA	2	9						0.100	1.000	0.100
10	MNEA	2	9	0.000	0				0.100	1.000	0.094
11	DMFS										
12	BLNK	Sequential merit function: RMS wavefront centroid GQ 20 rings 12 arms									
13	BLNK	No air or glass constraints.									
14	BLNK	Operands for field 1.									
15	OPDX		1	0.000	0.000	0.059	0.000		0.000	1.537E...	0.190
16	OPDX		1	0.000	0.000	0.134	0.000		0.000	3.543E...	0.175

Figure 13.16 Merit function for optimizing the microscope objective.

The CTF and image distortion curves for this objective are plotted in Fig. 13.17. All of the CTF curves stick together, which means that the CTF is uniform across the entire field and is why a weight of 1 is selected for every field (this is often the case if the object field is small). The CTF ≥0.3, up to about

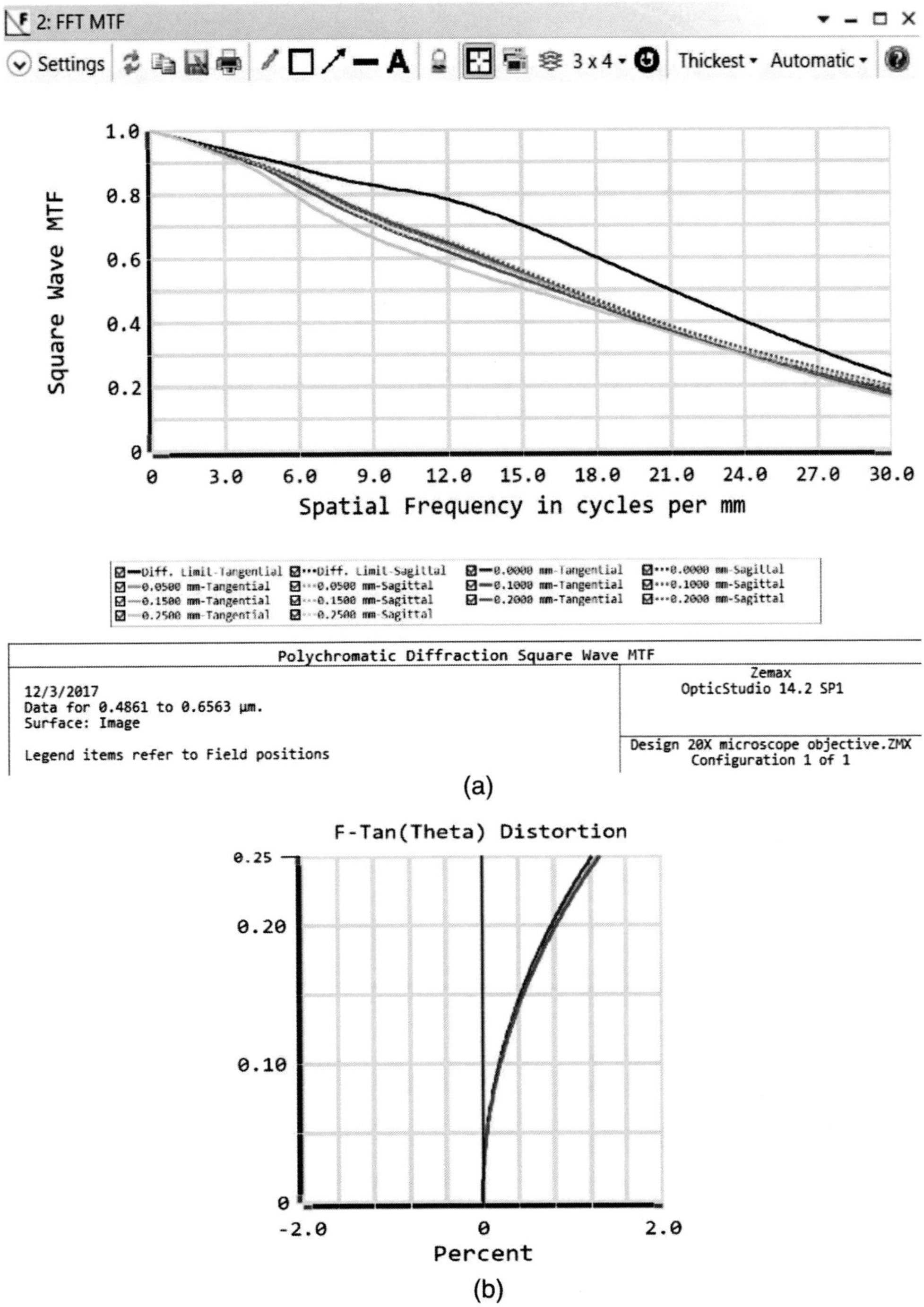

Figure 13.17 Performance of the objective: (a) CTF curves, and (b) the distortion diagram.

24 line pairs/mm. Since the whole image size is 20 mm, this objective can provide a total of $24 \times 20 = 480$ line resolution. Such a resolution is low because the small NA of 0.5 results in a low CTF ceiling (the diffraction-limited CTF).

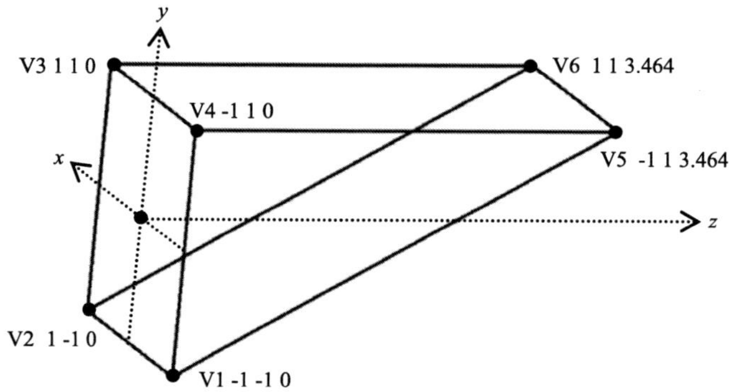

Figure 13.18 Definition of the six vertices of a right-angle prism.

Table 13.1 The Notepad file for a 30-60-90° right-angle prism.

Notepad Contents	Explanations
! Front-face vertices	In Notepad, all lines that start with ! are notes. The following four vertices are the front vertices.
V 1 −1 −1 0	Place vertex 1 at location $x = -1$, $y = -1$, and $z = 0$.
V 2 1 −1 0	Place vertex 2 at location $x = 1$, $y = -1$, and $z = 0$.
V 3 1 1 0	Place vertex 3 at location $x = 1$, $y = 1$, and $z = 0$.
V 4 −1 1 0	Place vertex 4 at location $x = -1$, $y = 1$, and $z = 0$.
! Back-face vertices	Define the two back vertices.
V 5 −1 1 3.464	Place vertex 5 at location $x = -1$, $y = 1$, and $z = 3.464$.
V 6 1 1 3.464	Place vertex 6 at location $x = 1$, $y = 1$, and $z = 3.464$.
! Front-face rectangle	The following line forms the front-face rectangle.
R 1 2 3 4 0 0	Symbol R means to form a rectangle. Link vertices 1, 2, 3, and 4 to form the front rectangular face. The second-last number defines the face type: −1 for absorptive, 0 for refractive, and 1 for reflective. The last number is assigned to a face. More than one face can share this number.
! Top-face rectangle	The following line forms the top-face rectangle.
R 4 3 6 5 0 0	Link vertices 4, 3, 6 and 5 to form the bottom rectangular face.
! Bottom rectangle	The following line forms the top-face rectangle.
R 1 2 6 5 0 1	Link vertices 1, 2, 6, and 5 to form the bottom rectangular face. Number 1 is assigned to this face for the convenience of selecting this face for further work, such as applying a certain type of coating. The numbers for all other faces are 0.
! Left-side triangle	The following line forms the left-side triangle.
T 1 4 5 0 0	Symbol T means to form a triangle. Link vertices 1, 4, and 5 to form the left-side triangle.
! Right-side triangle	The following line forms the right-side triangle.
T 3 2 6 0 0	Link vertices 3, 2, and 6 to form the left-side triangle.

13.6 Build a Polygon

All four designs discussed earlier are sequential designs for imaging optics. This section discusses a non-sequential topic: how to build a polygonal object using a 30-60-90° right-angle prism as an example.

All polygons in Zemax are designed using Notepad. These Notepad files can be accessed by opening the program first and then navigating to *Document/Zemax/Objects/Polygon Objects*. All of the files there with an extension. pob are for polygons.

The 30-60-90° right-angle prism is plotted in Fig. 13.18. The Notepad file is reproduced in the left column in Table 13.1, and the explanations are in the right column.

Three rectangles and two triangles form the right-angle prism. Other shapes of polygons can be built in the same way. After building a polygon, the Notepad file must be saved in the *Document/Zemax/Objects/Polygon Objects* folder with the extension .pob.

Chapter 14
Lens Fabrication, Tolerances, Procurement, Inspection, and Mounting

14.1 Optical Specifications and Standards

A complete set of specifications and standards covers every aspect in detail of the optical manufacturing process, from drawings to fabrication and inspection. These specifications and standards are important for manufacturers to maintain high and stable qualities for their products. However, for optical-component users, they appear to have too many details and are not very relevant. For example, users very rarely need to inspect the quality of the glass used in their lenses against the specifications; they may occasionally measure the performance of their lens, such as the focal length or the image resolution. Therefore, this chapter mainly discusses the principles of specifications and standards, not the manufacturing details. Manufacturers will usually take care of the details of specifications and standards for users if the latter can convey their needs to the former. See Hausner[1] for more details regarding specifications and standards.

14.2 Lens Fabrication Methods

The fabrication of optical components is mainly about creating optical surfaces, either lens surfaces or mirror surfaces. Spherical surfaces are the most commonly used surfaces since flat surfaces can be considered as spherical surfaces with infinitely large radii of curvature. Hausner[2] provides more information about the methods of producing optical components.

14.2.1 Spindle grinding and polishing

Spindle grinding and polishing is the traditional method used to fabricate spherical lenses as follows:

1. Cut glass blanks into pieces slightly larger than the final lens.

2. Rough grinding (shaping). Tools with a spherical working surface and large abrasive grain are used to grind the glass blanks to the shape of the final lens but a couple of millimeters larger. The spherical surface radius of curvature of the tool should equal the surface radius of curvature of the optical components under fabrication.
3. Precision grinding (lapping). Tools with a spherical working surface and small grain are used to finely grind the lens surfaces. The spherical surface radius of curvature of the tool should equal the surface radius of curvature of the optical components under fabrication.
4. Polishing. Tools with a spherical working surface and special abrasive are used to finalize the lens shape to meet the specifications and to reduce the micro-roughness of the lens surface caused by previous grindings. Again, the spherical surface radius of curvature of the tool should equal the surface radius of curvature of the optical components under fabrication.
5. Cleaning and inspection. The need for cleaning before inspection is obvious. The lens surface radius and central thickness are inspected during this step. For windows, the parallelism of the two surfaces is checked.
6. Centering. This step is unique to spherical lenses and is explained in Section 14.3.2. The lens edge is trimmed to make the two surfaces centered within the specifications.
7. Inspection. This step inspects the centering of the two surfaces of the lens.
8. Beveling. A bevel is created by grinding the lens edge. A bevel at the lens edge will facilitate lens mounting and reduce the likelihood of chipping or cracking the edge, potentially cutting a handler's finger.
9. Final cleaning for further handling (e.g., bonding, coating, or shipping).

14.2.2 Diamond turning

When the lens surface cannot be grinded, diamond turning is a common alternative method. Such surfaces include aspheric surfaces or the spherical surface of certain materials. For example, a plastic-lens spherical surface cannot be grinded because of the viscosity of plastic.

A schematic of diamond turning is drawn in Fig. 14.1(a). The lens under fabrication is mounted on a rotor with a rotation axis that coincides with the lens optical axis. The lens is being rotated, and a cutting tool with a small diamond tip is pressed against the lens surface. As the lens rotates, the cutting tool cuts a groove symmetric about the rotating axis on the lens. The cutting tool starts at the lens center and gradually moves toward the lens edge to cut into the lens. By properly controlling the cutting-tool position, a desired surface profile can be created.

Modern diamond-turning machines are computer controlled. The machine operator types in the data of the desired surface profile, and the computer controls the tool to make the correct cuts.

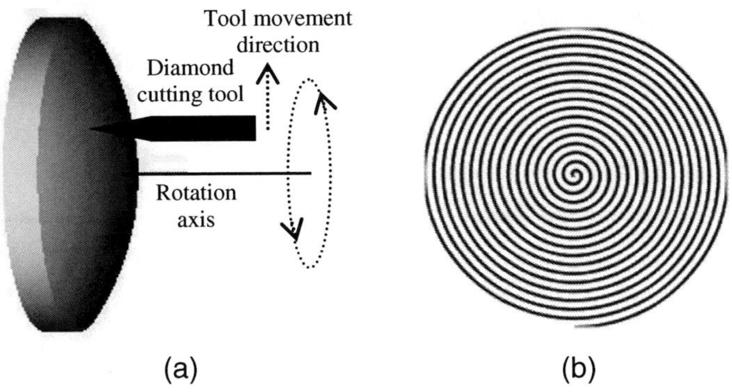

Figure 14.1 (a) Illustration of the diamond turning method for creating a lens surface. (b) A dramatically exaggerated scheme of the spiral track on a lens surface created by diamond turning.

The cutting grooves created by diamond turning has a spiral track, as illustrated in Fig. 14.1(b), with a certain shape and depth. Sometimes the grooves can be seen by the human eye under certain illumination. Generally speaking, the quality of the lens surface created by diamond turning is lower than the quality of a surface created by polishing, although ultrafine diamond turning can cut a groove with a depth less than 5 nm. The cutting grooves are usually not a problem for IR optics, which handle a relative longer wavelength. The cutting grooves are usually not a problem for optics in the visible range, either, so long as the light or beam is not focused or nearly focused on the surface created by diamond turning.

When using this method to create an aspheric surface on glass, the traditional spindle method should first be used to create a spherical surface that is close to the desired aspheric surface. The diamond turning tool must cut off much less material by working on this spherical surface. For a given aspheric profile, optical software (such as Zemax) is capable of providing the radius of a sphere that will minimize the amount of material removed by the diamond turning tool.

The cost of diamond turning a surface is about three times the cost of grinding a surface of the same material and similar size and quality. Diamond turning plastic materials is much easier since plastic materials are much softer than glasses.

14.2.3 Molding

Molding is another method of producing both spherical and aspheric lenses. Molding method has a high non-recurring cost for making the mold at the beginning. For large-quantity production, i.e., at least 100 pieces or more, the per lens production cost of the molding method can be lower than the per lens cost of the spindle grinding method and much lower than the per lens cost of the diamond turning method.

Molds are often produced by diamond turning and therefore have spiral grooves. Lenses produced by molding usually have the same-shape grooves on their surfaces. As noted in the previous section, these grooves do not cause apparent problems for most applications in the IR or visible ranges.

To produce glass spherical lenses, the molding method is less widely used than traditional spindle grinding because

1. The production quantities are often not large enough to justify molding, which can have a much longer production time.
2. Most glasses can be grinded but not be effectively molded because of a too-high melting temperature and other issues.
3. The optical quality of molded lenses is usually lower than the optical quality of ground lenses.

To produce large-quantity plastic lenses, molding is the best choice since plastic materials cannot be ground. To produce large-quantity glass aspheric lenses, molding is the often the best choice because diamond turning is much more expensive and the individual production process is very slow. However, there are several unique issues associated with the molding process:

1. Material melting temperature. Most glasses have too-high melting temperatures. The thermal contraction and even the thermal refractive index change of glasses during the cooling process can be too large. Before starting to design aspheric glass lenses for molding, the lens designer should contact molding lens vendors for a list of glasses that the vendors feel comfortable to mold. All plastic materials have a melting temperature low enough for molding.
2. The molding method cannot produce surfaces with strong aspheric profiles, which is not a problem for the diamond turning method, because of the uneven thermal expansion/contraction during the molding process. Figure 14.2 illustrates this issue. After selecting the moldable

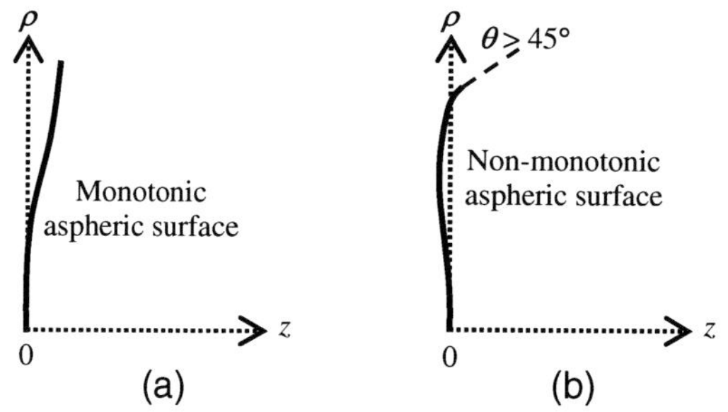

Figure 14.2 ρ is the lens surface radial coordinator, *z* is the surface sag, and 0 marks the origin of the coordinator. (a) A moldable monotonic glass aspheric surface. (b) A non-monotonic aspheric surface with a large slope angle. This aspheric surface is not suitable for glass molding but can be molded for plastic materials. This surface can be produced by diamond turning glass.

glasses, lens designers should show molding lens vendors their design to confirm that such a lens can be molded.

3. There are other unique limitations to the molding method, such as maximum size, maximum flank angle, radius of curvature range, minimum edge thickness, etc. Different vendors may have slightly different molding limitations, and the limitations may be relaxed as the technology progresses. Reference 3 provides further details.

14.3 Fabrication Tolerances

Nothing is perfect in the world. Every lens fabricated has certain tolerances, some of which are unique to lenses. Lens fabrication tolerances affect lens performances and are discussed in this section.

14.3.1 Tolerances of the surface curvature radius and surface sag difference

The tolerance of a lens surface curvature radius is one of several tolerances that must be specified. Since the surface radius of curvature tolerance and sag difference are equivalent parameters, radius tolerance is often specified as sag difference. However, the relations between these two parameters may not be clear at first glance, may cause confusion, and is thus discussed here.

From Fig. 14.3, the following relations can be written

$$2A + B = \pi \rightarrow \frac{B}{2} = \frac{\pi}{2} - A, \tag{14.1}$$

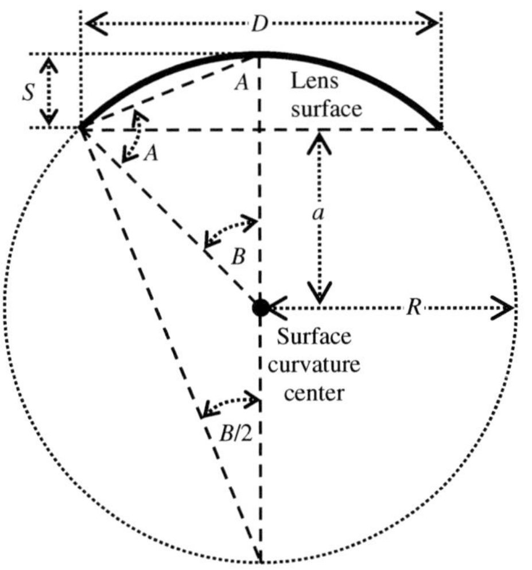

Figure 14.3 Illustration of the relation between lens diameter *D*, surface radius of curvature *R* and surface sag *S*. The thick solid curve represents the lens surface. The dot line circle is the extension of the lens surface to facilitate the illustration.

$$\frac{S}{D/2} = \cot(A) = \tan\left(\frac{\pi}{2} - A\right) = \tan\left(\frac{B}{2}\right), \tag{14.2}$$

$$\frac{D}{2(R+a)} = \tan\left(\frac{B}{2}\right), \tag{14.3}$$

$$a = \left(R^2 - \frac{D^2}{4}\right)^{0.5}, \tag{14.4}$$

where D, R, and S are the lens diameter, surface curvature radius, and sag, respectively; and a, A, and B are some length and angles, as marked in Fig. 14.3. Equations (14.1)–(14.4) are combined to eliminate a, A, and B, and S can be expressed in terms of R and D as

$$S = \frac{D^2}{4R\left[1 + \left(1 - \frac{D^2}{4R^2}\right)^{0.5}\right]}. \tag{14.5}$$

Differentiation of Eq. (14.5) results in

$$\Delta S = -\frac{D^2}{4R}\left(\frac{\Delta R}{R}\right)\frac{1 + \left(1 - \frac{D^2}{4R^2}\right)^{0.5}}{\left[1 + \left(1 - \frac{D^2}{4R^2}\right)^{0.5}\right]^2\left(1 - \frac{D^2}{4R^2}\right)^{0.5}}, \tag{14.6}$$

where ΔR is the radius of curvature tolerance, and ΔS is the sag tolerance. Note that ΔS is approximately proportional to $\Delta R/R^2$, not to $\Delta R/R$. Equation (14.6) is an accurate expression, but it may not be convenient to use. Simplifications are used for different $D^2/(4R^2)$ values.

For $D^2/(4R^2) \ll 1$, Eq. (14.6) reduces to

$$\Delta S \approx -\frac{D^2}{8R}\left(\frac{\Delta R}{R}\right). \tag{14.7}$$

Equation (14.7) is widely cited but often without clarification that it is valid only for $R \gg D$.

In Fig. 14.3, $D^2/(4R^2) \approx 0.5$. Inserting $D^2/(4R^2) \approx 0.5$ into Eq. (14.6) results in

$$\Delta S \approx -\frac{D^2}{4.83R}\left(\frac{\Delta R}{R}\right). \tag{14.8}$$

Equations (14.7) and (14.8) have a difference of about 66%, which means that Eq. (14.7) has a quite a large error for small-F-number lenses.

If taking the first-order approximation $[1 - D^2/(4R^2)]^{0.5} \approx [1 - D^2/(8R^2)]$, then after inserting it into Eq. (14.5) and differentiating it, the result is

$$\Delta S \approx -\frac{D^2}{8R}\left(\frac{\Delta R}{R}\right)\frac{1+\frac{D^2}{16R^2}}{\left(1-\frac{D^2}{16R^2}\right)^2}. \tag{14.9}$$

For $D^2/(4R^2) = 0$, Eq. (14.9) reduces to Eq. (14.7). For $D^2/(4R^2) \approx 0.5$, Eq. (14.9) results in

$$\Delta S \approx -\frac{D^2}{5.44R}\left(\frac{\Delta R}{R}\right), \tag{14.10}$$

which is about 13% smaller than the accurate result of Eq. (14.8), which may be acceptable from a practical point of view. Equation (14.9) is not very complex and can be used to replace Eq. (14.6).

Figure 14.4 provides an intuitive view of the errors contained in Eqs. (14.7) and (14.9), respectively. The errors can be rapidly estimated for any given values of $D^2/(4R^2)$. The figure shows that the widely used Eq. (14.7) contains a large error in most cases.

The surface curvature radius of a lens can be tested by comparing the surface with a standard surface under illumination of a coherent light. The reflections from the two surfaces form interference fringes. The fringe number N is related to the sag difference between the two surfaces by

$$N = \frac{2\Delta S}{\lambda}. \tag{14.11}$$

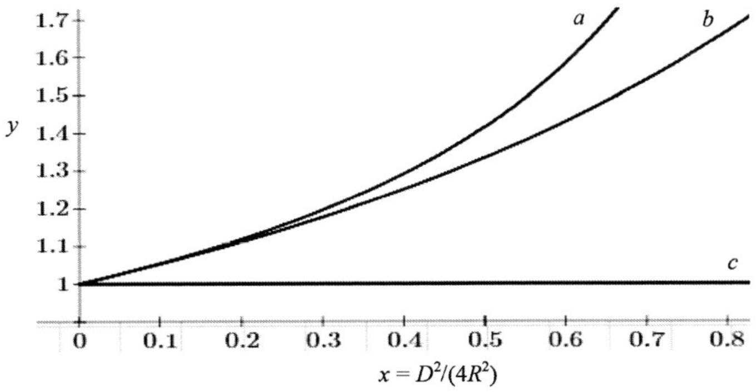

Figure 14.4 Illustration of the errors contained in Eqs. (14.7) and (14.9). Curve *a*: $y = (1 - x)^{-0.5}$ represents the accurate result from Eq. (14.6). Curve *b*: $y = (1 - 0.5x)^{-1}$ represents the first-order approximation result from Eq. (14.9). Curve *c*: $y = 1$ represents the inaccurate result from Eq. (14.7).

By counting the fringe number, the difference of sags and thereby the difference of the radii of curvature between the two surfaces can be found. Section 14.5 will further discuss this topic.

14.3.2 Decenter, tilt, and wedge between the two surfaces of a lens

The two surfaces of a real lens will always have certain amount of decenter d that equals the wedge angle θ or the sag difference ΔS between the two surfaces, as illustrated in Fig. 14.5.

With a decenter magnitude d, the left- and right-side sags of one surface are different given by Eqs. (14.12) and (14.13), respectively:

$$S_1 = \frac{(D-d)^2}{4R\left[1 + \left[1 - \frac{(D-d)^2}{4R^2}\right]^{0.5}\right]}, \tag{14.12}$$

$$S_2 = \frac{(D+d)^2}{4R\left[1 + \left[1 - \frac{(D+d)^2}{4R^2}\right]^{0.5}\right]}. \tag{14.13}$$

Current lens fabrication technology can guarantee $d \ll D$. Then, d in the denominators of Eqs. (14.12) and (14.13) can be neglected, and the sag difference can be calculated as

$$\delta S = S_2 - S_1 = \frac{Dd}{R\left[1 + \left(1 - \frac{D^2}{4R^2}\right)^{0.5}\right]}. \tag{14.14}$$

Equation (14.14) is accurate enough and simple; there is no need to further simplify it.

Note that δS is different from ΔS discussed previously. The latter is the difference between the two sags of one decentered surface, and the former is the sag difference between a reference surface with an accurate radius of curvature and the surface with an inaccurate radius of curvature.

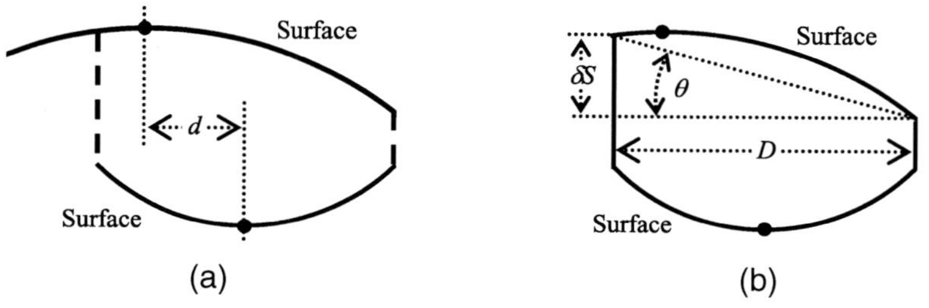

Figure 14.5 (a) The two surfaces of a lens have a decenter d between them. (b) After the lens edge is trimmed around the center of the lower surface, as marked by dash lines in (a), the decenter appears as a wedge angle θ or a sag difference δS between the two surfaces.

Also note that R in Eq. (14.14) is the radius of curvature of the decentered surface, the lens is trimmed about the other surface of the lens during the centering process. For example, in the case shown in Fig. 14.5, R is the radius of curvature of the upper surface, and the lens is trimmed about the lower surface.

The wedge angle and the sag difference is simply related by

$$\theta = \frac{\delta S}{D} = \frac{d}{R\left[1 + \left(1 - \frac{D^2}{4R^2}\right)^{0.5}\right]}. \tag{14.15}$$

14.3.3 Optical fabrication tolerance chart

Different optical component manufacturers have slightly different fabrication tolerance charts, if not the same. These differences are insignificant. Generally speaking, there are three grades of tolerances: commercial, precision, and high precision. Table 14.1 is a typical optical fabrication tolerance chart that includes all of the commonly seen tolerances. The chart includes the following notes:

A. The glass index and Abbe number of every production run can be slightly different. The "melt data" approach adjusts the lens surface curvature radius and all other optical parameters based on the index and Abbe number of a specific production run to minimize the negative effects of the index and Abbe number tolerance. The optics involved in the melt data approach means that every product run will be slightly different, which will create a lot work and therefore should not be used for products that will be continuously produced in the future.

Table 14.1 A typical optical fabrication tolerance chart.

Parameters	Commercial	Precision	High Precision
Glass index, Abbe number	±0.001, ±0.8%	±0.0005, ±0.5%	Melt data[A]
Diameter (mm)[B]	+0.00/−0.10	+0.000/−0.025	+0.000/−0.015
Center thickness (mm)	±0.15	±0.050	±0.025
Sag (mm)	±0.050	±0.025	±0.015
Clear aperture[C]	80%	90%	See note
Radius (larger of two)[D]	±0.2% or 5 fringes	±0.1% or 3 fringes	±0.05% or 1 fringe
Surface irregularity (RMS)[E]	2 fringes	0.5 fringe	0.2 fringe
Lens wedge (edge thickness difference, mm)	±0.05	±0.01	±0.005
Prism wedge (TIA, arc min)	5	1	0.5
Bevel (mm, face width at 45°)	<1.0	<0.5	<0.5
Scratch (μm)-dig (10 μm)	80-50	60-40	20-10
AR coating (residual reflectance)	Single layer <1.5%	Broadband <0.5%	V-coat <0.2%
Multilayer coatings[F]		Consult coating vendors	

B. The diameter tolerance usually has the form of +0/−x.xxx; the +0 is used to avoid the possibility that the lens is too large to fit inside the lens housing.

C. The clear aperture diameter is a certain percentage of the whole diameter. The part of the lens outside the clear aperture is for holding it during the fabrication and/or coating process, so its optical quality can be reduced. The percentage definition used in Table 14.1 is a little too simple. For example, a 1-mm edge outside the clear aperture is about the minimum for properly holding or mounting the lens. For a small lens with a 3-mm radius, the largest clear aperture is only $(3 - 1)/3 = 67\%$. For a lens with a 5-m radius, the largest clear aperture is 80%. For a larger lens with a 50-mm radius, the minimum edge should be increased to ≥3 mm. Then, the largest clear aperture becomes $(50 - 3)/50 = 94\%$. Therefore, for the same tolerance grade, smaller lenses have a smaller-percentage clear aperture.

D. Tolerance of surface radius of curvature can be specified as the percentage radius tolerance $\Delta R/R$ or fringe number N, as discussed in Section 14.3.1 and will be discussed further in Section 14.5. Since N and $\Delta R/R$ are not equivalent, the larger of the two is used in the chart.

E. Surface irregularity can be specified using the RMS value or P-V value. Generally speaking, the RMS value is about 1/5 of the P-V value. Surface irregularity can be measured by using either interferometers (the result is "fringes") or profilometers (the result is microns). Because the test light, which is usually a He-Ne laser beam, must make one round trip to interfere, 1 fringe $= \lambda/2$ or 0.5 wave ≈ 0.3 μm. Section 11.2 explains more details about how to use interferometers to measure the wavefront (optical surface) quality.

F. Multilayer thick coatings are sometime used to push the limits of the performance. These coatings have low production yields, and producing these coatings is more difficult than designing them on paper. Coating customers/users should anticipate this in advance.

14.4 Purchasing Optical Components

14.4.1 Off-the-shelf optical component vendors

The business of optical industry appears to be relatively stable compared with other sections of the high-tech industry. The US-based, off-the-shelf optical component vendors listed in Table 14.2 have been in business for more than twenty years. Their websites provide information and prices of thousands of various types of optical components. Most of these components are available for immediate shipping to customs. These vendors also provide opto-mechanical components. The list in Table 14.2 is not complete; there are tens of smaller vendors to choose from. The prices and qualities of the optical

Table 14.2 Major US-based vendors for off-the-shelf optical components.

Number	Name	Website
1	Edmund Optics	www.edmundoptics.com
2	Thorlabs	www.thorlabs.com
3	Melles Griot	www.mellesgriot.com
4	Newport	www.newport.com
5	CVI Laser Optics	www.cvilaseroptics.com
6	OptoSigma	www.global-optosigma.com

components provided by these vendors are similar. The main issue is to find the right components to meet the requirements.

14.4.2 Custom component vendors

If the off-the-shelf optical components do not meet special requirements, custom optical components should be used, which will involve a lot of additional work, e.g., designing the components with all of the tolerances specified, preparing data sheets for vendors, obtaining a price and delivery quote, and waiting a long time (eight weeks or so) for delivery. If the custom optics has tight tolerances or the custom coatings have many layers, the production yield can be low. As a result, the vendor must rework the parts that fail to meet the specifications and suffer a delayed delivery. Late deliveries are bound to happen occasionally, and so one should try to avoid very tight schedules when planning. Table 14.3 is an incomplete list of vendors who are actively pursuing custom optical component business.

14.4.3 Custom component price estimation

The price structure of custom optical components is complex:

1. Tolerance associated price. For a given lens, each step up in tolerance grade raises the cost of fabrication by about 50%. However, a lens can

Table 14.3 Major US-based vendors for custom optical components. A: Specialized in fast delivery of prototypes at a higher price. B: Good at coatings. C: Good at making large-quantity custom optical components.

Number	Name	Website
1	Optimax Systems^{A}	www.optimaxsi.com
2	Rainbow Research Optics	www.rr-optics.com
3	Rocky Mountain Instruments	www.rmico.com
4	Melles Griot	www.mellesgriot.com
5	Newport	www.newport.com
6	CVI Laser Optics^{B}	www.rmico.com
7	OptoSigma	www.global-optosigma.com
8	Edmund Optics^{C}	www.edmundoptics.com
9	Thorlabs^{C}	www.thorlabs.com

have a mixed tolerance. For example, the central thickness is ±0.15 mm in the commercial grade, whereas the surface irregularity is 0.2 fringe in the high-precision grade, and the price of such a lens is between the commercial and precision grades.

2. Quantity breaks. These are no simple rules for quantity price breaks. It depends on the sizes of the lens, the polish machine, and/or the coating chamber used to produce the lens. Any quantities that can be fit into one production run will likely have a better per-piece price.
3. Roughly speaking, the first quantity price break is at 10–20 pc. The second quantity price break is at 50–100 pc. The per-piece price for a large quantity (hundreds) will drop to about half of the per-piece price of a few pieces. For large quantities, the custom optical component price is close to the price of off-the-shelf optical components with similar specifications.
4. Optical material. Off-the-shelf optical components usually use inexpensive, widely used optical materials. Custom optical components may use expensive optical materials. The material price difference can be more than tenfold. The material cost is a small fraction of the total cost. However, with a tenfold higher material cost, the total price can be doubled.
5. There is often a nonrecurring tooling cost and setup cost added to the sale price.
6. Fast delivery cost. Nowadays, the standard delivery of custom optical components is about eight weeks. This is relatively long compared with the delivery of custom made metal parts. Vendors, particularly small vendors, may be willing to charge a premium of 20% or so to speed up the delivery by a few weeks. Some vendors do not want to charge extra for fast delivery because it may complicate their production schedules.

A very rough estimate of the custom-component price for small quantities is about three times higher than the price for similar off-the-shelf optical components.

When ordering custom optical components, it is common to order an extra one in case the user damages one while handling the components. A custom optical component vendor usually produces a couple extra in case one is damaged during handling. If the user has damaged more than one custom component, call the vendor and ask if there are any extra parts in stock.

14.4.4 Lens drawings and data sheet for custom components

It is nice to prepare a formal lens drawing for a vendor to make a custom lens, but it is not necessary to spend a lot of time and effort to do so in order to meet the industrial specifications and standards. A simple lens drawing generated by Zemax and a list of all specifications, such as the one shown in

Table 14.4 A typical specification sheet for a custom lens. A: The tolerance of the lens-surface radius of curvature can also be specified by the "lens unit," or millimeters in this case.

Names	Specifications
Lens type	Single lens
Lens shape	Spherical convex-convex
Glass	CDGH H-F2
Index, Abbe number	$\leq \pm 0.0005$, $\leq \pm 0.8\%$
Center thickness	7 ±0.05 mm
Outside diameter	25 +0/−0.1 mm
Front-surface clear aperture	≥22 mm
Back-surface clear aperture	≥20 mm
Front-surface radius of curvature[A]	100.000 mm ± 0.5%
Back-surface radius of curvature[A]	67.330 mm ± 0.5%
Wedge	<0.01 mm
Surface irregularity	<0.5 wave
Scratch and dig	60-40
Bevel	1 mm @ 45°
AR coating	Average <1% for 400–700 nm

Table 14.4, are sufficient. Even lens drawings are not necessary—several lines of text that describe each parameter and the tolerance of the lens are acceptable by most custom lens vendors.

A Zemax lens drawing can be created by opening the Zemax file of the lens to be drawn and then clicking *Analyze/System Viewers/Zemax Element Drawing* so that a lens drawing will appear. Figure 14.6(a) shows a drawing of the first element of the double Gauss lens, which is plotted several times in the book such as in Fig. 12.9(b). Most lens parameters will be automatically displayed in the drawing, such as radius of surface curvature, central thickness, glass type etc. If tolerance analysis has already been performed, there will be tolerance data saved in *Tolerance/Tolerance Data Editor* box. Zemax will copy the relevant tolerance data to this drawing, such as radius of surface curvature tolerance, surface irregularity tolerance, central thickness tolerance etc.

The drawing can be completely edited by clicking the *Setting* button in the upper-left corner of the drawing. A new box will appear, as shown in Fig. 14.6(b), where one can select which element to draw, manually change or add any parameter and tolerance numbers, and revise or add to the note by clicking the *Edit Note File.*

14.4.5 Review inspection report

Few users have the capability to test their custom lenses, and so they must trust the vendors. Fortunately, the vast majority of them live up to the expectations.

As a common practice, vendors will inspect a couple of randomly selected sample lenses during every production run and provide an inspection report for all of the parameters that are requested by the customer and agreed to be

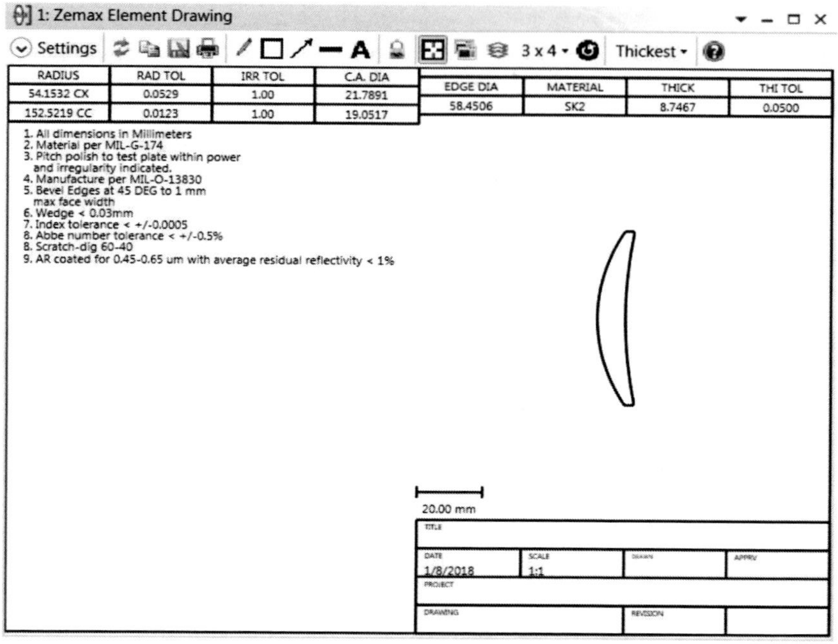

(a)

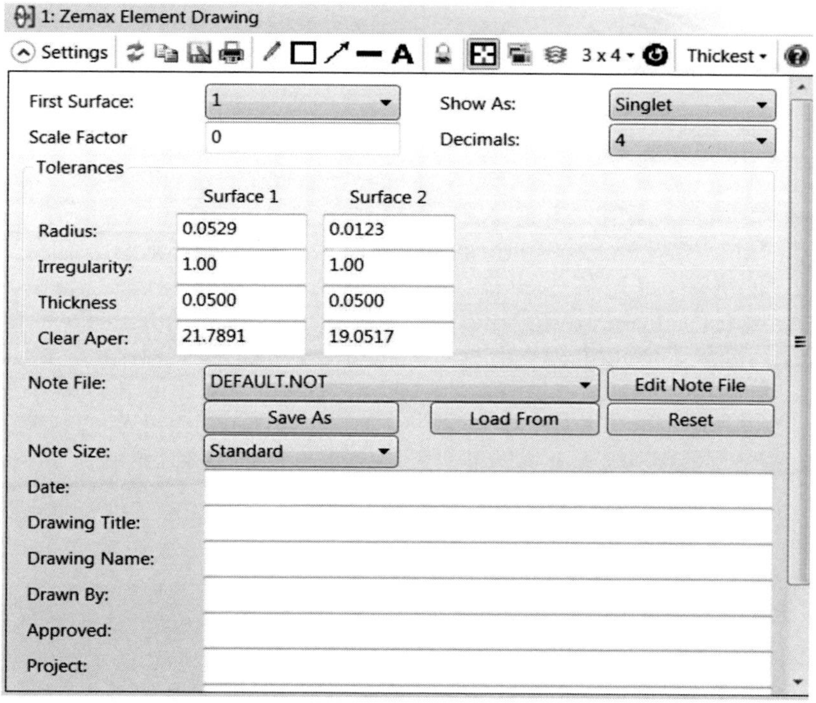

(b)

Figure 14.6 (a) Zemax-generated lens drawings of the first element of the double Gauss lens. (b) The lens drawing can be completely edited.

met by the vendor. The report should include real measured numbers for all applicable parameters, such as thickness, diameter, etc.; a scan (reflectance or transmittance) of the coating over the specified wavelength range; an interferogram for surface quality; the index and Abbe number values of the glass used; and some check marks to confirm that the lens passes inspection for all applicable parameters, such as scratch and dig.

The inspected lenses will be numbered for tracing back when needed. Custom-lens users can pay a little extra to have the vendor inspect each lens.

14.4.6 Incoming inspection

All off-the-shelf optical components are fabricated in large quantities. The fabrication and inspection processes are well set, and the quality is stable. Most custom optical components are made for small quantities, and the quality varies. Most users of custom optics do not have the equipment and expertise to perform a serious inspection of the incoming custom optical components. All they can do is a visual inspection, mainly of the cosmetic aspects, such as scratches, digs, chips, bubbles in the glass, and other defects.

Scratches, digs, etc. can be inspected using an eyepiece with a paddle on it or a measuring microscope with a reticle on it to estimate the size and length of the scratches and digs. Such an inspection is often not accurate but rather an estimate. Arkens[4] provides an overview of scratches and digs.

Coatings can be visually inspected for the following:

1. Color, if the correct color of the coating is known. In many cases, the color should be "complementary," e.g., a longpass coating with a cutting wavelength at red. The reflection color of the coating would then look bluish, and the transmitted light would look red.
2. Pinholes, voids, dust, and stains. Commonly used specifications do not cover these defects, so common sense is required. Some traces of these defects are normal and acceptable.
3. Cracks and peel-off. Many-layers-thick coatings often have tensions and stresses among the layers that can cause cracks and peel-off in the coatings. These defects can be found by using an eyepiece or microscope to visually inspect the coating. These defects can occur after the vendor inspected the coating since such coatings are fragile, so the discovery of these defects does not necessarily mean the vendor did a poor job.

14.5 Measuring the Surface Radius of Curvature, Surface Quality, and Lens Physical Size

The inspection and test of optical components are mainly about inspecting and testing optical surfaces, either lens surfaces or mirror surfaces. Spherical surfaces are the most commonly used surfaces since flat surfaces can be considered as spherical surfaces with infinitely large radii of curvature. References 5 and 6 are recommended for a comprehensive review.

14.5.1 Profilometers

There are two types of profilometer: traditional, contact-based mechanical profilometers, and non-contact optical profilometers.[7]

A contact profilometer has a stylus with an indicator connected to it. When measuring, the stylus gently touches and scans the optical surface under measurement. The indicator indicates the 2D position of the stylus along the lens surface. The surface radius of curvature is thereby measured. The scan is usually made along two orthogonal diameters across the surface to reduce the possible measurement errors and offer a more complete profile of the surface.

Profilometers have a stylus with a tip radius from 20 nm to 50 μm and can measure surface features from 10 nm to 1 mm. The advantage of contact profilometers is that their vertical (perpendicular to the lens surface) resolution is higher than non-contact profilometers. The disadvantages of contact profilometers are that the entire surface cannot be profiled, and the surface (particularly polymer surfaces) may be scratched.

Non-contact profilometers utilize various optical techniques, such as triangulation sensors, confocal microscopy (to profile very small objects), interferometry, and digital holography. (All of these techniques are very specialized and beyond the scope of this book.) The advantages of non-contact profilometers are that they can easily profile the entire lens surface, their lateral (along the lens surface) resolution can be as high as submicrons, and they will not damage the surface under measurement. The disadvantages are that non-contact profilometers are usually more expensive, more complicated to operate, and less flcxible.

14.5.2 Test plates

Every lens manufacturer has kept many test plates, which are high-quality lenses. The radii of curvature of these test plates are precisely known and spread over a large range from a few millimeters to over one meter. After the design of a custom lens is completed, the two surface radii of curvature should be adjusted to be the same as the radii of curvature of the two test plates, respectively, of the selected lens manufacturer.

After the custom lens is fabricated, its real radii of curvature will be slightly off the designed values because of fabrication tolerances. The lens can be tested by contacting the lens surface under test against the surface of the test plate and let a collimated coherent light pass through the two optical components, as illustrated in Fig. 14.7(a). Interference rings (Newton's ring) will appear. The relation between the ring number and the delta of the two surface radii of curvature is discussed in detail in Section 14.3.1. Based on the

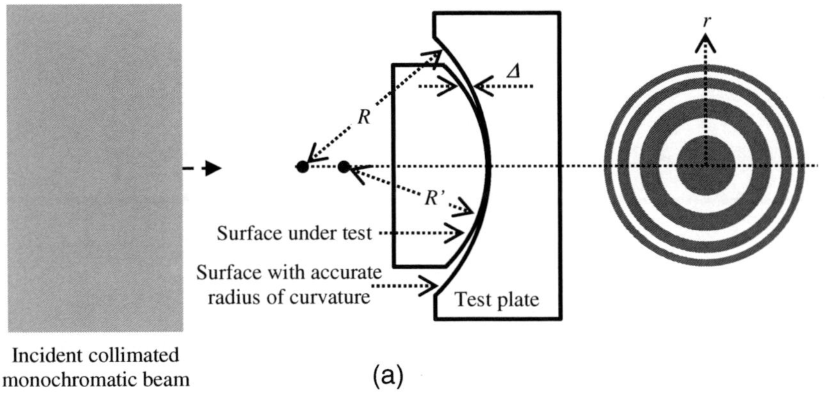

(a)

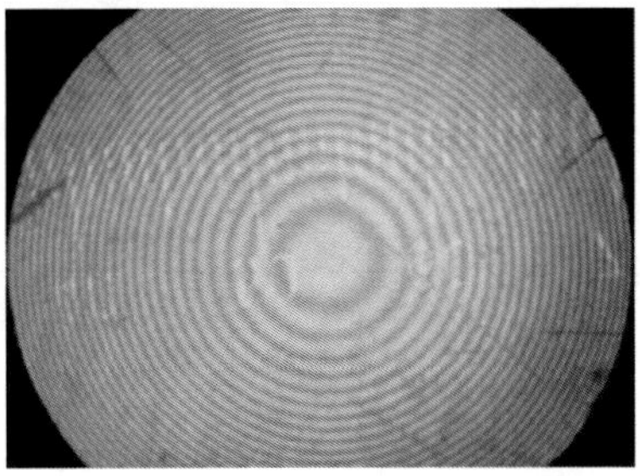

(b)

Figure 14.7 (a) Illustration of a test plate used to test the surface radius of curvature of a lens. The difference between the two radii is exaggerated here for clarity. A collimated coherent illumination light passes through the two components. Interference fringes can be observed from another direction. (b) A photograph of a real Newton's ring (reprinted courtesy of Warrencarpani (own work) [CC0], via Wikimedia Commons).

interference ring number, the surface radius of curvature of the custom made lens can be calculated. Figure 14.7(b) shows a real photo of a Newton's ring.

14.5.3 Interferometer methods

Interferometers are widely used to test the surface radius of curvature as well as surface quality of lenses and mirrors. Various types of interferometers have been developed. Figure 14.8(a) shows a simple schematic of a setup that uses a Twyman–Green interferometer to measure a spherical surface. A collimated coherent beam is split in two. One beam serves as the reference beam and is not shown; another beam serves as the test beam and is focused by a high-quality lens onto the optical surface under test. The focal point of the lens and the center of the surface under test should coincide. The test beam is reflected by the surface under test, traces its way back and is collimated by the focusing lens, and then is combined with the reference beam to interfere. The radius curvature and other qualities of the surface under test can be found in the interferogram.

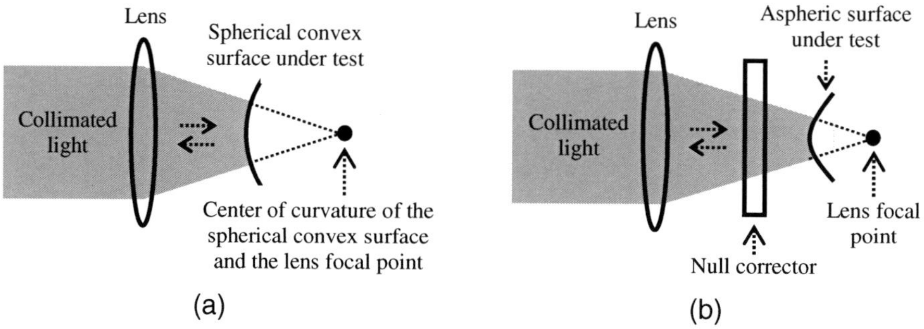

Figure 14.8 (a) The setup uses a Twyman–Green interferometer to test a spherical convex surface. (b) With a properly designed null corrector being placed in the optical path, the setup can be used to test an aspheric surface.

Note that Fig. 11.19(a) provides a more detailed description of this technique, and Section 11.2 briefly introduces a few commonly used interferometers that can be used to test optical surfaces. References 7–11 in Chapter 11 provide more detailed descriptions about interferometers.

14.5.4 Test aspheric surfaces

The immediate choice for the testing of aspheric surfaces is a profilometer, as discussed in Section 14.5.1. Interferometry, such as a Twyman–Green interferometer, can be used with a null corrector to test an aspheric surface, as shown in Fig. 14.8(b). A null corrector is a multi-element lens. The spherical convergent wave passes through the null corrector and is incident on the aspheric surface under test. The wave reflected by the aspheric surface passes through the null corrector again and interferes with a reference wave. The null corrector is designed so that the effects of passing twice through the null will correctly transform the wavefront reflected by the aspheric surface back to a spherical wave with the same radius of curvature. If the aspheric surface profile is not exactly as it should be, there will be interference fringes. The interpretation of such fringes is very complex and is a special topic. One null corrector works only for one specific aspheric surface.

14.5.5 Measuring surface quality

Optical surface quality is specified in different ways by different standards,[8] but for optics users, quality issues refer to all surface imperfections, and there is no need to dig into these standards.

Optical surface quality is commonly defined as the RMS or the P-V deviation of a real optical surface from the ideal surface. Generally speaking, the RMS value equals 1/5 of the P-V value.

Optical surface quality can be measured by using interferometers or profilometers. When an interferometer is used, an interferogram will be generated, and the measurement result is in terms of fringes or waves. 1 fringe = 0.5 wave because the test light must make one round trip to interfere. The light used to produce interference fringes is usually He-Ne laser light with a wavelength of 632.8 μm. A visual review of the interferogram can only find the P-V values of the quality of the surface under measurement. The computer and software associated with the interferometer should be able to calculate the corresponding RMS value based on the fringe shapes.

When a profilometer is used, the measurement result is also the P-V value in terms of microns. Since a profilometer can only measure the surface quality in a very limited area, converting such a measured P-V value to a RMS value does not make much sense. Also, most profilometers do not have suitable software for the task.

A common optical surface quality ranges from about 0.2 fringes for high-precision-grade optics to about 2 fringes for commercial-grade optics.

14.5.6 Measuring physical size

The physical size of an optical component, such as a lens, includes its diameter, central thickness, and sometime edge thickness. All of these parameters can be measured using a proper caliper or gauge. The process is simple and straightforward.

The wedge of a lens or glass plate can be considered in the physical size category, and can be measured using a caliper/gauge or using an interferometer. When a caliper/gauge is used to measure wedge, the edge thickness along the circumference is measured. The largest edge thickness difference divided by the diameter of the component is the wedge.

When an interferometer is used, the measurement principle and method are the same as measuring the surface quality. A collimated light wave is incident on the optical component in a direction perpendicular to one surface of the component. The reflected waves from the two surfaces of the optical component has a small angle between them because of the wedge. The two reflected waves interfere each other and form straight fringes. The thickness difference Δ of the two edges of the component can be determined by $\Delta = m\lambda/(2n)$, where m is the fringe number, λ is the test light wavelength, and n is the refractive index of the component under test. The wedge under measurement can then be calculated by the ratio of the thickness difference over component size.

14.6 Methods for Measuring the Lens Focal Length

Several methods can be used to measure the lens focal length. For example, focal length can be indirectly measured by measuring the radii of two surface curvatures of the lens using an interferometer or a profilometer as described in Section 14.5 and calculating the focal length using Eq. (1.10).

Morel[9] provides an overview about lens focal length measurement. The section titled "Nodal Testing" in Ref. 10 provides an overview of measuring the most important properties of lenses, including the length focal length. This section discusses two methods of measuring the lens focal length.

14.6.1 Transverse magnification method

The setup of this measurement method is illustrated in Fig. 14.9. An object (reticle) is placed at the focal plane of a high-quality collimation lens and is illuminated by a single-wavelength light source. The lens under measurement forms the image of the reticle. In such a setup, the object and image distances are f and f', respectively, where f and f' are the focal lengths of the collimating lens and the lens under measurement, respectively. f' can be found from the lateral magnification relation

$$f' = f\frac{h'}{h}, \tag{14.16}$$

where h is the size of the reticle, h' is the size of the reticle image, and h'/h is the lateral magnification. Both f and h are known before measurement, and h' can be measured by a microscope objective.

This method is effective only under paraxial condition $h \ll f$. Otherwise, the aberration and distortion will introduce large errors to the measurement.

14.6.2 Nodal slide method

The measurement setup is illustrated in Fig. 14.10(a). A pinhole is illuminated by a single-wavelength light source and is the object point. A high-quality lens collimates the light from the pinhole. The lens under measurement focuses the light on its focal point. All of the parts are mounted on a rail with the optical axis along the rail. The back focal length BF of the lens is easy to measure, but the true focal length is measured from the focal point to the nodal point, which is usually somewhere inside the lens.

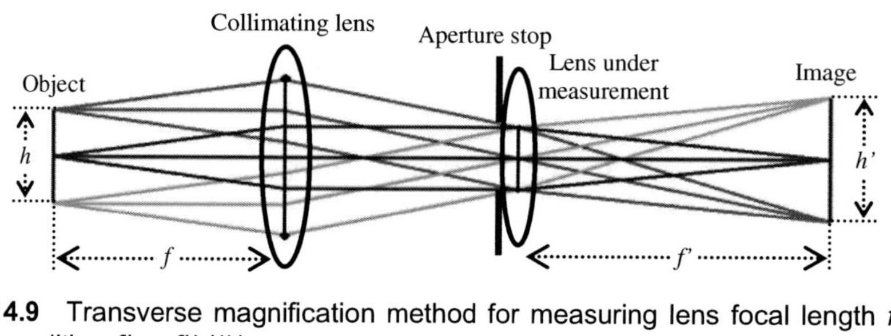

Figure 14.9 Transverse magnification method for measuring lens focal length f'. Under paraxial condition $f' = f(h'/h)$.

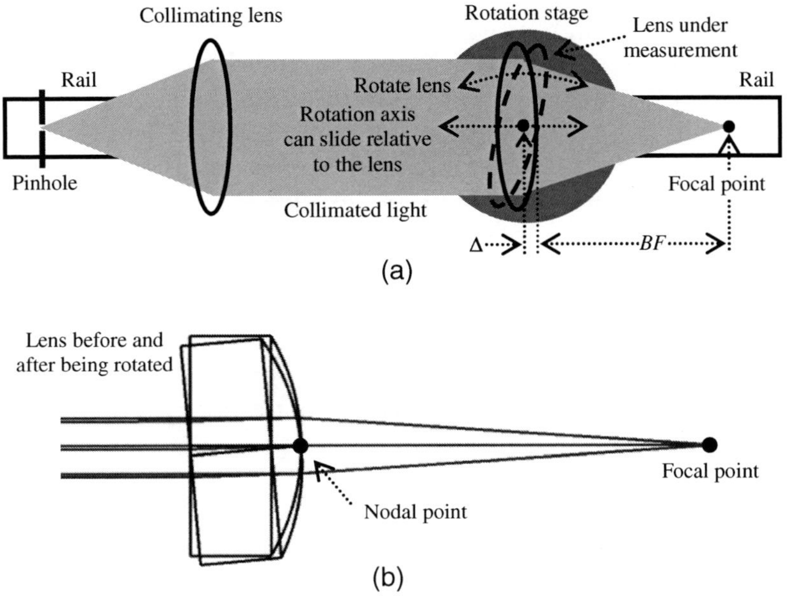

Figure 14.10 (a) Nodal slide method for measuring the focal length f' of a positive lens. Once the rotation axis coincides with the nodal point, rotating the lens does not move the focal point position, and $f' = \Delta + BF$. (b) Two Zemax-generated raytracing diagrams plotted for a positive lens before and after being rotated about the nodal point. The focal point stays at the same location.

The lens under measurement is mounted on a rotating stage. The rotation axis position relative to the lens can vary. The measurement process is as follows:

1. Find the focused spot and measure *BF*.
2. Slightly rotate the lens and check if the focused spot moves.
3. If the focused spot moves, then slightly change the rotation axis position relative to the lens, and repeat step 2.
4. Repeat steps 2 and 3 until a rotation axis position is found where slightly rotating the lens does not change the focused spot position, i.e., the rotation axis coincides with the lens nodal point.
5. The rotation axis position can be measured as the distance Δ to the lens back surface. The focal length f' under measurement is given by $f' = \Delta + BF$.

Based on the shape of the lens under measurement, the nodal point position can be roughly determined before the measurement, so it does not take long to find the nodal point position.

Figure 14.10(b) shows accurate raytracings for a positive lens before and after rotating about its nodal point, which is at the vertex of its back surface.

14.6.3 Measuring the focal length of a negative lens

The two measurement methods described previously measure the focal length of positive lenses. Little of the literature discusses the method for measuring the focal length of negative lens, although negative lenses are of equal importance to the positive lenses. A possible reason for this situation is that the measurement of the focal length of negative lenses is more complex than measuring the focal length of positive lenses.

The negative lens focal length can also be indirectly measured by measuring the radii of two surface curvatures of the lens using an interferometer or a profilometer, as described in Section 14.5, and calculating the focal length using Eq. (1.10).

There are simpler methods to measure the focal length of a negative lens. Figure 14.11(a) shows one of them. A high-quality positive lens focuses light from a single-wavelength point source. The location of the focal point is recorded, and then the negative lens under measurement is placed in the optical path. The light is still focused since the focal length of the positive lens is selected to be shorter than the focal length of the negative lens. The distances between the back surface of the negative lens and the two focal points can be measured as o and i, respectively, as marked in Fig. 14.11(a). The negative lens is mounted on a rotation stage that is the same as the one used to measure the focal length of positive lenses, as shown in Fig. 14.10(a). The rotation axis position is varied, and the lens is slightly rotated until the nodal point position Δ is found. The focal length f' under measurement can be calculated using the thin lens equation as

$$\frac{1}{f'} = -\frac{1}{o+\Delta} + \frac{1}{i+\Delta}, \tag{14.17}$$

where $o + \Delta$ and $i + \Delta$ are the object and image distances, respectively, and the negative sign in front of the $o + \Delta$ term appears because the object point is at the right side of the negative lens and is a virtual object.

14.7 Methods for Measuring the Image Resolution of a Lens

The concepts of the contrast transfer function (CTF) and modulation transfer function (MTF) were discussed in Section 1.13. Reference 10 provides a general discussion about how to measure them. Reference 11 provides a general discussion of the physics involved. Figure 14.12 provides an illustration of resolution measurement targets, images formed by a lens, and the CTF or MTF of the image. Several resolution charts are used to measure the CTF or MTF of an image formed by lens, three of which are described in this section.

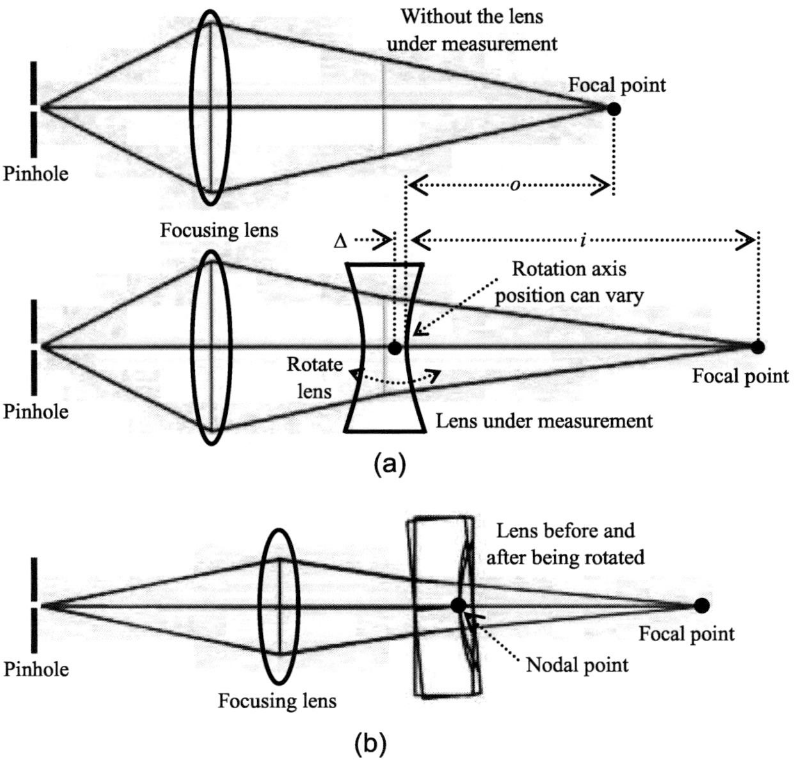

Figure 14.11 (a) Nodal slide method for measuring the focal length f' of a negative lens. A high-quality positive lens focuses a point source, and then the negative lens under measurement is placed in the optical path. The negative lens is mounted on a rotation stage with a slide axis the same as that shown in Fig. 14.10(a). (b) Two Zemax-generated raytracing diagrams plotted together for a negative lens before and after being rotated about its nodal point. The focal point stays at the same location.

14.7.1 USAF resolution test chart

A USAF resolution test chart was plotted in Fig 1.31. When measuring, the lens under measurement forms the image of the test chart onto a 2D detector array, and software or computer code process the image and calculate the CTF of the image.

The target brightness of the black-and-white bars heavily depends on the illumination level, which affects the brightness of the bars in the image. The sensor background noise and the possible scattering light in the lens under measurement also affect the brightness of the bars. Thus, the black bars may not be very black, and the white bars may not look very white. The measurement accuracy will be reduced. When processing the image, the software takes the central portion of the big black square (see Fig. 1.31) as the black reference and takes the central portion of a big white area as the white reference. Then, the software goes through all the black-and-white bars and compares them with

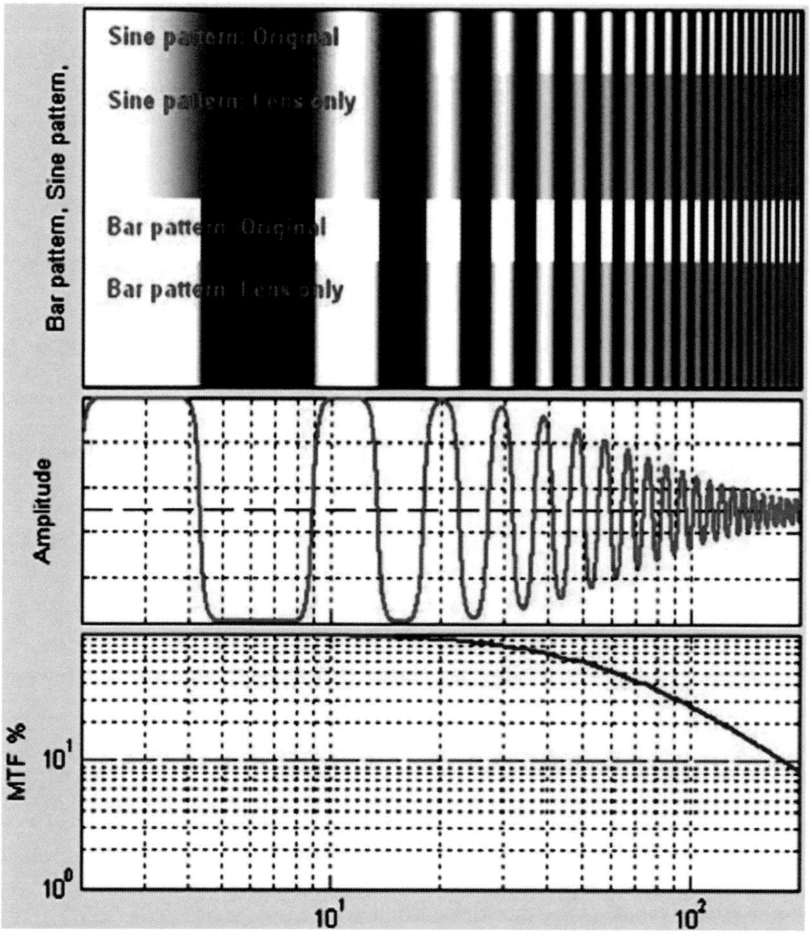

Figure 14.12 Illustration of measuring CTF and MTF (image reprinted courtesy of Imatest LLC). The top row (sine pattern: original) is a resolution test chart with sinusoidal intensity variation and varying spatial frequency. The second row (sine pattern: lens only) is the image of the top row formed by the lens under test. The sharpness of the chart is partially lost in the image. The modulation value of the pattern shown in the second row is the MTF. The third row (bar pattern: original) is a black-and-white-bar resolution test chart with varying spatial frequency. The fourth row (bar pattern: lens only) is the image of the third row formed by a lens under test. The sharpness of the chart is partially lost in the image. The modulation value of the pattern shown in the fourth row is the CTF. The second-last row "Amplitude" plots the amplitude of the image of the fourth row (CTF). At a high spatial frequency, CTF > MTF. The last row "MTF %" appears to plot the modulation depth of the amplitude shown in the second-last row.

the black-and-white references to calculate the CTF. Thus, the aforementioned error effects will be minimized. The software used for this task is not complex. People sometime write their own computer code to calculate the CTF.

If white light or RGB wavelengths are used to illuminate the chart, the CTF or MTF measured is the average for these spectra.

One problem with this method is that when the bar width is close to the size of the detector pixels, the alignment of the pixels to the bars will significantly affect the measurement result, as illustrated in Figs. 14.13(a) and (b). In such a case, several measurements must be taken. For every measurement, the chart position is shifted by a distance corresponding to a subpixel distance in the image and the best CTF measured is used.

14.7.2 Siemens star chart

The second method of measuring the CTF of a lens uses a Siemens star chart.[12] Figure 14.13(c) shows a Siemens star chart with 80 black-and-white bar pairs. The bar pair width D at radius distance r from the center of the chart is given by

$$D = r\theta, \tag{14.18}$$

where $\theta = 2\pi/N$, and N is the total bar pair number. The commercially available Siemens star charts contain tens of bars.

Most imaging lenses are symmetric about their optical axes; most of these lenses have their peak CTF/MTF value at their optical axes, and the CTF/MTF value drops along the radial direction. When using a Siemens star chart, the chart must be centered with the optical axis of the lens. A 2D detector array takes the image of the star chart, and a piece of software or code calculates the CTF/MTF along several or many circles centered with the chart, as marked in Fig. 14.13(c). The white reference is obtained by taking a "flat field" (background image, without the star chart) with the illumination on it. The black reference is obtained by taking a flat field image without the star chart and the illumination.

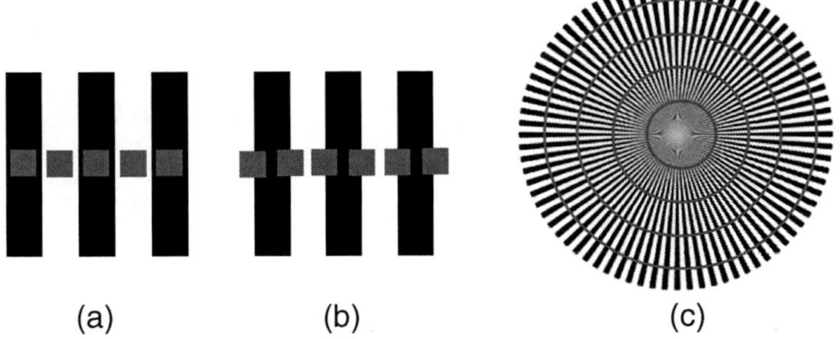

Figure 14.13 (a) Sensor pixels (grey squares) are well aligned with the black-and-white bars. The measured CTF is relatively the highest and is the correct result. (b) Sensor pixels (grey squares) are completely out of alignment with the black-and-white bars. The output of each pixel is the same. The measured CTF is zero. Such a result is not correct. (c) A Siemens star chart with 80 bar pairs. The CTF is measured along several or many co-centered circles as marked.

Compared with the USAF resolution charts, Siemens star charts have their advantages and disadvantages. The advantage is to be able to provide a complete CTF over the entire field. The disadvantage is more complex procedure of taking black/white references.

Just like the USAF resolution charts, star charts also have an issue with aligning the sensor pixels to the black-and-white bars.

14.7.3 Slant edge

The third method uses a slant-edge target to measure the MTF of lenses.[13] A slant-edge target has an "L" shape and is tilted to the sensor array by a few degrees, as shown in Fig. 14.14(a).

Because of the diffraction and aberration, the image of the target formed by a lens does not have a clear black-and-white cutting edge; rather, it has a gradually varying grey area, as shown in Fig. 1.32. The profile of this grey area is called the "line spread function" (LSF) or edge spread function in this case. The slant target lets sensor pixels measure the LSF, and the Fourier transform of the LSF is the MTF (Morel[9] gives an overview of a variety of functions). The two arms of the L-shaped target enable the measurement of the horizontal and vertical MTF simultaneously. The tilted black-and-white edges of the target cut through a series of pixels, as shown in Fig. 14.14(b). There are always some pixels that are well aligned with the edges of the target. Thereby, the pixel and target alignment problem is solved.

When measuring, the MTFs along many horizontal and vertical pixel lines are measured. The best MTFs are taken as the final measurement results. The measured MTF can be converted to the CTF as described in Section 1.13.5.

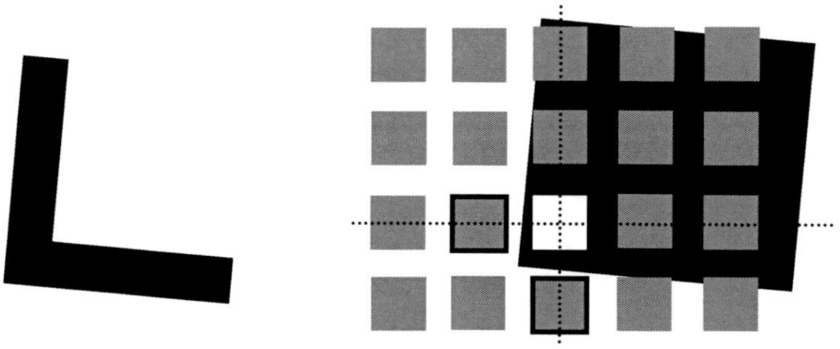

(a) (b)

Figure 14.14 (a) An "L"-shaped slant-edge target with a 5° tilt. (b) Illustration of the edges of the target cutting through sensor pixels in both the horizontal and vertical directions. The small grey (white) squares represent the pixels. The big black rectangle represents the slant target. The white pixel completely samples the target. The two grey pixels with black frame next to the white pixel completely sample the background. So, the pixels along the horizontal and vertical dot lines can avoid the "partial" sampling problem and provide the correct measurement results.

The shortcoming of the slant-edge method is that in order to measure the MTF of a certain location in the image, the target must be moved there so that the edges of the target can cut through this location.

14.8 Principles of Mounting Lenses

The performance of a lens can be significantly lowered by the mounting errors of the optical elements inside the lens, in terms of centering, tilt, and axial positioning. Lens mounting is a standalone and important technology that overlaps with mechanics and optics. Generally speaking, there are two approaches: passive mounting and active mounting.

Yoder[14] provides a comprehensive review of lens mounting techniques. Some companies, such as Opto Alignment[15] and Ruda Cardinal,[16] provide custom services.

14.8.1 Passive mounting

Passive mounting is a straightforward approach. The inner diameter (ID) of the lens housing, the outer diameter (OD) of the lens, and the centering of the two surfaces of the lens are made within the specified tolerances so that directly dropping the lens into the housing will result in an accurate-enough lens centering, as illustrated in Fig. 14.15(a). To avoid the lens being stuck midway as the lens is moved along the housing tube to the position, the ID of the housing must be slightly larger than the OD of the lens by at least ten microns or so. Therefore, no matter how precise the housing and the lens are fabricated, there is always room for decentering.

The two surfaces of a lens always have a certain amount of decentering or wedge, as illustrated in Fig. 14.5(b). The physical center and optical center of the lens are not along the same line. Minimizing the lens-surface decentering error is more difficult than making a metal housing with a precise ID or making a lens with precise OD. Passive mounting will likely produce a result like that shown in Fig. 14.16(b). Whether or not such a result is sufficient depends on the lens performance requirements.

Passive mounting is relatively simple and effective only for lenses that require relaxed mounting tolerance, but it is not accurate enough for lenses with a tight mounting tolerance.

14.8.2 Total indicator of runout

Active mounting is required for lenses that require a tight mounting tolerance. One key concept used in active mounting is the total indicator of runout (TIR). When a lens is being mounted, there is more or less physical centering error. In addition, there is lens fabrication error in terms of wedge, which moves the optical axis of the lens away from its physical axis. The widely used concept of TIR includes the contributions of both of these errors to the

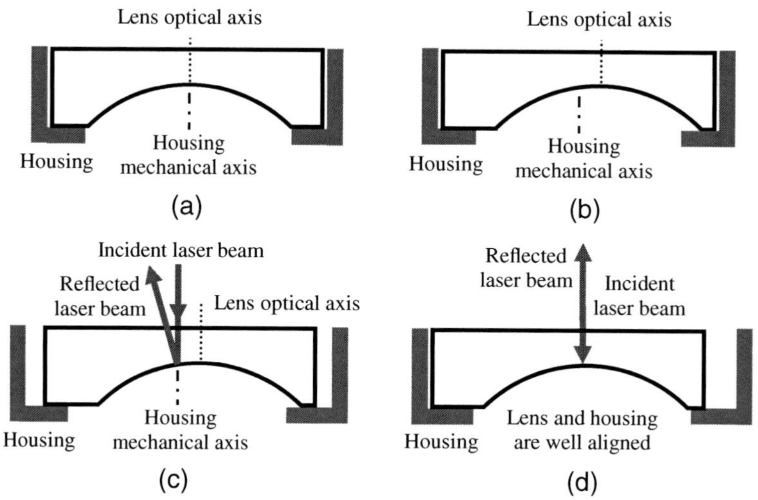

Figure 14.15 (a) Passive approach mounting a lens. Both the lens OD and the housing ID are precisely made so that simply dropping the lens in the housing will result in good lens positioning. (b) The lower surface of the lens has large decentering. The passive approach cannot well align the optical and mechanical axes. (c) The lens housing ID is intentionally made much larger than the lens OD, so that the lens has enough room for lateral displacement. (d) Using the active approach, the lens in (b) can be optically centered.

centering, as illustrated in Fig. 14.16. Every optical surface, when being mounted, has a TIR in units of degrees.

When designing a lens, one must perform tolerance analysis, specify the maximum acceptable TIR for every optical surface, and properly split a TIR

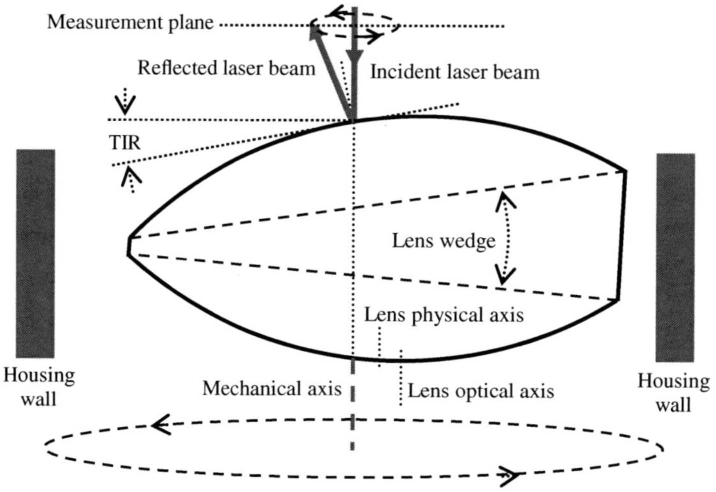

Figure 14.16 Illustration of total indicator of runout. All errors are exaggerated for clarity.

to the mounting decentering the fabrication wedge. When mounting a lens, all one needs to do is control the TIR for every surface to be within the tolerancing range. It is not necessary to determine how much the error, mounting decentering, or fabrication wedge contribute to the TIR.

14.8.3 Active mounting

The active mounting of a lens is a process that reduces its TIR. The following is a common technique to reduce the TIR while mounting a lens:

1. The lens housing ID is intentionally made large enough so that the lens has enough room for lateral displacement inside the housing, as illustrated in Figs. 14.15(c) and 14.16.
2. The housing is mounted on a rotatable stage. The housing mechanical axis is aligned with the rotation axis.
3. A laser beam is propagating along the mechanical/rotation axis through the lens. The reflected beam spot from the lens surface will mark a circle on the measurement plane as the housing is being rotated because of the TIR, as shown in Fig. 14.16. The TIR can be calculated by specially developed software using the circle radius and the distance between the measurement plane and the lens surface.
4. As the housing is being rotated, the lens assembler gently touches the lens to reduce the radius of the circle marked by the reflected beam spot until the TIR falls within the acceptable range, as illustrated in Fig. 14.15(d).
5. The lens position is then fixed, usually first by applying three spots of UV fast-curing adhesive for temporary fixation, followed by applying some other adhesive for permanent fixation.

For a multi-element lens, the lens at the deepest position is mounted first. Since every element has two surfaces, multiple elements will produce multiple reflection spots on the measurement plane. The lens assembler must input the optical prescription of the lens into specially developed software. The software will indicate which spot is the reflection from which surface. Opto Alignment[15] is the main manufacturer for the active lens alignment stage.

14.9 Techniques for Mounting Lenses

Some engineers or scientists may occasionally need to mount small quantity of lenses for lab experiment or prototype test. The mounting techniques involved can be different from large quantity production. In this section some commonly used mounting techniques are described

14.9.1 Mounting a laser diode beam-collimating lens

Since laser diodes are the most commonly used lasers and their beams are difficult to manipulate, they are used as an example here to illustrate the technique for mounting a beam-collimating lens.

The working principle is illustrated in Fig. 14.17(a). The lens is mounted in a threaded lens holder. The lens holder is screwed onto the laser module housing with the matching thread. The axial spacing between the lens and the laser is adjusted by rotating the lens holder. The lateral position of the lens is adjusted by three plastic set screws, while the far-field beam pattern is monitored, until the far-field beam pattern looks like something similar to that shown in Fig. 14.17(b). The lens is then well centered. The plastic screws not only center the lens but also hold the lens in position. There is certain pressure between the lens and the screws. The reason for using plastic rather than metal screws is to avoid placing too much pressure on the lens and cracking it. If the lens has $NA < 0.5$ or so, diffraction rings will appear in the fast axis direction, but the ring pattern should be vertically symmetric. This technique can be used for align collimating lens for any other lasers.

14.9.2 Passive method of centering a lens

Aligning a laser beam collimating lens is relatively easy because the far-field laser beam pattern can be used as feedback to examine the alignment quality. Passive alignment of an imaging lens without any feedback is more difficult.

The main difficulty in passively centering a lens is that the flat surface of a lens cannot be set on a flat seat, as shown in Fig. 14.15, because such a mounting tends to have larger centering error unless an active mounting technique is used as described in Section 14.8.3.

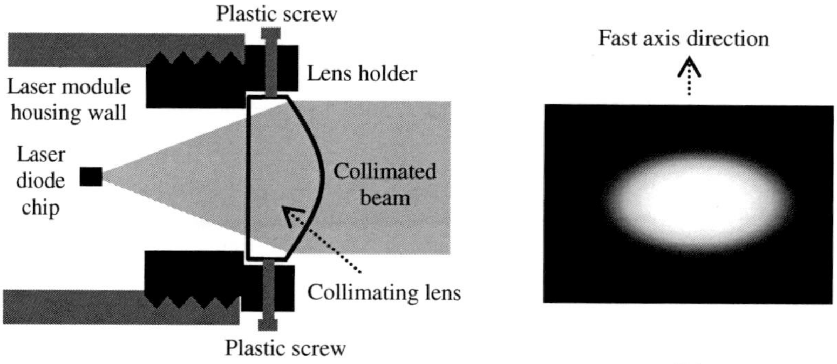

(a) (b)

Figure 14.17 (a) Illustration of aligning a laser diode beam-collimating lens. The lens is mounted in a threaded lens housing. The axial position of the lens is adjusted by rotating the lens holder. The lateral position of the lens is adjusted by three plastic set screws. (b) If the lens is well centered, the far-field beam pattern should look like something shown here, clean and symmetric.

One-contact-point seats are often used to mount a lens, as illustrated in Figs. 14.18(a) and (b). The surface curvature and the weight of a lens naturally center the lens. When mounting, the lens housing must be gently tapped to let the lens overcome the friction between the lens surface and the seat to slide to the right position.

The lens seat tip that contacts the lens surface should have a round shape, as shown in Figs. 14.18(c) and (d), to avoid possibly scratching the lens surface or cracking the lens if too much pressure is applied between the lens and the seat.

If it is difficult to machine a one-contact-point seat on the lens housing for a concave surface, as shown in Fig. 14.18(a), the seat can be machined on a retaining ring and is threaded to the right position in the lens housing.

14.9.3 Setting the axial spacing between two lenses

In most cases, the axial alignment is less sensitive than lateral alignment. A spacer can be placed between two lenses to set the axial distance, as illustrated in Fig. 14.19(a). The spacer can have a one-point-contact seat or a flat seat, depending on what seat the other surface of the lens sits.

If the axial distance between two lenses must have high accuracy, a few spacers with slightly different thicknesses can be made to compensate for the lens-thickness variation.

14.9.4 Fixing the lens position

Once a lens is well positioned, the lens position must be fixed. A retaining ring is commonly used to hold a lens in position. The ring can have either a flat surface or a one-contact-point surface. A rubber ring can be placed between

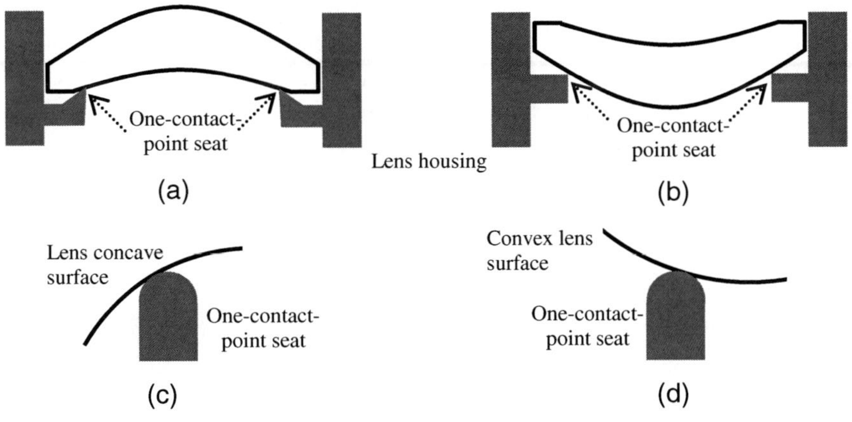

Figure 14.18 Illustration of mounting a lens on one-contact-point seat for better centering: (a) Mounting on a concave surface. (b) Mounting on a convex surface. (c) and (d) One-contact-point seat for concave and convex surfaces, respectively.

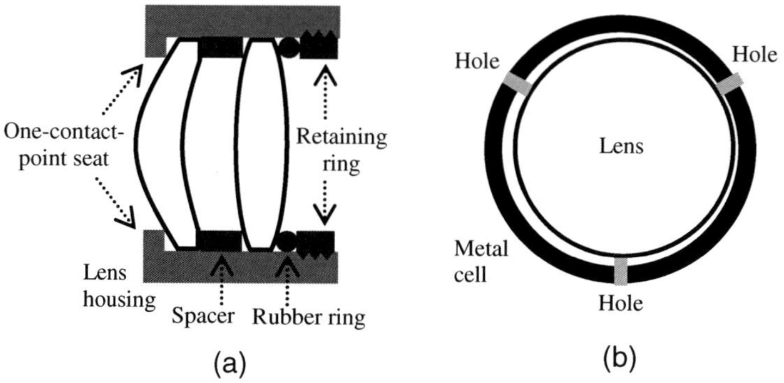

Figure 14.19 (a) A spacer is used to set the axial distance between two lenses. (b) The lens is not physically aligned with the metal cell, but it is assumed to be optically aligned with the cell. The lens position relative to the cell can be fixed by injecting elastomer through three holes previously drilled into the cell.

the lens and the retaining ring, as illustrated in Fig. 14.19(a), so that the pressure on the lens can be gentle and even.

One other way of fixing a lens in position is illustrated in Fig. 14.19(b). Three holes are made in the lens housing. After the lens is optically centered, some type of elastomer, such as silicone RTV sealant, is injected through the holes to hold the lens in position.

14.9.5 Aligning a lens to a cell

The alignment of two metal parts involves aligning two physical shapes and is relatively easy, whereas the alignment of a lens to a housing means aligning the optical axis of the lens (which often does not coincide with the physical axis of the lens) to the housing and is more difficult. Figure 14.19(b) illustrates a case in an exaggerated way. The lens can be optically aligned to the cell, while is physically not aligned. Since lens housings can be large and heavy, and may contain certain proprietary information, it's common to send a lens and a precisely made metal cell to alignment service providers to let them align the lens to the cell. The lens user, then, needs align the cell with the lens in it to the housing, which is a relatively easier task.

14.10 Handling and Cleaning Optical Components

14.10.1 Handling components

Optical components are usually picked up by using a tweezer or fingers with clean finger cots to hold the edge or non-optical part of the component. Caution should be taken to avoid directly touching an optical surface with bare fingers because even one brief, gentle touch will almost guarantee a fingerprint on the optical surface. It is acceptable to pick up optical components that are wrapped in clean optical tissue with bare fingers.

Optical components should be handled in a clean environment. When packaging, one should first wrap optical components in optical tissue, followed by a layer of plastic foam, and finally placed in boxes.

14.10.2 Cleaning components

Dust particles on optical surfaces can be blown off with an air puffer or a can of clean, compressed air. If air flow is not enough to remove the dust, use a cotton swab or optical tissue to gently wipe the surface. Wiping has more power than air flow and can gently remove stuck dust.

If these techniques do not work, then wet cleaning may be necessary. A cotton swab or a folded optical tissue soaked with a certain optical cleaning fluid, such as acetone or ethyl alcohol, can be used to wipe the optical surface under cleaning. The fluid will help resolve the pollution and make it easy to remove.

The method of wiping optical surfaces is not trivial. Two ways are suggested,[17] as illustrated in Fig. 14.20, but there appears to be no clear best way. The principle is that every wipe should end outside the optical surface so that the pollution—if it cannot be picked up by the cotton swab/optical tissue—will be pushed off of the surface. The wipe movement should be gentle and stable.

When cleaning, the cotton swab/optical tissue should be soaked with an amount of fluid that is enough to cover the wiped area of the optical surface but not so much that the fluid flows over the surface and drips off.

Optical components can also be cleaned with ultrasonic cleaners. The ultrasonic wave can remove the pollution without damage the lens.[18]

Hausner[17] provides more information about handling and clean optical components. Several vendors, such as Thorlabs,[17] sell optical cleaning sets that include a compressed air can, cotton tip, clean fluid, and wipe issue.

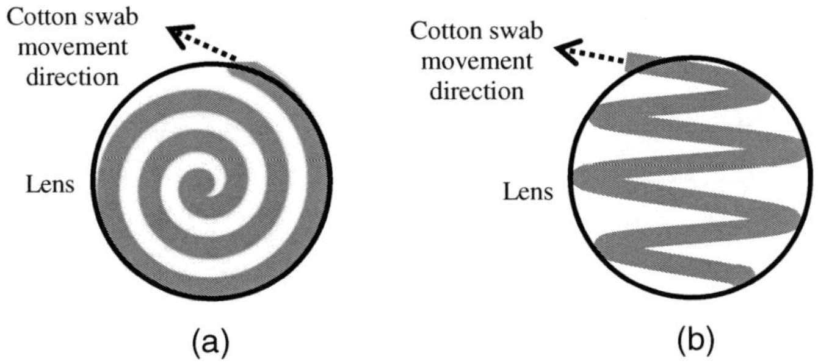

Figure 14.20 Two suggested ways of wiping to clean an optical surface. (a) Spiral: start at the center of the lens and end at the outside of the lens. (b) Spring: start at one edge of the lens and end at the outside of the lens.

References

1. M. Hausner, "Optical Standards and General Technical Specifications," Chapter 8 in *Optics Inspections and Tests: A Guide for Optics Inspectors and Designers*, SPIE Press, Bellingham, WA (2017) [doi: 10.1117/3.2237066.ch8].
2. M. Hausner, "Methods for Producing Optical Components," Chapter 5 in *Optics Inspections and Tests: A Guide for Optics Inspectors and Designers*, SPIE Press, Bellingham, WA (2017) [doi: 10.1117/3.2237066.ch5].
3. Wikipedia, "Precision glass molding," wikipedia.org/wiki/Precision_glass_molding
4. D. M. Arkens, *The Truth about Scratch and Dig*, www.lambdaphoto.co.uk/pdfs/Savvy/The%20Truth%20About%20Scratch%20and%20Dig.pdf
5. D. Maracara, *Optical Shop Testing*, 3rd ed., Wiley-Interscience, Hoboken, NJ (2007).
6. M. Hausner, *Optics Inspections and Tests: A Guide for Optics Inspectors and Designers*, SPIE Press, Bellingham, WA (2017) [doi: 10.1117/3.2237066].
7. Wikipedia, "Profilometer," wikipedia.org/wiki/Profilometer
8. J. C. Smith, "Understanding Surface Quality: A Practical Guide," www.photonics.com/a57165/Understanding_Surface_Quality_A_Practical_Guide
9. S. Morel, *Methods for Measuring a Lens Focal Length*, wp.optics.arizona.edu/optomech/wp-content/uploads/sites/53/2016/10/Tutorial_MorelSophie.pdf
10. *How to Measure MTF and Other Properties of Lenses*, www.optikos.com/wp-content/uploads/2013/11/How-to-Measure-MTF.pdf
11. Wikipedia, "Optical transfer function," wikipedia.org/wiki/Optical_transfer_function
12. *How to Measure Modulation Transfer Function (4)*, harvestimaging.com/blog/?p=1328
13. D. A. Kerr, *Determining MTF with a Slant Edge Target*, dougkerr.net/Pumpkin/articles/MTF_Slant_Edge.pdf
14. P. R. Yoder, Jr., *Mounting Optics in Optical Instruments*, 2nd ed., SPIE Press, Bellingham, WA (2008) [doi: 10.1117/3.785236.]
15. Opto Alignment Corp, optoalignment.com
16. Ruda Cardinal, www.ruda.com
17. M. Hausner, "Handling Optical Components," Chapter 19 in *Optics Inspections and Tests: A Guide for Optics Inspectors and Designers*, SPIE Press, Bellingham, WA (2017) [doi: 10.1117/3.2237066.ch19].
18. Thorlabs, "Ultrasonic Cleaner," www.thorlabs.com/newgrouppage9.cfm?objectgroup_id=10165
19. Thorlabs, "Optical Clean Suppliers," www.thorlabs.com/newgrouppage9.cfm?objectgroup_id=330

Index

Haiyin Sun has over thirty years of industrial and academic experience in the US, China, and Germany. He holds senior engineering and management positions at L-3 Communications, Coherent, Oplink, etc., and is currently the leading optical engineer with ChemImage Corp. working on hyperpectral imaging systems. His main expertise includes optical design and engineering, and laser-diode applications and physics. He is the sole author of four books and one book chapter published by Springer, CRC Press, and SPIE Press, respectively, and over thirty research articles published by peer-reviewed journals.

Sun has been the leading optical engineer and sole optical designer of several major optical projects, including XM25 (Counter Defilade Target Engagement System), CT2WS (Cognitive Technology Threat Warning System), NGCD (Next Generation Chemical Detector), VeroVision Mail Screener™, VeroVision™ Portable Standoff Threat Detection (recipient of the Platinum 2016 Technology Innovation Award from *Military & Aerospace Electronics* and *Intelligent Aerospace*), as well as numerous complex optical and laser systems. His technical works have been reported by *Photonics Spectra* and cited by the Melles Griot catalog.

Sun is an SPIE fellow, serving on several SPIE committees and teaching a short course at Photonics West. He was an editorial board member of the *Journal of Optical Communications*, the adviser of the optics and photonics book series published by River Publishers, and an adjunct faculty member of applied science at the University of Arkansas–Little Rock. His Ph.D., M.S., and B.S. are in applied science, optics, and physics, respectively.